65th Annual Safety in Ammonia Plants and Related Facilities Symposium 2021

AIChE Technical Manual 2021 Volume 61

Online
30 August – 3 September 2021

ISBN: 978-1-7138-4502-7

Printed from e-media with permission by:

Curran Associates, Inc.
57 Morehouse Lane
Red Hook, NY 12571

Some format issues inherent in the e-media version may also appear in this print version.

AIChE
120 Wall Street, FL 23
New York, NY 10005-4020

Phone: (800) 242-4363
Fax: (203) 775-5177

www.aiche.org

Additional copies of this publication are available from:

Curran Associates, Inc.
57 Morehouse Lane
Red Hook, NY 12571 USA
Phone: 845-758-0400
Fax: 845-758-2633
Email: curran@proceedings.com
Web: www.proceedings.com

AIChE® TECHNICAL MANUAL 2021 VOL 61

SAFETY

IN AMMONIA PLANTS & RELATED FACILITIES SYMPOSIUM

21 SAFETY IN AMMONIA PLANTS & RELATED FACILITIES SYMPOSIUM

AIChE

The Global Home of Chemical Engineers

CLARIANT
A New Era of
Ammonia Synthesis Catalyst
AMOMAX® 10 PLUS
AmoMax® 10 Plus is Clariant's latest generation of **wustite-based ammonia synthesis catalysts**. The new catalyst is built on the successful and well proven AmoMax® 10 formulation with a further **improved promoter set**. The optimized promoter concentrations lead to higher activity, higher stability and a faster startup of AmoMax® 10 Plus compared to its predecessor.
PRODUCT BENEFITS
HIGHER PRODUCTIVITY
8 - 12 % more productive than AmoMax® 10
IMPROVED WATER RESISTANCE
of AmoMax® 10 Plus
LOWER REDUCTION AND LIGHT-OFF TEMPERATURE
18 - 27 °F lower than for AmoMax® 10
CUSTOMER BENEFITS
ENERGY SAVINGS AND LESS CO_2 EMISSIONS
Lower loop pressure: Reduced by up to 145 psi
Lower recycle ratio: Ammonia concentration at converter outlet increased by up to 1 % point
INCREASED AMMONIA PRODUCTION BY UP TO 3 %
(If not limited by make-up gas and/or compressor curve)
LONGER LIFETIME
Improved poison and water resistance
QUICKER STARTUP
Lower reduction and light-off temperature
CLARIANT.COM/CATALYSTS
what is precious to you?

Introduction

Welcome to the 65th Safety in Ammonia Plants and Related Facilities Symposium. The COVID-19 pandemic has affected all of us in many ways, and this conference is no different. The 2020 symposium was cancelled due to the travel and other logistic challenges created by the virus. Many of these conditions remain in place, so this symposium will be held in a virtual format for the first time in our history.

This symposium focuses on improving the safety, reliability, and operability of ammonia plants, ammonia storage facilities, and other fertilizer and hydrogen based operations. The papers you read and presentations you participate in were prepared and presented by individuals who strongly feel these topics will improve the safety of our facilities. I urge all of you to take the lessons learned and apply them to improve the safety of our industry! Besides the proceedings from the 65th Symposium, you will also receive electronic access to the proceedings of the previous sixty-four symposiums, which is an invaluable resource.

On behalf of the members of the AIChE Ammonia Safety Committee, I wish to thank the authors, sponsors, and all the attendees for making this year's symposium a success. The participation and commitment of the authors throughout the years has made this symposium an ideal of the value in shared learnings. We also thank the AIChE staff for their excellent support in the organization and preparation of this event.

Eugene Britton, 2021 Program Chairman

2021 AICHE Ammonia Safety Committee

Front: Taylor Archer (Clariant), Michel Warzee (Yara, Retired from Committee), John Mason (Nutrien), Venkat Pattabathula (Becht), AK Singh (IFFCO, Retired from Committee), Dorothy Shaffer (BakerRisk), Ahmed Esmael Rahimi (Qatar Fertiliser Company), Eugene Britton (CF Industries)

Back: John Brightling (Johnson Matthey), Klaus Noelker (thyssenkrupp Industrial Solutions AG), Ian Welch (Consultant), Scott Rodrigue (CF Industries), Bob Collins (KBR, Retired from Committee), Neal Barkley (Coffeyville Resources, Retired from Committee), Harrie Duisters (OCI NV)

Not Pictured: Marc Gilbertson (East Dubuque Nitrogen Fertilizer), Umesh Jain (KBR), Svend Erik Nielsen (Haldor Topsoe A/S), Mohamad Noueiri (Yara SA/NV)

TABLE OF CONTENTS

MONDAY, August 30, 2021
8:00 AM - 4:00 PM US Central Standard Time

TUESDAY, August 31, 2021
8:00 AM - 4:30 PM US Central Standard Time

TABLE OF CONTENTS

WEDNESDAY, SEPTEMBER 1, 2020
8:00 AM - 4:00 PM US Central Standard Time

TABLE OF CONTENTS

Improving Safety and Reliability of Ammonia Plants with Focus on Boiler Water Treatment

In this paper, incidents reported in the past AIChE Ammonia Safety Symposiums related to boiler water systems are recapped. Comparisons between utility boilers and Ammonia plant waste heat boilers are provided. Typical challenges in managing boiler water systems are elaborated. Finally, the international standards & guidelines applied and recommended best practices in boiler water systems for the ammonia industry are discussed to improve waste heat boilers' safety and reliability and process coolers.

Venkat Pattabathula
SVP Chemical Plant Services

Dr Robert Terrell
Principal Consultant
Industrial Water Consultants (IWC)

Introduction

Water plays an important role in ammonia plants' safety and reliability, both as cooling water and steam raising/process cooling. In modern ammonia plants, about 3 to 4 tonnes of steam are produced for every tonne of ammonia, making ammonia the byproduct of a steam-raising process.

Over the last 60+ years, waste heat boiler (WHB) damage has been a regular point of discussion at the AIChE Ammonia Safety Symposiums, as well as in other industry symposia. Are these recent incidents of new problems or a reluctance to learn the lessons from past failures? Where can we look to find better answers, both for design and operation, to prevent future failures?

The power sector provides the most obvious comparison and has made real progress in the last 30 years in improving reliability. Besides, as modern utility plants operate at significantly higher pressures (up to 2500 psig (17.2 MPa) and higher) than the ammonia plant WHBs (900 to 1900 psig, 6.2 to 13.1 MPa), it also provides the opportunity to understand the operating constraints better.

This paper intends not to describe in detail the corrosion mechanisms that were well covered by Duisters and Savelkoul [1] but to focus on the conditions that allow them to occur.

Comparison of Utility Boilers and Waste Heat Boilers

Utility Boilers

For utility boiler plants using recirculating drum boilers, raising steam is recognized as the primary function, and close attention is given to producing make-up water of the required quality, ensuring that the internal boiler water treatment is accurately maintained, recovering as much condensate as possible for reuse, and polishing it to the required quality as necessary.

Utility plants commonly have more than one boiler available, allowing regular outages and inspections without affecting power generation. The boilers are designed to have an even heat flux at all points and good water circulation. If considered necessary, forced circulation is introduced, for example, in Heat Recovery Steam Generators (HRSG) used for gas-fired power generation. Older oil or coal-fired utility boilers are physically larger per steam production unit than WHBs - although newer gas-fired boilers tend to be more compact, allowing faster start-up. It is not uncommon for utility boiler tubes to last for 30 years or more without replacement with good maintenance.

Consistently supplying the best possible feedwater quality is seen as a vital aspect of the operation. With well over ninety per cent designed condensate recovery, make-up water plants, whether demineralization or reverse osmosis, are relatively small. At normal condensate recovery and blowdown rates, the impact on the boiler water quality of any impurities in the make-up water is close to 1:1, e.g. 20 µg/l of chloride the make-up water would be 20 µg/l in the boiler water. Condensate polishers have historically been oversized to handle the larger water volumes and accommodate potential cooling water contamination introduced by leaks in surface condensers. Recognizing that the condensate polishing plant's faulty regeneration presents a greater risk to the feedwater quality than condenser tube leaks, ex-situ regeneration, in which the ion exchange resins are removed from the polisher vessel before regeneration is carried out, is used to minimize the risk of regenerant contamination. Newer plants using thin-walled titanium tubed surface condensers can operate without condensate polishing, recognizing that the risk of condensate contamination through leaks is lower than the risk from poor regeneration.

The internal chemical treatment for utility boilers is now well defined and better understood. The high-quality feedwater allows the application of simple, minimal water treatment programs because internal treatments are contaminants that can cause damage which should only be used where necessary and then dosed at the minimum concentration needed. This approach is only sustainable with consistently excellent feedwater quality.

Understanding the key control parameters and key performance indicators has allowed monitoring to become more precise. Online monitoring plays a vital role through the continuous measurement of dissolved oxygen, Oxidation Reduction Potential (ORP), iron in boiler feedwater and cation conductivity. Better, more reliable instrumentation has been a key factor in this change, allowing a reduction in off-line sampling and analysis schedules. This has allowed a zero-tolerance approach to out-of-specification results with rapid response to correct issues before they develop into problems.

While it would be incorrect to conclude that all the water issues in utility boilers have been resolved, the progress made in the last 30 years has been substantial. It is reflected in the improved guidance available in the International Standards from the Electric Power Research Institute (EPRI) [2], VGB PowerTech (VGB) [3], and the International Association for the Properties of Water and Steam (IAPWS) [4], among others. Has this improvement been mirrored in Ammonia plants?

Waste Heat Boilers (WHB)

Waste heat recovery is an essential competitive element of a modern Ammonia plant. This is achieved through a series of dedicated "process coolers" – or WHBs and boiler feed water preheaters, integral to the ammonia manufacturing process. There are no installed spares.

The boilers are designed primarily for the process duty, resulting in poor water circulation and very high and unevenly distributed heat fluxes. Over the last five decades, steam pressures have been raised from 3 MPa to 13 MPa (435 psig to 1885 psig), with limited basic design changes.

The water quality operating standards employed are based on the manufacturers' instructions and international standards, which prevailed when the plant was built, often ignoring the changes in standards and operating practices that have since been introduced. The standards are based upon steam pressures rather than heat flux, thus making the implicit assumption that heat flux in the boiler is evenly distributed.

External water treatment is a mixed bag, often driven by cost pressures. With typical make-up rates in the range of 30 to 40%, any make-up water quality excursions have a rapid impact throughout the system. Recovered steam condensate may or may not be polished, together with recovered process condensate containing ammonia-based compounds and carbon dioxide, which can rapidly overwhelm the polisher. Also, methanol and ethanol in the process condensate would pass through the polisher and, depending on operating pressure, resulting in the formation of weak organic acids such as formic acid and acetic acid in the boiler. Polishing may be carried out independently on the three streams, or two streams in combination, or not at all. The result is feedwater quality that could be excellent, but more often is not.

Various internal chemical treatment programs are used, sometimes following the international standards' advice, but often using a proprietary blend of chemicals as recommended by the water treatment supplier. These are intended to compensate for the inconsistent feedwater quality rather than the demands of the boiler itself. The application of an unsuitable chemical treatment program can lead to problems both at the lower steam pressures and the high pressures employed in newer plants. This is often exacerbated by using incorrect and inconsistent control limits.

With the focus on ammonia production rather than steam raising, it is not surprising that the monitoring and control applied is below the standard applied in the utility and power boiler plants. Online instrumentation is often limited and poorly serviced, while off-line analysis may be hampered by inadequate sampling and analytical techniques, inadequately qualified analysts, and a general lack of priority. It is not uncommon to find out-of-specification results running at more than 20% - even up to 50% - for some parameters over extended periods. Rather than precision dosing and control, non-compliances are tolerated on the basis, presumably, that "if it hasn't failed, the boiler must be okay".

Those plants that have not suffered WHB failures yet may consider this description unduly pessimistic, while those who have experienced failures would recognize several of these issues. The challenge for all plants is how to achieve the required operating standards and how to know that they are being maintained. In our industry, the best performing boilers may last for 20 to 25 years without tube failures. But, in the worst cases, the boiler's life may be as little as two years if the water quality is poor, resulting in extended outages and severe damage to the business.

What lessons can be learned from the power utilities and the best ammonia plant boilers to improve all ammonia plants' operation and reliability? Are the failures due to a complex, challenging design or just poor control? These issues are examined in more detail in the following section.

Understanding the challenges

Focusing only on the WHB, there is an inevitable tension between the boiler design and duty, water quality, and the selection and management of the internal treatment at the heart of the problem.

Boiler Design

The WHB designs adopted by different ammonia plant technology licensors follow the same design principle – how to get the maximum performance for the least cost. The design is dictated by the process conditions and the need to maximize heat recovery rather than the water side's requirements. As steam pressures have been raised from 3 MPa to 13 MPa (435 psig to 1885 psig), design changes have been relatively limited.

The waterside design bears superficial similarities to standard "shell" boilers, the workhorse of many industrial processes requiring low-pressure steam. Hot gases pass through tubes while the water is retained in the shell. However, the high-pressure operation demands several additional features, with external downcomers to improve water circulation, a separate (generally standard) steam drum, and much stiffer tubesheets. Further, learning from early tube / tubesheet failures, the hot gases are transported into the boiler and away from the tubesheet through the use of ferrules.

This design has the following consequences for the waterside:

1. The water circulation pattern within the boiler is dominated by the annular flow, which maybe 10 to 12 times larger than the flow out of the boiler. The axial view in Figure 1 shows how the annular flow provides the bulk of the circulating water.

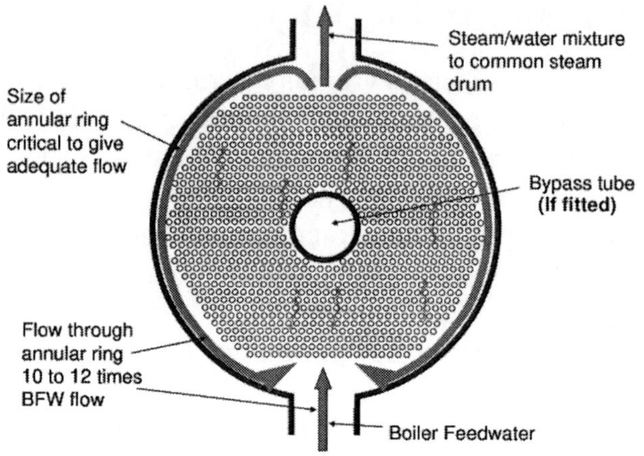

Figure 1. Illustration of axial water flow pattern

The lateral view in Figure 2 shows that the water flow needs to move laterally and through the annular ring.

The water flow through the compartment would not be even with minor pressure drop differences dictating the relative flows in different boiler areas from these illustrations. Inevitably, the water flow close to baffles or the tubesheets at the end of the section would be lower than the water flow in the section's center. Also, the process gas would be hotter at the inlet end than at the outlet end of the section, leading to higher water flows at the hot end.

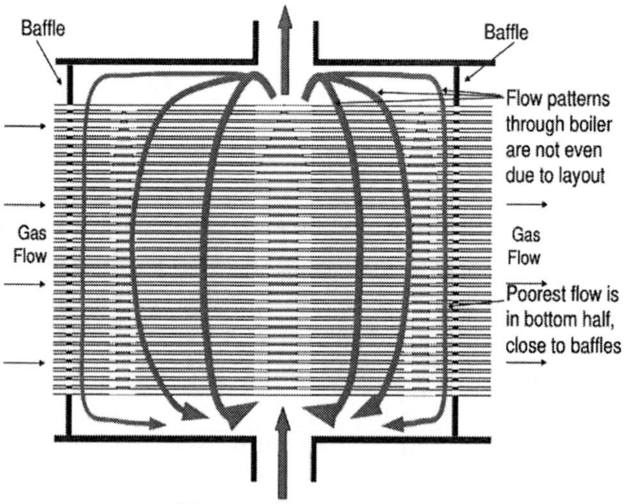

Figure 2. Illustration of lateral water flow pattern

Furthermore, the water at the bottom of the section would be marginally colder and denser than the water higher in the section, containing more steam. This is important in driving the circulation around the annular ring. However, it also results in the water flow in the bottom of the section being marginally lower than in the top of the section. Because of the natural circulation from the bottom to the top of the section, the steam/water flow on the top of each boiler tube would be lower than the flow around the sides of the tubes.

Thus the "worst" conditions in terms of water flow will be on the top of the boiler tubes at the inlet end of the section, close to the baffle walls and in the lower half of the section. This is where any waterside deposits, if present, would be expected to accumulate.

2. The stresses on the hot end tubesheet imposed by the high temperature / high-pressure operation, together with the desire for more rapid start-up and shut down, have driven the designers to follow several different paths to avoid the need for very thick tubesheets – including making physically smaller boilers operated in series or parallel, and the use of thinner tubesheets strengthened by internal structures such as vertical plates between the tubes and extending into the boiler as far as the end of the ferrules or beyond. Where used, these plates have the potential to affect the water/steam flow patterns and allow waterside deposits to collect.

3. The use of ferrules results in the heat flux being maximum in a narrow section of the tube bundle, maybe 30 to 50 cm (12 to 20 in) long, immediately beyond the ferrules' length. It is beneficial that it is away from the tubesheet for water circulation reasons, but it is no coincidence that the WHB failures reported due to damage caused while online is very limited, mainly to this section of the boiler.

The inlet gas temperatures at this point in the process are similar regardless of the steam pressure, so the ability of the water/steam to transport heat away from the tubes would be a function of the relative boiling pattern close to the tubes.

Boiler designers aim to ensure that any boiling on the heat transfer surfaces is short-lived and transient, with a microscopic steam bubble forming and collapsing continuously. Departure from Nucleate Boiling (DNB) is the term given when these steam bubbles remain on the heat transfer surfaces and coalesce into larger steam bubbles insulating by nature soluble and insoluble impurities present or chemical treatment to dry out and deposit on the heat transfer surfaces. The extreme version of DNB is "thin-film boiling", in which there is a constant steam bubble formed on the heat transfer surface with the likelihood of overheating, deposition and corrosion occurring.

The boiler manufacturers conduct extensive modelling of their designs to ensure that the water circulation in this section is enough to avoid stable steam bubble formation or (DNB), but this would assume that the tubes are clean and well protected by a thin magnetite film. The boiler designer's skill is to ensure that there is adequate water flow at all points in a boiler and at all boiler loads to guard against overheating, DNB, and scale deposition. In simple terms, the higher the heat flux, the more critical the flow and the higher the flow required to avoid DNB.

Unfortunately, cleanliness is challenging to maintain because of the inevitable variations due to operational changes, plant outages, water quality, and chemical treatment. While the flow differences are minor, they may be enough to allow DNB or deposition to occur in a specific location, while the boiler tubes elsewhere in the section are unaffected.

4. The design makes the boiler very difficult to inspect and, if necessary, to clean. The limited access makes detailed inspection almost impossible, even with the judicious location of inspection ports and the use of borescopes, so it is challenging to identify if and where any deposits are collecting. If the tubes close to an access point appear clean, this is generally considered good enough, despite the circulation challenges identified above.

   For the same reasons, cleaning is also complicated. Most owners are reluctant to introduce acid into their boilers if they cannot guarantee their complete removal or neutralization. But the reality is that it is difficult to achieve the necessary conditions of circulation, temperature, and disturbance to ensure that the cleaning is effective without very carefully planned and detailed procedures executed by highly skilled and experienced cleaning professionals. The associated high cost and the fear of leaving acid-rich deposits on the tops of boiler tubes in the high heat flux zones are enough to dissuade many owners unless there is no alternative.

The typical two to five-year turnaround cycles limit inspection opportunities, adding to the challenge.

This is not an indictment of the boiler designs or the designers. Many ammonia plants are using these designs, which operate reliably over extended periods without boiler tube failures. Existing plants have no option but to continue with the boiler design they have. The boilers are not impossible to operate, but they do demand an understanding of the challenges they present and close attention to how they are managed. This not only applies to a newly designed and constructed plant; it is also essential to consider the long-term impact resulting from accumulated damage over the plant's life.

For such high heat flux conditions and circulation constraints, high purity water is needed to prevent corrosion damage, and the internal chemical treatment needs to be minimal and carefully managed. For technical and economic reasons, including:

a. The proportion of make-up required to replace unreturned condensate
b. Cost of water purification (both Capital and Operating Costs)
c. Size of the water treatment plant
d. Impact on cost of ammonia production
e. Ability/willingness of ammonia plant operators to manage high purity water production, etc.

Ammonia plants can't mirror the conditions used in utility boilers, but that does not mean that lessons cannot be learned. Investigations into boiler tube failures conducted by Structural Integrity Associates Inc. [5] on over 250 boiler and HRSG plants worldwide have identified Repeat Cycle Chemistry Situations (RCCS) as instrumental in many of the failures, i.e. frequent repetition of small quality excursions which, in isolation, are not sufficient to result in tube failures. Reducing the number and magnitude of such excursions would have a significant impact on boiler reliability.

The rest of this paper would focus on water treatment aspects.

Feedwater Quality

The better the feedwater quality, i.e. purer, less contaminated, and more consistent, the more reliable the boiler would be, and the simpler the internal chemical treatment can be. But that is easier said than done. With the high make-up water percentage and low steam condensate recovery, there would always be an enormous impact from make-up water impurities. Typically, in ammonia plants, the recovery of process condensate containing high concentrations of ammonia, carbon dioxide, methanol, ethanol, CO_2 removal system solvent degradation byproducts depending on the applied technology (Potassium carbonate, vanadium, aMDEA, silicone antifoams, etc.) add to the confusion.

Water Purification

The demineralization process is remarkably reliable and effective, as long as the regeneration process is carried out correctly. Although it is common for the ion exchange resin to be held responsible for poor water quality, most issues are associated with poorly executed regeneration. These systems operate in cyclical service, and the large number of valves required to operate faultlessly in the proper sequence and for the correct duration means that problems are inevitable. Installed instruments are often not maintained, and controllers on older plants were often allowed to become obsolete. Above all, the lack of understanding of the demineralization process, its works, and its failure to monitor and troubleshoot difficult. Gone are dedicated water treatment plant operators for many plants, leading to the reliance on outside service companies.

Reverse Osmosis (RO) has some excellent advantages over demineralization in its ability to remove dissolved organic materials, which can otherwise be degraded in the boiler to form organic acids. It also operates continuously, offering fewer opportunities for contamination. But for many installations, it is more expensive to operate than demineralization and, unless two-pass RO is installed, will produce poorer quality water. The successful operation of RO is also critically dependent on the pretreatment applied, which may double the plant's cost; a membrane that can remove impurities down to ionic size is itself very vulnerable to becoming fouled.

The final stage in the treatment process is polishing, where the already "pure" water is further purified. Although ion exchange polishing, either for demineralization or condensate treatment, looks the same as demineralization, it typically operates at higher throughput rates, allowing much shorter contact times to remove ions. To produce consistently very high quality, the inlet water quality must also be high. Sending poor quality water to a polisher, e.g. due to a poorly operated demineralization plant, single-pass RO, or poorly stripped process condensate, would result in poor quality polished water using more than 50% of the nominal polisher capacity.

Maybe the most prominent issue lies with designing the water purification stage, which receives too little attention yet is subject to rigorous "value analysis" to reduce the cost. Good knowledge of the raw water quality, good design specifications, correct sizing for the duty and foreseeable upset conditions (raw water quality changes, condensate contamination, mechanical failures of the purification plant, and media changes), the use of top quality valves, and the inclusion of sufficient online instrumentation and a diagnostic controller sound expensive but would be repaid many times over in improved reliability of the boilers. This requires an understanding of the processes, how they work and how they fail, and the potential impact on the ammonia plant and merits both the EPC contractor and the end customer's close attention. "Cheap is good, but good is not necessarily cheap".

Water Quality Monitoring

The only way to confirm that the water quality is good throughout the water/steam/condensate system is through monitoring, preferably online. Accurate sampling for off-line testing is fraught with challenges unless conducted from continuously flowing, suitably located, and adequately cooled sample points. In a utility plant, these are generally gathered to a single point; in an ammonia plant, collecting samples may involve a significant hike and climb up and down many levels. Analysis of these samples is a specialist task requiring expensive analytical equipment, especially when looking for parts per billion (μg/kg) concentrations of impurities. Even the smallest speck of contamination of the sample can produce wildly misleading results.

Online instrumentation provides several advantages – it is continuous, it can identify variable and cyclical events, and it less susceptible to sampling errors or sample contamination (if correctly installed). It is more expensive and requires routine attention, but these are again repaid by future plant reliability.

Key online instruments for most water/steam/condensate systems are dissolved oxygen (exit the deaerator), pH and cation conductivity (also known as post cation conductivity or hydrogen ion conductivity), and DOC (Dissolved Organic Carbon). ORP (Oxidation Reduction Potential) is also becoming more common, as is online iron measurement.

The best measurement to identify contamination in the BFW and steam/condensate systems is degassed "Cation Conductivity", which is the conductivity of the BFW or condensate after passing it through a column of cation ion exchange resin in the hydrogen form to exchange any cations (most notably sodium and amines) with hydrogen ions. The cation conductivity is a measure of the residual acidic anions (e.g. chloride, sulphate) in the water. Exchanging the cations for hydrogen ions enhances the conductivity, so the cation conductivity measurement is very sensitive.

Given the challenging boiler design issues, as discussed, and the multi-year interval between inspections, having good feedwater quality and applying best practice monitoring of the water quality is the minimum standard required to avoid unexpected and expensive boiler tube failures.

Internal Chemical Treatment

It is necessary to understand what is needed and only add what is necessary to control water quality within tight limits.

Internal chemical treatments are used to reduce the corrosion and fouling resulting from residual impurities in the feedwater. Higher concentrations of impurities demand more complex chemical treatment programs, use more internal treatment, and make control much more difficult. Yet the chemical treatments themselves have the potential to cause damage, including the following damage mechanisms:

- Flow Assisted Corrosion (FAC) of the feedwater system due to over-addition of oxygen scavenger,
- Acid corrosion of tubes in boilers using phosphate control programs,
- Caustic corrosion of boiler tubes following phosphate hideout,
- Condensate system corrosion from the degradation of complex oxygen scavengers and amines.

Having spent time and effort purifying the feedwater to the required standard, it would be a shame to add treatments that themselves cause more damage than the residual impurities.

Dissolved Oxygen (DO)

Most modern deaerators should achieve lower than 10 μg/kg dissolved oxygen (DO) in the feedwater during normal operation, and any excursions should be identifiable through online DO analysis. This is good enough for most applications using high purity feedwater. Despite this, plants are often encouraged by their water treatment suppliers to continue to add an oxygen scavenger (sometimes gross over-addition) to reduce the DO concentration further. This creates a reducing environment and causes FAC in high-temperature feedwater systems, with water temperatures over 150°C (300°F) and transporting iron into the boilers to deposit on heat transfer surfaces.

There is a trade-off between dissolved oxygen, feedwater impurities, and oxygen scavenger addition which needs to be observed, and this is better resolved by improving the feedwater quality than through the excessive addition of oxygen scavenger. Adding oxygen scavenger during start-up and upset conditions may be necessary, but where possible, it should be minimized during the long term, steady operation.

Corrosion Control

The most critical corrosion control parameter for the water/steam/condensate system is pH. This is commonly controlled by adding caustic, phosphate-based chemicals, amines, and/or ammonia, with the choice determined by the boiler operating pressure and feedwater quality. (The tables in the international standards are based on operating pressure rather than heat flux.)

Lower pressure boilers ($\leq$ 900 psig (6.2 MPa)) typically allow lower feedwater quality but require stronger alkalis (caustic, phosphate) to counteract the corrosivity of acidic ions such as chloride.

For higher boiler pressures, a range of treatments is available, typically based on phosphate (coordinated phosphate, congruent phosphate, equilibrium phosphate – now more generally referred to as "phosphate treatment") or neutralizing amines (All Volatile Treatment or AVT). (Control ranges for pH and phosphate concentration are given in the international standards, which should be consulted for specific applications.) The choice of treatment is again dependent on operating pressure and feedwater quality, although it would be correct to allow for high heat flux as found in ammonia plant WHBs.

The phosphate treatment methods are more challenging to operate well but can accommodate feedwater quality excursions.

AVT is much simpler to manage, with pH being the critical control parameter, although it has minimal tolerance for impurities as it is not essential enough to cope with acidic ions and therefore demands better feedwater quality. Some high-pressure WHBs have successfully used polyamine treatments as a version of AVT, which similarly requires high-quality feedwater.

It is also important to note an optimum pH range for each treatment program to minimize corrosion, rather than just a minimum pH value. Operating at pH values above the specified control range can itself cause damage. As the feedwater quality improves, the amount of chemical treatment required is reduced, and the pH control range can be lowered as the impurity-induced corrosion rates fall.

The volatility of neutralizing amines means that they can be used to protect all parts of the water/steam/condensate system, while the non-volatile caustic and phosphate are only used in the boiler itself. In utility boilers with minimal steam circuits (boiler – turbine – condenser), ammonia is the most commonly used amine, but in more extensive steam circuits with different steam pressures, it is common to use a blend of amines to protect all parts of the circuit from corrosion. The presence of carbon dioxide in the steam, from degrading organic molecules and process condensate, can lead to low pH values at different points in the steam/condensate system and should be investigated through structured sampling and analysis at multiple points throughout the steam/condensate system to confirm that the blend of amines is suitable.

Fouling Control

Iron is the most common foulant in boiler systems fed with demineralized water quality. This occurs every time the water quality is poor or the chemical treatment is out of specification and continues due to the spalling of loose and fluffy magnetite formed during these upsets. [1] It also occurs due to FAC of the feedwater system and following any shutdown due to thermal shock.

The power sector identified over 30 years ago that, in a tightly managed system operated continuously at baseload, the amount of iron deposited was minimal, leading to a significant reduction in acid washing frequency. For ammonia plants, for reasons already discussed, this is more difficult to achieve.

Iron would deposit throughout the WHB system but is most damaging on horizontal heat transfer surfaces. It tends to accumulate, forming the basis for wick boiling, the extreme concentration of impurities in the feedwater over time and under deposit corrosion leading to tube failures.

"Wick boiling" is the term used to describe the process of drawing boiler water into a porous deposit by a "chimney effect", driven by the boiling action on the hotter surfaces beneath the deposits. Once there, the boiler water boils to form steam, leaving behind the deposit's contaminant ions. Once established, this wick boiling process becomes self-sustaining, resulting in ever-higher concentrations of contaminants within and underneath the deposits leading to accelerated corrosion. Studies have indicated that the concentrations of contaminant ions can exceed 10,000 mg/l, or 1%.

Iron deposited on non-heat transfer surfaces does little harm, but the disturbance created by intermittent blowdown is insufficient to remove it. The only way iron is removed if it remains in suspension and removed in the continuous blowdown line is where dispersants may provide some assistance.

The presence of hardness in the water, either knowingly through softened water (low-pressure boilers) or heat exchanger tubes leaks (unpolished condensate), creates very similar iron issues when it forms deposits. There are dispersants available to cope with hardness and keep it in solution, but their ability to work in the highest heat flux conditions should always be questioned. As with iron, prevention is preferable to treatment.

Reducing the build-up of deposits on heat transfer surfaces is critically important in preventing tube failures in WHBs, and the aim should always be to minimize the formation of iron deposits through the correct management of the water quality and chemical treatment to prevent deposits from being formed. However, this may require significant investment and be beyond the capacity of many ammonia plants. In these circumstances, regular high quality, correctly executed acid washing may be the only option to remove the deposits; the frequency is determined by an assessment of how well the water quality and chemical treatment are managed.

The Application of International Standards and Guidelines

As stated earlier, different standards and guidelines are available, depending on where the boilers are designed. The boiler manufacturers are members and contributors to the bodies which produce these standards, and there is significant commonality between them.

While the guidance provided is largely experiential, based on knowledge gained from previous boiler failures, the last 30 years have seen an increasingly theoretical approach to the internal chemical treatments and the detrimental impact of low concentrations of impurities. [2] [4]

There is also an increased recognition that control upsets would occur but can be tolerated if not too severe, too long, or too frequent. The VGB Action Levels provide good guidance for operators. [3]

However, it is important to recognize that the Standards and Guidelines are primarily written for utility boilers rather than WHBs. This implies evenly distributed heat flux, good circulation, condensate recovery, feedwater quality, internal chemical treatment, and monitoring. The tables of control limits are presented mainly by steam pressure, with limited reference to heat flux. When applying this advice to ammonia plant WHBs, due account must be taken of the very demanding duty as described in this paper, such as applying the advice in the standards and guidelines given for boilers operating at significantly higher pressure. This may mean that some internal chemical treatments are not suitable for use even though the steam pressure alone might suggest otherwise. It may also mean that the feedwater quality needs to be much better.

As the international standards make clear, creating hard and fast rules for every situation is impossible, and local conditions need to be considered. In addition, it is unwise to "cherry-pick" values for different parameters from different standards to suit local conditions.

In this paper, we have avoided being too specific on control limits but instead encourage all owners to review and revise where necessary their standards, working with their boiler manufacturer to make sure they are following the best and most recent standards.

Examples of Failures

A list of boiler water-related papers that have been presented over the years at the past AIChE Ammonia Safety Symposiums are provided in Table 1. These incidents highlight the challenges in managing boiler water systems and the focus required to operate modern-day high-pressure boilers in ammonia plants.

Boiler failures have been a nightmare in the fertilizer industry in general, as any failure of this equipment seriously hampers the production for at least eight to nine days, besides the consequential damages to other critical equipment, machinery, and catalysts.

Based on reported incidents, the following common lessons can be drawn:

For existing plant

1. Highly reliable and safe utility service facilities are essential because of their significant impact on plant operations.
2. Ensure that all modifications to the plant and its operation are correctly managed throughout the plant's life, no matter how small they are (MOC, Risk Analysis, HAZOP, etc.).
3. Ensure that boiler feedwater oxygen concentration is consistently less than 10 µg/kg.
4. Monitor the copper concentration in boiler feedwater and other areas of the circuit routinely. The combination of ammonia and oxygen is highly corrosive towards copper and its alloys at pH values over 9.3.
5. All pH variations would damage the magnetite passivation layer and result in the re-depositing of a porous insulating magnetite layer, which would provide the substrate for future corrosion.
6. The primary cause of tube failures in boilers is sludge formation and accumulation due to low circulation rates.
   - The tubes' failures from overheating in the high heat flux zones may occur if there is a significant accumulation of deposits on the boiler tubes.
   - Deposits provide the opportunity for the collection and concentration of acidic ions and leading to corrosion failures.
7. The boiler system must be adequately protected when the plant is not operating. A nitrogen purge must be placed on the boiler system as soon as it is secured and mechanically intact and maintain this nitrogen purge until start-up.
8. The feedwater pipework should be inspected during every major turnaround for signs of corrosion.

For new boiler equipment

1. During repair or manufacturing of new bundles, care should be taken to procure all components from proven manufacturers under third party inspection from a reputed agency.
2. The vendor must ensure that a nitrogen blanket has been established on the shell and tube side before dispatch. When on-site, ensure that spare tube bundles are always kept under a positive nitrogen atmosphere to prevent the formation of rust and corrosion products at

tube surfaces. Just filling and sealing the bundles with nitrogen may allow "sweating" to occur.

3. Be careful while inserting the tube bundle into the pressure shell as the clearance between the shell shroud and baffle is very low. If the bundles get tilted while pushing into the shell, the peripheral tubes get rubbed or weakened because of the bundle weight.
4. Take extreme care while carrying out any repairs in the boiler drum to ensure that no foreign materials like electrodes, nuts, bolts, or other particles fall into the boiler tubes. It is evident from the failure history that two failures occurred while starting the plant after the shutdown.
5. When plugging the tubes, make sure that the plug length covers the tube sheet thickness and the top thermal shield baffle so that the hot process gases do not come in contact with the tube sheet.
6. Inadequate water circulation due to uneven flow in risers causes tube overheating, leading to rupture.
7. Use the best water quality for hydrotesting (feedwater quality) and ensure that it is removed from inside the tubes after hydrotesting. The tubes should be blown, dried, and cleaned thoroughly before boxing-up.

Conclusions

It is not possible in this paper to provide specific recommendations to cover all possible boiler designs, operating pressures, and internal chemical treatments. These need to be considered for each plant based on the specific conditions – design, age, operating pressure, water quality, etc.

However, some key conclusions relate to the safe and reliable operation of Ammonia plant WHB systems that can be drawn by comparison with utility boiler operation, which may reduce failures.

1. The process WHB design is not inherently flawed, but it does demand very good attention to the water quality, the internal chemical treatment, and rapid response to upset conditions.
2. The design of the boiler may lead to poor water circulation within the tube bundles. The use of internal stiffening of the tubesheet may exacerbate this.
3. Most online tube failures occur in a very narrow section of the boiler immediately downstream of the ferrules, where the heat flux is at its highest. For the boiler to be reliable, the tubes in this section of the boiler must be kept very clean with no deposits accumulation.
4. The higher the percentage of make-up water and recovery of process condensate, the more difficult it is to produce the consistently high-quality boiler feedwater required. Changes to the gas supply or process condensate stripping would add to these challenges.
5. The internal chemical treatment is not designed to compensate for poor feedwater quality. It should be used in the minimum quantities capable of meeting the specified control limits. Improving the feedwater quality would allow the use of simpler internal chemical treatments.
6. Existing international standards and guidance give good direction on setting control limits but do not specifically address the high local heat fluxes found in ammonia plant WHBs. If possible, operators should work toward better water quality and internal chemical treatment specification than indicated by the operating pressure.
7. Monitoring and response are the keys to good results. This starts with representative samples, an accurate analysis conducted by skilled analysts using appropriate analytical techniques, but also requires an accurate and timely response to upset conditions. In too many cases, this activity is not important or urgent, resulting in compliances that fall well

below the 95% limit, which should be considered the minimum standard required. Damage to boilers should be considered to start as soon as the upset condition occurs, be it pH, phosphate, chloride, etc. A policy of zero tolerance to upsets is preferred.

8. Dissolved oxygen is often seen as the major enemy of feedwater systems, and the concentration must be monitored, ideally online. Yet, in our desire to protect the feedwater system from dissolved oxygen, it has become common practice to overdose chemical oxygen scavengers, leading to an excessively reducing environment that causes FAC, which is worst at water temperatures 150°C (300°F) and when the feedwater pH is low. FAC has likely been a significant but unseen and unmeasured contributor to higher pressure boilers' deposits leading to many tube failures. The best practice is to add oxygen scavenger when needed because of poor feedwater quality and then only add the minimum necessary.
9. The boiler's most significant risk comes from accumulating deposits on the boiler tubes, which may lead to overheating failures or corrosion failures due to the concentration of trace impurities underneath deposits. It does not require out of specification feedwater for these failure mechanisms to occur.
10. While prevention of deposition is the preferred option, it may be necessary for installed boilers to adopt a routine cleaning policy to remove existing deposits. This must be carried out to a very high specification to ensure the complete removal of any deposits. The cleaning frequency would be determined by an assessment of the feedwater quality and compliance to control limits.

Finally, it should be considered that boilers have long memories and are unforgiving. Damage may be caused over an extended period before boiler tubes fail and are the cumulative impact of poor control. Just because the boiler has not failed (yet) should not be taken to indicate that the present operation is good enough. All boilers look okay until they fail.

References

[1] Duisters and Savelkoul, DSM Agro, "Corrosion damage in Waste Heat Boilers; Major root causes and remediation AIChE Ammonia Symposium 2005, Paper 006.

[2] EPRI (Electric Power Research Institute) Comprehensive Cycle Chemistry Guidelines for Fossil Plants 2011. Product ID 1021767

[3] VGB (VGB Powertech e.V.) Feed Water, Boiler Water and Steam Quality for Power Plants / Industrial Plants (Formerly R 450 Le) ISBN: 978-3-86875-381-3

[4] IAPWS (International Association for the Properties of Water and Steam) Phosphate and Caustic Treatments 2015; Volatile Treatments 2015

[5] Dooley, Shields and Shulder, Structural Integrity Associates Inc., PowerPlant Chemistry 2008, 10(10)

Table 1

AIChE Ammonia Safety Symposium: Incidents related to boiler water failures

These were included in papers printed in the AIChE Ammonia Technical Manuals, and using the "year reported", the source paper can be located in the "Ammonia Technical Manual."

Item	Year Reported	Incident	Root Cause	Mitigation
1	1973	Steam generating tube failure	Accumulation of boiler chemicals & low circulation	Increased circulation rate, changeover of the water treatment program
2	1985	Failure & repair of primary WHB.	Thermally-induced stresses due to liquid level fluctuations.	Coating the upper exposed area of the shell and use of oxygen scavengers in BFW.
3	1988	Problems in a process gas cooler	Poor BFW quality.	Cooler leaks were repaired.
4	1989	Magnetite layer in primary WHB and auxiliary boiler	The boiler leaked due to corrosion.	Boiler water pH lowered, blowdown rates increased.
5	1991	Failure of secondary reformer WHB	High thermal stress between tube-to-tube sheet joints.	Routine intermittent blowdown restored.
6	1991	Failure of two 100 bar BFW lines	High flow velocities caused erosion in BFW piping.	BFW piping was repaired.
7	1992	An incident on motive steam raising gas cooler	Crevice corrosion promoted by chlorides.	New design cooler installed.
8	1993	Retubing of reformed gas boiler	Boiler water deposits caused build-up on tubes.	Done in situ replacement of all tubes.
9	1994	Tube failure in Synloop WHB	Lack of water circulation.	Leaking tubes plugged, repaired.
10	1994	Ammonia WHB replacement	The boiler ran dry.	A new boiler of floating head type was installed.
11	1996	Damage of reformed gas fire tube boiler	High pH boiler water.	The boiler was repaired.
12	1998	Plugging of bayonet/scabbard tube WHB	A power failure caused damage to boilers.	Leaking tubes plugged.
13	2002	Auxiliary boiler failures	Poor water quality. Not adequately cleaned steam system.	All the BFW and steam system chemically cleaned.
14	2003	Secondary reformer WHB failure	Sludge deposits and corrosion under sludge.	Exchanger design modified to allow intermittent blowdown.
15	2005	Corrosion in WHBs	The insulation layer of deposits	Changed to polyamine dosing as an alkaline filming agent.
16	2005	Failure of 110 bar WHB	Poor BFW quality	
17	2005	Failure of RG reboiler	Blocked BFW sample point	Sample points cleared off
18	2007	Failure of WHB at Profertil	High O2 in BFW and low pH excursion	Leaking tubes plugged and improved BFW treatment
19	2012	WHB Failure	Deposits of corrosive products and rust	Leaking tubes were plugged and provided an air gap
20	2014	High-Pressure Process Boiler Failure	Corrosion associated with the deposits caused the thinning, which led to tube failure	The tube bundle was replaced with a spare one
21	2015	Synloop WHB failure	Chloride SCC	Replaced the equipment with improved materials

Failure of a Hot Walled Ammonia Synthesis Converter at Yara Sluiskil

Ammonia Plant C at Yara's Sluiskil site was shut down for a turnaround on October 17, 2019. As part of the Risk-Based Inspection (RBI) program, an inspection was performed in and around the shell and nozzle welds of the second synloop converter (R306) and crack indications of Hydrogen Assisted Cracking (HAC) were found between the man-way nozzle top head and bottom inlet/outlet nozzle. The root cause investigation concluded that the cracking phenomenon is the result of a combination of nitriding at the inner surface and hydrogen-induced embrittlement. The nitrided inner layer is prone to cracking when taking the equipment in and out of service and acts as a crack initiation point. Moreover, the hydrogen charged material exhibits a lower threshold stress resistance factor as well as a higher crack propagation rate with respect to uncharged material. The usage of a high strength material (3Cr type compared to 2.25Cr) also played a role in increasing the stress level at the cracked areas. Based on these results, a plan to develop criticality rankings the reactors in all Yara plants as well as a detection and monitoring strategy using NDE is under preparation.

R. Bosselaar, S. Cavalleri, M. Noueiri, J. Thoelen, F. Vaes
Yara International ASA

Introduction

This paper describes the Root Cause Analysis (RCA) performed on one of the ammonia converters at Yara Sluiskil as a follow-up to inspection activities performed during a 2019 turnaround. The ammonia plant was shut down as planned on October 17, 2019, for a scheduled turnaround. According to the Risk Based Inspection (RBI) program [1], the second synloop converter was inspected for internal corrosion, nitriding [2, 3] and hydrogen assisted cracking (HAC) [2, 6]. From previous experience and considering the combination of operating conditions and materials of constriction all were expected as degradation mechanisms. The inspection, utilizing Ultrasonic Time of Flight Diffraction (UT TOFD) found several cracks, mainly at the welds between the manway nozzle and the top head and also at the bottom inlet/outlet nozzle. Maximum crack depth was about 40% of the total thickness. The cracks, perpendicular to the weld line, were uniformly distributed on the entire circumference with a distance between them of about 100 mm (3.9 in). Inspections performed during the previous turnaround in 2016 did not find any cracks, suggesting the cracks grew entirely during the last three years of operation.

Repair procedures were discussed, but the severity of the cracks and the lack of available welding material could not guarantee a reliable repair. Therefore, the converter was taken out of service and bypassed. The plant has been running stable at 80% of Maximum Design Rate (MDR) since then and the converter is scheduled to be replaced during the next turnaround in 2022. Although HAC has been assigned as the failure mode and, based on previous experiences and internal models for susceptibility assignment (this vanadium modified alloy material is known to be more prone to embrittlement and cracking phenomena than standard P22 material), the extent and the speed of cracking was greater than expected.

With the converter out of service, some material was available to perform a failure analysis and begin Root Cause Problem Solving (RCPS). The root cause of the failure was assessed, critical aspects that can affect similar equipment have been identified, and preventive/monitoring actions have been defined to mitigate the risk of similar failures for the other Yara reactors.

Background and Equipment Data

Yara International ASA is a leading chemical company that converts energy and nitrogen from the air into essential products for farmers and industrial customers. As a global supplier of mineral fertilizers and agronomic solutions, Yara has offices and production sites all over the world. One of the largest sites is Yara Sluiskil, located in the Netherlands. Three of the plants located in Sluiskil are ammonia plants, Unit C, D, and E. All three plants use the Brown & Root purifier design. The total capacity of the three plants is 5.385 MTPD (5940 STPD). The subject reactor of this analysis is the second inline within the ammonia synthesis loop of Unit C. The main equipment data is noted in Table 1.

Installation year	1996
Shell material	SA-336 Gr. F3V (3Cr-1Mo-0.25V)
Medium	Ammonia synthesis gas: H_2:63%, N_2:19%, NH_3:14%
Operating Pressure	170 barg (2466 psig)
Design Pressure	190 barg (2756 psig)
Operating Temperature	380°C (716°F)
Design Temperature	425°C (797°F)

Table 1. R306 Equipment Main Data

In Sluiskil, there are six comparable converters, all built by the same manufacturer, Japan Steel Works. The oldest convertors have been in service since 1983, without any damage. The cracked convertor is the newest one, put in service in 1996. It is the only one made with a Vanadium enhanced material.

After finding the cracks on R306 several solutions were discussed. Repair of the convertor was not feasible because of uncertainty of the quality and duration of repair. Also, no adequate spare or mothballed convertor was found on short notice. Therefore, it was decided to bypass R306 while a new convertor was fabricated.

As indicated by process simulations, operating the plant while bypassing the R306 was feasible (see PFD of the hot loop in Figure 1). However, this results in lower energy efficiency and reduced production capacity. The main effect is less conversion per pass, from 21% to 16%, which is compensated somewhat by a higher loop circulation rate.

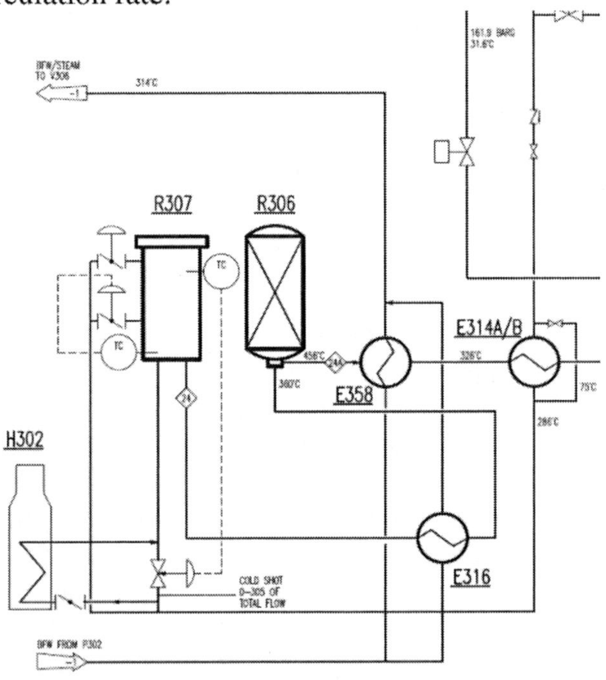

Figure 1. PFD Extract of the Hot Loop Portion of the Synloop.

The easiest and quickest way to bypass the R306 would be the simultaneous bypass of the reactor and its effluent Waste Heat Boiler (WHB), E358, as only piping work is required. However, this option would recover far less energy from the ammonia synthesis reaction, leading to lower HP steam production, and further reducing production capacity and energy efficiency. Therefore, it made sense to keep E358 WHB in service.

Keeping E358 in service created some technical complications. The connection between R306 and E358 had to be cut and capped. This connection is relatively short with limited space for workers, as it is obstructed by the concrete support of R306. Furthermore, a new complex suspension was needed to support the E316 as it is normally suspended under the reactor. Some of the E358 internals were also removed. A new hemisphere cap on the E358 would take a long time to fabricate, so a manhole lid from a different spare convertor was reused (see Figure 2. for a picture of E358 after R306 removal).

Figure 2. Picture of E358 after R306 removal.

A HAZOP (HAZard and OPerability analysis) was performed to validate all the modifications. Some of the identified actions were:

- Verify the new conditions with the Original Equipment Manufacturer (OEM).
- Prevent condensation in the long medium pressure (MP) steam export/import line.
- Recalculate heat exchanger vibrations.
- Update operating procedures and retrain operators.
- Update the RBI program.
- Calculate the remaining strength of R306 to ensure it could be safely lifted.

The activities to bypass R306 took more than five weeks, leading to a two-week overrun of the planned turnaround duration. After the plant start-up, a plant load of 80% of MDR was achieved at a higher energy factor, as expected. The steam balance changed, with MP steam continuously imported rather than exported. The main bottleneck was synloop pressure. While optimizing the R307 inlet temperature, the first bed temperature dropped too low and the convertor reaction died. After a couple of hours ammonia production was reestablished.

The new R306 is being fabricated with cold wall design and will be installed during turn around at the end of 2022.

Inspection Findings

According to the RBI program, the following inspection actions are recommended for R306 to cover the HAC degradation mechanism:

- Time of Flight Diffraction (TOFD): 100% of circumferential welds (shell/heads)
- Phased Array Ultrasonic Testing (PAUT): 50% of all the nozzle insert welds.

During the turnaround of 2016 no indications were found, i.e. no detectable defects were reported with the same NDE techniques. However, during the 2019 turnaround several transverse cracks were detected during the standard HAC inspections.

With reference to Figure 3, the majority of the cracks (29) were found in the insert weld of the top manway (M1). In addition, nine cracks were found at the bottom outlet nozzle (N1) insert to weld. The orientation of the cracks discovered in the insert welds is transverse to the weld seams, starting in the weld material and then propagating into the base metal. The maximum crack depth was approximately 65 mm (2.6 in) in a nominal wall thickness of 134 mm (5.3 in), starting from the inside of the vessel. The normal NDE scope (TOFD, PAUT, and Manual UT) is mainly focused on longitudinal cracks in the weld/ heat affected zone. In this case; however, transversal (radial) inspection paths were also used.

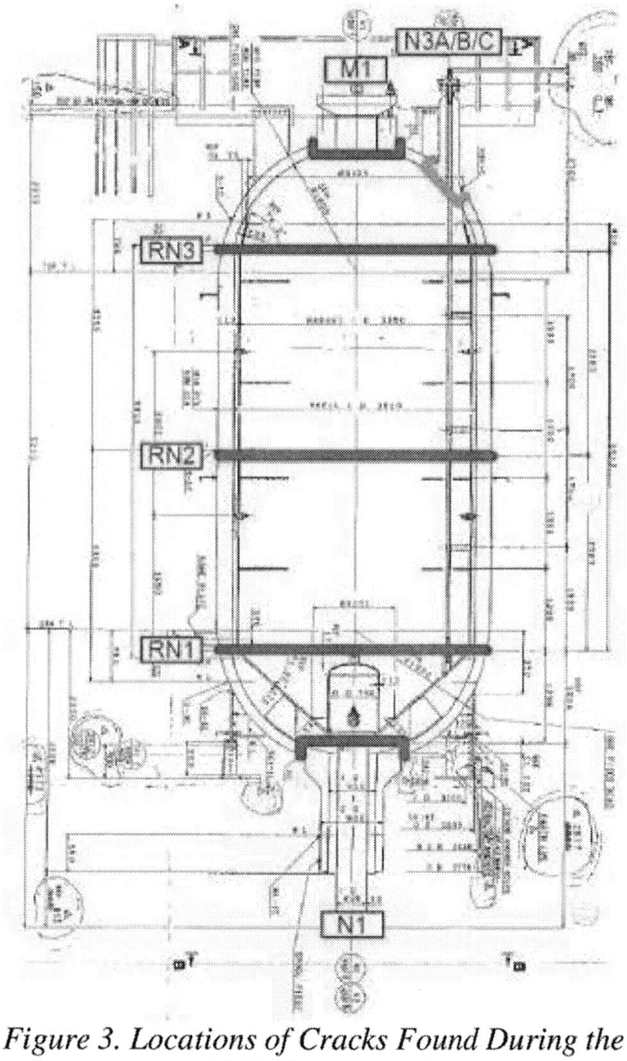

Figure 3. Locations of Cracks Found During the 2019 Turnaround

In Figure 4, the resulting images of TOFD and PAUT (parallel and non-parallel scanning direction) of the M1 weld are shown. In the areas with high noise levels (found with TOFD and PAUT at same locations), the velocity ratio, backscatter, and spectral analyses have also been executed.

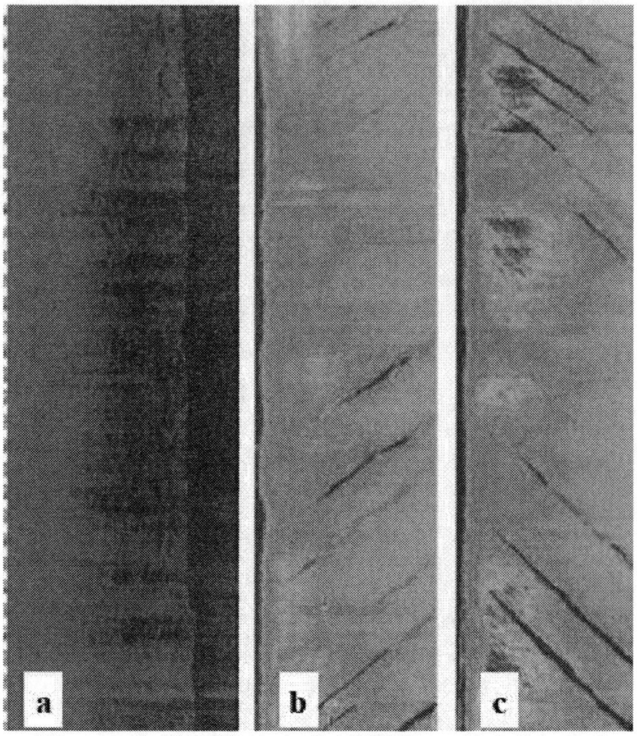

Figure 4. TOFD (a), PAUT Parallel (b) and Transversal (c) Image of M1 Weld

Metallurgical Investigation

After the converter was taken out of service, some material containing cracks was cut out from a section of the top manway for a failure analysis and metallurgical investigation. Tensile and Charpy-V impact test specimens were machined out to evaluate the mechanical properties and extract transition curves. Specimens were submitted to Element Materials Technology Rotterdam b.v. Results of the investigation are summarized in following paragraphs.

Visual and stereomicroscopic examination

The characteristics found at the crack surfaces point to brittle cracking initiated at the inside of the convertor. Figure 5 shows the section of the convertor used for failure analysis with indication of three cracks found with a distance between them of about 100 mm (3.9 in).

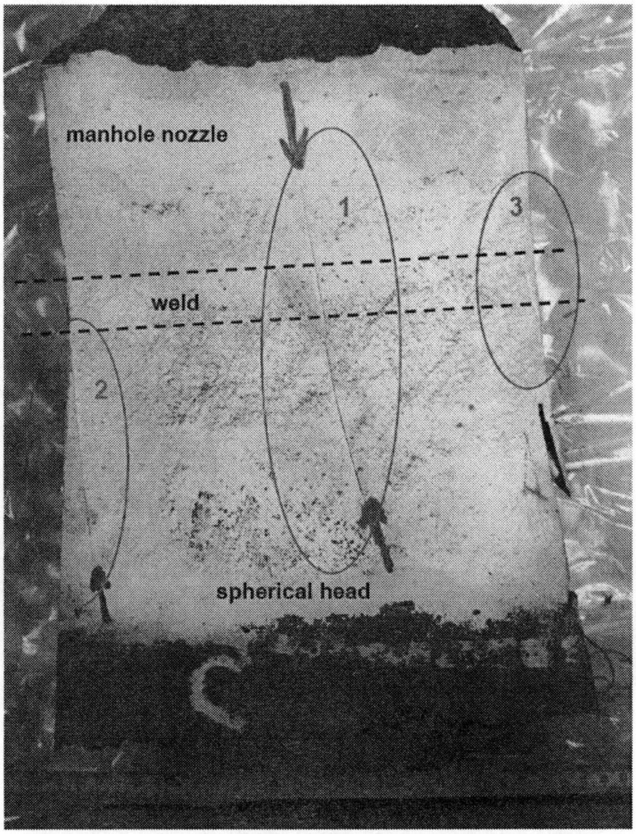

Figure 5. Portion of the convertor with Indications of the Cracks

Scanning Electron Microscopy (SEM) examination

SEM examination revealed no oxides or corrosion products covering the crack surface. Energy Dispersive X-Ray Analysis (EDAX) examination revealed that only the elements composing the convertor material type F3V were detected, though an increased nitrogen (N) content was present. Some titanium (Ti) was also identified near the crack tip in the base spherical head material. It must be noted that boron (B) cannot be detected by EDAX.

Chemical analysis – Optical Emission Spectroscopy (OES)

The OES analysis concluded that the base material satisfied the requirements for F3V according to ASTM A336, as specified. The filler weld metal matched material type F3V (lower in carbon but higher in manganese). It also revealed that the carbon (C) content at the inner surface of the converter was significantly decreased (decarburization), and the nitrogen (N) content significantly increased. This points to chemical reactions with the reactor contents under operating conditions as expected, considering the combination of material and process fluid.

Mechanical testing

Tensile testing showed that the base materials, as well as the weld metal, met ASTM A336 code requirements with respect to yield and ultimate tensile strengths, as well as to elongation. Impact testing concluded that the weld metal of the convertor had ductile properties and was not sensitive to brittle cracking at +20°C (68°F) due to a sudden (over)load. However, considering the original material certificates, a comparison between the ductile-to-brittle transition curves of the "as-built" condition and current "aged" material highlighted an evident temperature shift. Figure 6 shows the transition curves extrapolated from impact test results of the original certificates and from the last tests performed on the "aged" material. Temperature shifts of about 40°C (72°F) and 60°C (108°F) have been observed respectively for the average Charpy V-Notch value (CVN) (about 100 J) and for the upper-shelf CVN at about 150 J.

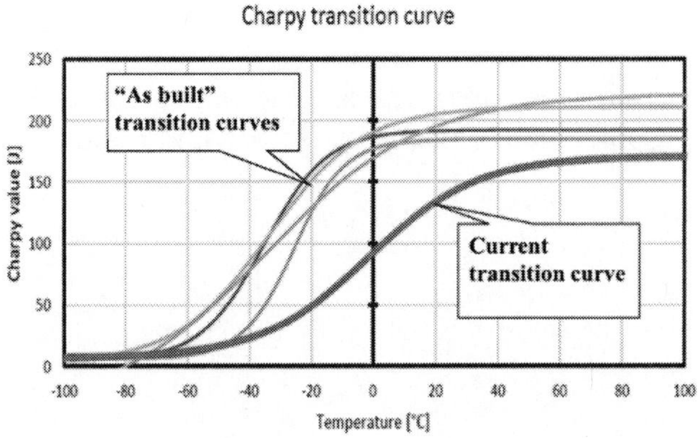

Figure 6. Plot of Charpy Transition Curves for Material in "As Built" and "Aged" Conditions

Microscopic examination

The cross section taken over the root of the weld (1R), corresponding with the inner surface of the convertor, revealed the presence of many fine cracks. All of these cracks were positioned in the root weld metal and heat affected zone up to a depth of approximately 2.0 mm (0.08 in). The cracks ran in an interdendritic/intergranular manner into the weld metal. The inner surface revealed small corrosion pits up to a depth of 60 μm and locally some intergranular cracks up to a depth of 50 μm.

Hardness measurements

ASTM A336 does not specify hardness for material type F3V. In general, the maximum hardness requirement, according to ISO 15614-1, for welded ferritic steels such as F3V is 350HV in the heat-treated condition. It is therefore concluded that the hardness of the base materials, heat affected zones, and the weld metal of the convertor satisfy the general requirement for the base and welded material type F3V. Hardness testing up to a depth of ~0.5 mm (0.02 in) along the inner surface shows that the hardness had increased significantly. This appears to relate to the high nitrogen (N) content in this zone, although the carbon (C) content had decreased. Furthermore, hardness testing along the crack surface indicated increased hardness over the entire crack depth. The hardness profiles through the thickness for the base metal and the weld metal in correspondence of the crack is shown in Figure 7.

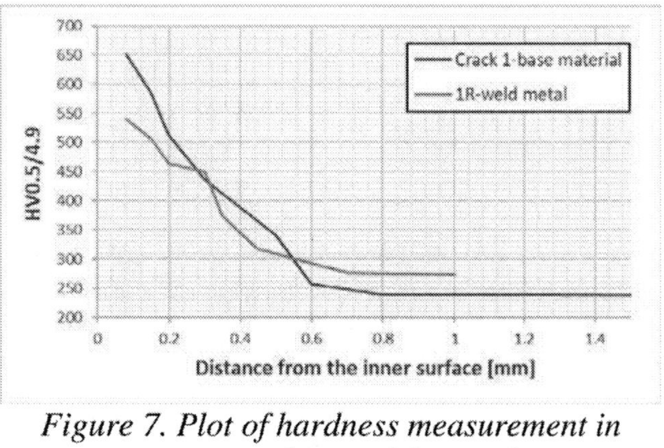

Figure 7. Plot of hardness measurement in correspondence of the crack

Mechanical Analysis

A mechanical analysis of the vessel was performed to evaluate the temperature and stress distribution for different operating conditions. This assessment was focused on the start-up and shutdown phases. Using the Fracture Mechanics approach, the fracture toughness at operating conditions was estimated, and a Fitness for Service (FFS) analysis completed to establish the criticality of the deepest crack with respect to brittle fracture. Finally, a crack propagation model for low alloy steels in hydrogen charged conditions was applied to estimate the critical flaw dimension for propagation. For the stress analysis, a typical start-up/shutdown curve has been extracted from data of the last five years of operations as shown in Figure 8, where, for convenience, the shutdown phase is plotted just after the start-up.

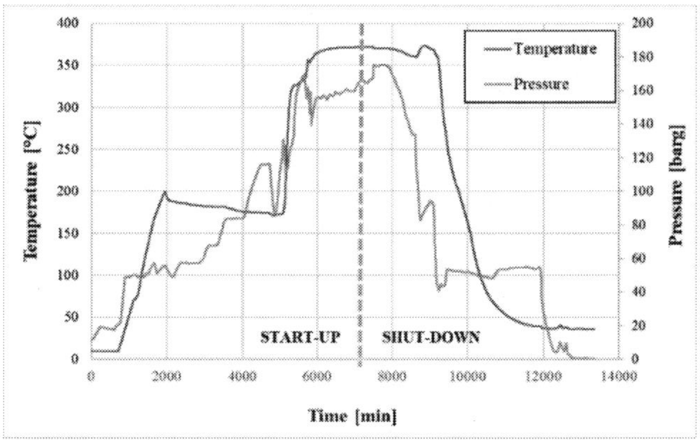

Figure 8. Typical Start-up/Shutdown Diagram

FFS analysis

The fracture toughness has been estimated according to the API 579 procedure using Charpy-V-Notch impact tests. Dynamic fracture toughness (K_{IR}) has been considered as it is a lower envelope to dynamic initiation and crack arrest toughness data. This curve represents a conservative estimate of the resistance of the material to rapid unstable crack propagation, and can be used to estimate the toughness of the steel containing dissolved hydrogen.

The top manway portion of the converter was considered for the calculation. The stress levels from both pressure and thermal loads were evaluated by means of a 2D axisymmetric Finite Element (FE) model. According to API 579, the residual stresses have been estimated as:

- 0.3 YS parallel to the weld
- 0.2 YS perpendicular to the weld

Having a yield stress at room temperature of 410 MPa, the residual stress parallel to the weld (which is the stress opening the cracks transversal to weld seam) is 123 MPa. A plot of the pressure thermal and residual stresses is shown in Figure 9, where it is evident that the weld residual stress plays a major role.

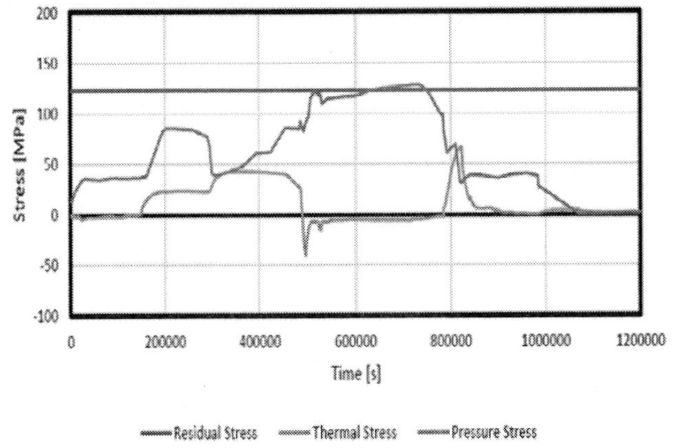

Figure 9. Plot of Primary and Secondary Stress Transverse to the Weld

Applying a FFS level 2 approach, as referenced in API 579), the failure assessment diagram (FAD) must be developed for all the operating points considering the start-up and shutdown. The stress intensity factor was calculated by applying the analytical solution for semielliptical internal cracks on spheres. The analytical solution has been also checked with a detailed FE analysis of the top manway section where a reference crack has been explicitly modeled. The assessment points plotted on the FAD are shown in Figure 10. Both a 1 mm (0.04 in) deep crack and the 65 mm (2.6 in) crack found in the reactor were considered for the assessment. The results pointed out that static failure is not expected during normal operating conditions as all the assessment points lay in the acceptability region: stress intensity ratio (K_r) lower than 1 and load ratio (L_r) lower than 1.2.

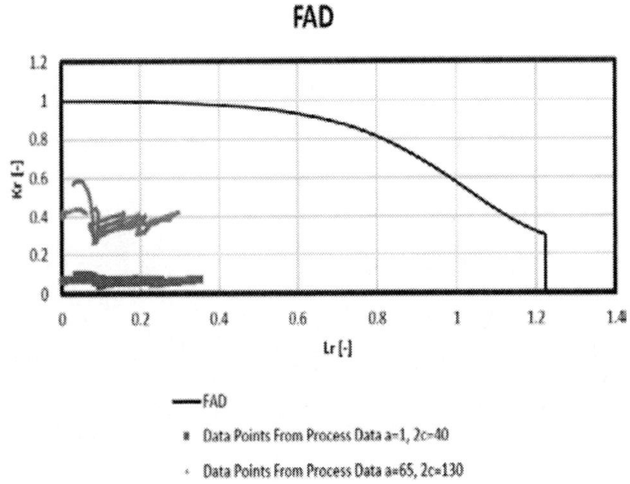

Figure 10. FAD for operating start-up and shutdown

Together with the static assessment, a crack growth analysis was also performed. The scope of the analysis was to establish the maximum crack depth below which no crack propagation is expected. The procedure reported in WRC 562 [8], the applicable welding bulletin, to assess the crack growth over time (da/dt) as the result of the hydrogen assisted cracking (also called "slow stable crack growth") has been applied. For Cr-Mo welded material, the threshold stress intensity factor (K_{th}) below which no crack propagation occurs is assumed to be 19 MPa$\sqrt{m}$ (600 MPa$\sqrt{mm}$). Because the available data is limited to P11 and P22 grades and no references are given for F3V, a conservative approach has been used.

Crack growth occurs if the combined primary and secondary stress intensity are above the threshold and the operating temperature is lower than a reference temperature (T_{IH}), which can be assumed equal to 110°C (230°F) for the material under investigation and a dissolved hydrogen content of 3.5 ppm. With such a procedure, a limited crack depth was calculated for different operating pressures as shown in Figure 11.

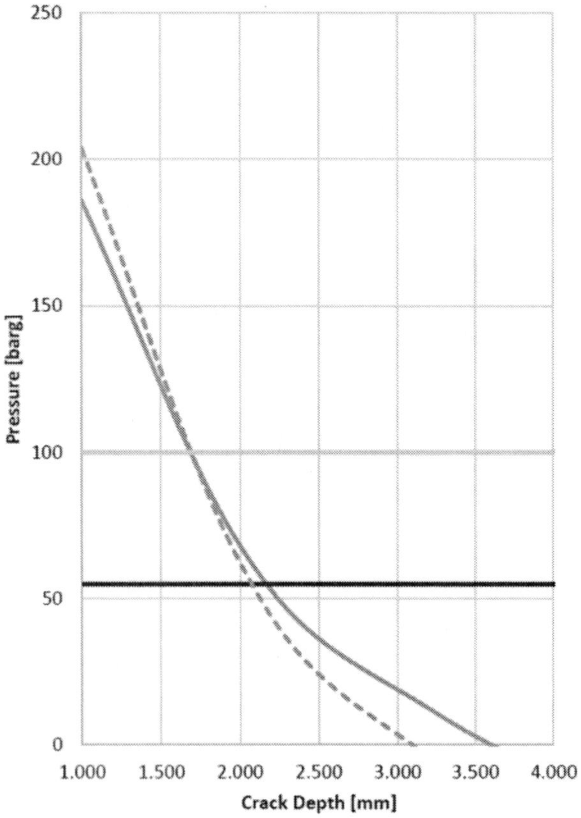

Figure 11. Limiting Flaw Size vs. Operating Pressure for Crack Growth

Lines at 55 barg (800 psig) and 110 barg (1450 psig) represent the maximum pressure levels occurring at start-up and shutdown with a correspondent temperature lower than 110 C (230°F). The calculated limiting crack sizes are summarized in Table 2. The analysis of historical data highlighted that during start-up/shutdown, the combination of temperature below 110°C (230°F) and pressure above 55 barg (800 psig) and 100 barg (1450 psig) respectively occurred more than incidentally.

Maximum Pressure (T < 110°C) (T < 230°F)	**Limiting Flaw Depth**
55 barg (800 psig)	≈ 2.1 mm (0.08 in)
100 barg (1450 psig)	≈ 1.7 mm (0.07 in)

Table 2. Limiting Flaw Depth at Maximum Pressure Levels Occurred at Startup/Shutdown with T <110 °C

Discussion

Based on the analysis performed, hydrogen assisted cracking (HAC) is considered the driving failure mode for the cracking. Under the given process conditions, ammonia dissociates into atomic N and atomic H, diffusing in the material. Since nitrogen is a larger atom, its diffusion is more restricted to the surface layer, whereas hydrogen diffuses more freely into the material. As a result, the formation of a very hard and brittle nitride surface layer leads to the development of small cracks during temperature changes. Such cracks then propagate due to embrittlement of the material at the crack tip, due to diffusion of H into the crack tip, which results in a crack propagation at temperatures below 110°C (230°F). These conditions often occur during start-up and shutdown conditions.

Based on the increased hardness of the crack flank, the crack must have propagated since the last start-up of the reactor in 2016. Furthermore, F3V material can allow diffusion of a higher amount of H compared to 2.25 Cr material and releases H less than the 2.25 Cr material, making this material much more susceptible to cracking.

The proposed sequence of events is as follows:

- Hydrogen (H_2) in process gas dissociates, and atomic-H diffuses into steel. The hydrogen concentration from the normal operating conditions (T=365°C (689°F), Hydrogen partial pressure, pH2= 108 bar (1570 psi)) for F3V-steel is calculated to be 2.1 ppm.

- Ammonia in process gas dissociates, and extra atomic-H diffuses into steel (total concentration is expected to be 4-5 ppm).
- The dissociated ammonia also allows atomic-N to diffuse into the steel, which causes hardness increase to about 0.5-1.0 mm depth (i.e. nitriding).
- Small cracks formed due to "over-nitriding" (most likely promoted by blistering $2H \rightarrow H2$).
- After ~20 years of service the nitriding depth and small cracks progressed to the extent that perpendicular flanks of these cracks are larger than the critical crack size.
- Both atomic-H and atomic-N are blocking dislocation movement (i.e. embrittlement) when T is below 110°C (230°F) during start-up and shutdown.
- Cracks propagate due to nitrogen-assisted Hydrogen Embrittlement.

Recommendations

As the decomposition of ammonia is intrinsically connected to the process condition, this will also occur in equipment made of 2.25 Cr material. Although less susceptible, HAC will be a failure mode and embrittlement of the material below 110°C (230°F) is present in any case. Therefore, the operating conditions must be controlled, meaning that the equipment may not be pressurized below 110°C (230°F). This can be refined by creating MPT (Minimum Pressurizing Temperature) curves for all equipment operating under similar conditions.

According to the conclusions above, the following preventive action have been planned;

- Adjust inspection planning for all converters in Yara making sure that TOFD and PAUT are performed in order to find cracks in all orientations.
- Establish MTP curves for all converters in Yara and adjust operating procedures if required.

Acknowledgments

The failure analysis process was performed with the support of Element Material Technology B.V, Sintra Engineers, and Materials and Corrosion Know-How.

References

[1] *Risk-Based inspection Methodology*, API RP 581.

[2] *Damage Mechanisms Affecting Fixed Equipment in the Refining Industry*, API RP 571.

[3] Marsch. "Nitriding of Steels After 5 Years in Ammonia Synthesis Service", AIChE Ammonia Symposium (1981).

[4] Alban , Harrison, and Jones. "Cracking in an Ammonia Converter Effluent / Boiler Feedwater Heat Exchanger", AIChE Ammonia Symposium (1990).

[5] Wagner, ,Heuser, and Heinke. "Hydrogen Attack in 2 ¼ Cr – 1 Mo Steel Below Nelson's Curve", AIChE Ammonia Symposium (1991).

[6] WRC Bulletin 535, "Toughness, Fracture and Fitness For Hydrogen Service" 2009.

[7] WRC Bulletin 562, "Recommendations for Establishing the Minimum Pressurization Temperature (MPT) for Equipment" (2017).

Prevention of Stress Relaxation Cracking in Ammonia Plants - The Haldor Topsøe Approach

Stress relaxation cracking is a failure mechanism of concern in the temperature range 500°C – 700°C (932°F – 1292°F), when welded joints or cold formed areas of precipitation strengthened austenitic stainless steels and nickel alloys cannot withstand the inelastic strain imposed upon the material. This paper presents the Haldor Topsøe approach to minimize residual stresses and the risk of SRC in terms of type of the alloy's chemical composition, grain size, design, welding and postweld heat treatment based on a thorough literature study and various plant experiences.

Maria J.L. Østergård Ph.D., Anette N. Hansson, Ph.D.
Haldor Topsøe A/S

Introduction

Stress relaxation cracking (SRC) is a failure mechanism of concern in the temperature range between 500°C and 700°C (932°F – 1292°F), when welded joints or cold formed areas of precipitation strengthened austenitic stainless steels and nickel alloys cannot withstand the inelastic strain imposed upon the material. In ammonia plants, SRC has been observed mainly in steam reformer inlet systems and at T-joint welds and the welds between straight pipes and elbows. It is often observed after approx. 0.5 to 2 years in service.

SRC is also known as reheat cracking, stress relief cracking, stress relaxation cracking, creep embrittlement cracking, post weld heat treatment cracking, stress induced cracking, and stress assisted grain boundary oxidation (SAGBO) cracking.

The fracture is often brittle in appearance and occurs in cold worked regions, frequently in the proximity of welds. The cracks are located along grain boundaries due to the formation of fine precipitates at elevated temperatures. This results in the loss of ductility and cracking when strained. Often, but not always, the cracks are oxidized and contain a metallic filament at the core of the crack.

Much research work and joint industrial projects have been published and are being conducted to get a better understanding not only of the failure mechanism itself, but especially, how to prevent it. This has resulted in multiple preventive guidelines for alloy selection and Post Weld Heat Treatments (PWHT), including API 942-B (2017), and the most recent version of API 571 (2020), which describe the factors affecting the mechanism, design considerations, and specific temperature ranges for both susceptibility and

PWHT of different austenitic stainless steels and nickel-based alloys.

However, there are still discrepancies in terms of which alloy to select for each condition, which parameters to control, the role of grain boundary oxidation in the mechanism and whether PWHT prevents SRC or damages the weld so it becomes more prone to SRC.

The Haldor Topsøe approach

As a world leading licensor of multiple technologies and provider of proprietary equipment, Topsøe has studied in depth critical factors affecting SRC, and has been able to utilize austenitic stainless steels and Ni-base alloys under conditions where there is a potential risk for SRC. For example, the transfer line between the hot reformer preheat coil and the reformer inlet in plants with a steam-methane-reformer (SMR) and similar high-temperature piping systems Topsøe has numerous running references of using 321H stainless steel with maximum pipe diameter of 24 in (0.6 m) and 41 mm (1.6 in) wall thickness dating back to 1997.

Furthermore, Topsøe participates actively in industrial partnership projects to continue developing knowledge about SRC.

This study, which is summarized in the following sections, has investigated the causes, mechanisms, morphology, and the preventive measures regarding the critical factors:

- type of material (chemical composition, impurity elements)
- large grain size
- high residual stresses from fabrication (cold working, welding)
- large pipe sizes
- heavy section thickness
- notches and stress concentrators
- weld metal and base metal strength; and
- welding and heat-treating conditions.

Causes, Mechanisms and Morphology of SRC Failures

Although SRC failures have been detected in low alloy steels (see [15] for description and preventive measures), the current paper deals with austenitic stainless steels and Ni-base alloys only.

According to literature, precipitation strengthened austenitic stainless steels and Ni-base alloys are susceptible to SCR. Most of the published work is on alloys 304H, 316H, 321/321H, 347/347H, 800/800H/800HT, 601, 602CA, 617, 625 and 825. And some publications establish a ranking of the susceptibility of the alloys to SRC. [13]

Cracks are located in zones with high level of residual stresses, which points as high residual stresses as probably the most important cause of SRC. [1, 2, 6, 7, 19, 20, 21, 22, 23, 24]

For example, in [1], the stress relaxation cracking of alloy 800HT was located in an overstressed reducer welded to a pigtail tube. It was found that the risk of stress relaxation cracking was larger than expected due to the high level of residual stresses from machining, contributed by a factor of large grained Alloy 800 HT material and sharp shoulders acting as stress riser.

In [6], cracks in alloy 800H outlet header were caused by high residual stresses from fabrication (cold working, welding) or by cyclic deformation during service at temperatures in the range 550°C to 760°C (1022°F – 1382°F), resulting in relaxation processes. In this temperature range, the critical nucleation radius for carbide precipitations is very small. Fine carbides, mainly Cr-carbides, are precipitated within the grains at dislocations after short periods of time leading to strengthening of the grains. At the grain boundaries, coarse Cr-carbides are present, leading to Cr-depletion in their vicinity. Consequently, precipitation free zones having low strength occur aside of the grain boundaries. Because of the relative difference in strength between the grains and the grain boundaries, relaxation will take place at the grain boundaries leading to crack formation at relatively low overall deformations by relaxation processes. The crack formation starts stepwise on both sides of the grain boundaries in the weakened zones, producing a metal filament in the middle, which possibly combined with strain induced oxidation processes becomes rich in Ni. It should be mentioned that gamma prime precipitation and minor element contents, leading to a weakening of the grain boundaries, are also discussed as reasons for the crack formation.

Welded structures are particularly susceptible because of the potential for high residual stresses (up to the yield strength of the base metal) resulting from welding operation that have not been post weld heat-treated. [7] Cracks are located on the grain boundary. The crack zone is usually hard, with a Vickers Hardness greater than 200 HV. [11, 14] The hardness increase is attributed to intermediate temperature-aging reactions often accelerated by prior cold work.

The same mechanism based on strain-induced precipitation is proposed in [10, 19, 20, 21, 22, 23, 24] for other alloys.

In addition to residual stresses and the microstructural changes during welding and service at high temperatures, non-optimal large grain size distribution has also been identified as a main contributor to SRC failures. [2, 9, 14]

As a summary, SRC failures in austenitic welded joints may occur during service when the weld joints:

- have high level of residual stresses
- have non-optimal grain size
- operate at metal temperature is between 500°C and 700°C (932°F – 1292°F)

Typical SRC fingerprints are:

- The cracks are present in the Heat Affected Zone, weld metal or cold deformed areas with Vickers Hardness higher than 200 HV.
- The cracks are always located on the grain boundaries, are very often oxidized, and might present a Ni-rich metallic filament (Figure 1).

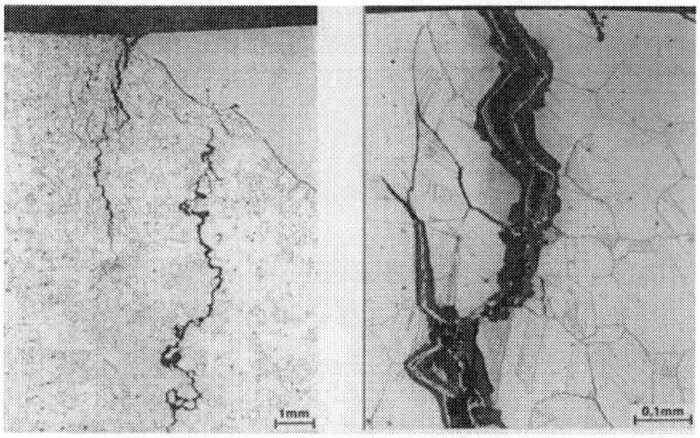

Figure 1. Typical SRC Cracks [24]

Prevention Measures Reported in Literature:

<u>Design measures</u>

Among the preventive design measures the following has been reported:

- Use uniform pipe thicknesses and diameters and avoid the use of reducers in inlet pigtail designs. [1]
- Limit the amount of bending of pigtails, limit thermal expansion and related bending loads of the whole system, and ensure that manifolds and pigtails have the appropriate support. [2]
- For heavy wall butt weld joints, thickness over 12 mm (0.5 in.), minimize restraint during welding, minimize stress concentration in joint designs and avoid metallurgical notches in welding operations (e.g. lack of penetration, lack of fusion, etc.), since such notches usually give rise to cracking of heat affected zones. [13]
- Avoid sharp changes in cross section, such as short radius fillets or undercuts that can give rise to stress concentrations, avoid long-seam welds which are particularly susceptible to mismatch caused by fit-up problems or by "peaking" due to the plate edges not being properly rolled. [15, 19]

Fabrication measures

Regarding alloy composition, it has been pointed out the importance for alloys 800/800H to limit the Al+Ti content to less than 0.7% in the base material and weld metal will minimize the risk of SRC. [13, 15] Special care shall be taken with "dual certified" 800H/800HT alloys, to ensure that the chemical composition meets the Al+Ti concentration constrain.

The grain size has an important influence on the high temperature ductility and on the SRC susceptibility, with coarse grain materials being more susceptible as larger grain sizes result in a less ductile Heat Affected Zone (HAZ). [13, 19] In practice it is recommended to specify a small grain size (smaller than ASTM 3.5) which can be achieved by hot rolling and forging practices.

For pigtails specifically, it has been recommended to limit or exclude cold working during fabrication to help control the grain size of pigtail material coarser than ASTM No. 5 but not larger than ASTM No. 1 and to control size variability to be no more than 2 or 3 ASTM grain size numbers. [2]

For welding, low heat-input welding techniques are recommended to minimize the risk of welding-induced grain growth. [13] Furthermore, for 18Cr-10Ni stainless steel types, the welding consumable E16.8.2-15 is recommended due to its superior SRC, creep-rupture ductility, and solidification cracking properties as well as less tendency for brittle sigma phase formation. [13, 15]

For alloy 800H specifically, small diameter filler metal rods, such as 2.5 mm (0.1 in), a maximum interpass temperature of 180°C (350°F) and the use of the stringer bead technique (rather than weaving) have been recommended. It is also recommended for this alloy, that the filler metal chemistry match the major elements of the base metal chemistry [13] such as AWS 5.14 ERNiCr-3 and AWS A5.11 ENiCrFe-2. Above 700°C (1300°F), AWS A5.11 ENiCrCoMo-1 and AWS A5.14 ERNiCrCoMo-1 can be used to match the creep strength of the base metal. For repair welding, the use of E16.8.2-15 [5] has also been recommended for this alloy.

Heat treatments

Heat treatments before and after welding have been discussed thoroughly in literature.

a) Heat treatments before welding

The advantages and disadvantages of a stabilizing heat treatment before welding have been discussed in literature.

A stabilization heat treatment creates benefits by generating coarse particles within the grains and on the grain boundaries. These carbides form by C reacting with age-hardening elements such as Ti, Al, Nb, etc. The result is that at operating temperature, less C is available for the formation of fine carbides within the grains. [13] This results in a more even strength between the grains and the grain boundaries. Hence, some authors claim that a stabilizing heat treatment of cold formed base material before or after cold forming is a very effective to avoid relaxation cracking during service. After this heat treatment and up to 15% cold deformation the material can withstand more than 2% relaxation strain. [4]

For alloy 800H, a stabilization heat treatment at 982°C (1800°F) for 3 hours is recommended after rolling, forming or other manufacturing steps [13, 17], while for the same alloy [19] recommends the stabilizing temperature to be 950°C (1742°F). However, stabilization heat treatments can result in sensitization of austenitic stainless steels. [8]

Pan, et al. [16] states that a stabilizing heat treatment at 850°C - 950°C (1562°F – 1742°F) for several hours is not beneficial for avoiding reheat cracking, in fact cracking may take place during the stabilizing process for thick components.

Messer, et al [18] also warns against this treatment for SS347 as it may lead to sigma phase formation and resulting embrittlement, reduction of impact toughness, and reduction in creep properties, as well as increasing the sensitivity to SRC. Instead they propose stress-relief at 593°C (1100°F) for 2 hours per inch, following by solution anneal at 1052°C (1925°F) for 2 hours per inch, then stabilization anneal at 954°C (1750°F) for 1 hours per inch.

b) Post Weld Heat Treatments (PWHT)

A test using the Gleeble® based test procedure has been presented in [12] to study the effect of PWHT temperature and cold working on SRC susceptibility of multiple alloys.

In general, for stainless steels and Alloy 800H, a solution treatment of thick austenitic weldments 1050°C (1922°F) for 1 h – 2 h significantly reduces the risk of delayed reheat cracking during service operation [19]. The rate of heating up to 430°C (806°F) should be slow, and rapid from 430°C to 1050°C (806°F to 1922°F). Components should preferably be forced air cooled. However, practical difficulties associated with solution treatment have been reported in literature obtaining a uniform temperature of 1050°C (1922°F) in thick section weldments and distortion problems during forced air cooling.

The ASME Boiler and Pressure Vessel Code recognized that the potential for SRC exists for pressure vessels and components to be exposed at temperatures over 538°C (1000°F). As a consequence, rule UNF-56 of Section VIII, Division 1 requires that pressure boundary welds and welds to pressure boundaries in alloy 800, 800H, and 800HT components with design temperatures above 538°C (1000°F) must be stress relieved prior to service. [3] The required heat treatment is at a minimum temperature of 885°C (1625°F) for 1.5 hr for a maximum section thickness of 1 in (25 mm) and for 1.5 hr + 1 hr/in (1 hour per 25 mm) for material of section thickness more than 1 in (25 mm). While the potential for stress relaxation cracking exists with essentially any stainless steel or nickel alloy used at temperatures in the 500°C to 700°C (930°F to 1292°F) range, it is interesting that ASME only requires this stress relief for the alloy 800-type products.

Requirements for alloy 800H are also included in [6], where it is indicated, that to avoid SRC a PWHT at 950°C (1742°F) or at least at 850°C (1562°C) can reduce the cracking problem.

For austenitic stainless steels, [4, 13] recommend PWHT in the temperature range 843°C to 899°C (1550°F to 1650°F) for most 300 series stainless steels. PWHT enables the material to accommodate relaxation strains of more than 2%, which is often sufficient to accommodate service-induced strains. For SS 321 and 347 specifically, the recommendation is to PWHT at 900°C (1650°F) for 1 hr 25 min per inch, though 3 hours minimum, followed by still air cooling. Furthermore, [13] indicates, that PWHT of SS 304, 304H, 316/316H, 317 304L, 316L and 317L is considered detrimental as it may cause sensitization.

For Ni-base alloys, [13] recommends different heat treatments and PWHT at different temperatures and times for alloys 625, 800, 800H, 800HT and 825. PWHT of Alloys 800H and 800HT is required by Code for pressure vessels for service temperatures above 540°C (1000°F). PWHT of Alloy 800H is considered the best prevention against in-service SRC (although it creates the risk of SRC during PWHT).

Heating rates at temperatures above 250°C (480°F) should generally comply with the limits provided by ASME BPVC Section VIII for PWHT. [15]

A multi-step optimized PWHT for stainless steels 347/347H/347L [18] has been reported to produce fewer but coarser – and thus less detrimental - Nb(C,N), prevent sigma phase embrittlement and provide a microstructural morphology that reduces future precipitation caused by creep during long-term, high temperature operation, thus providing substantially greater immunity to SRC (Figure 2).

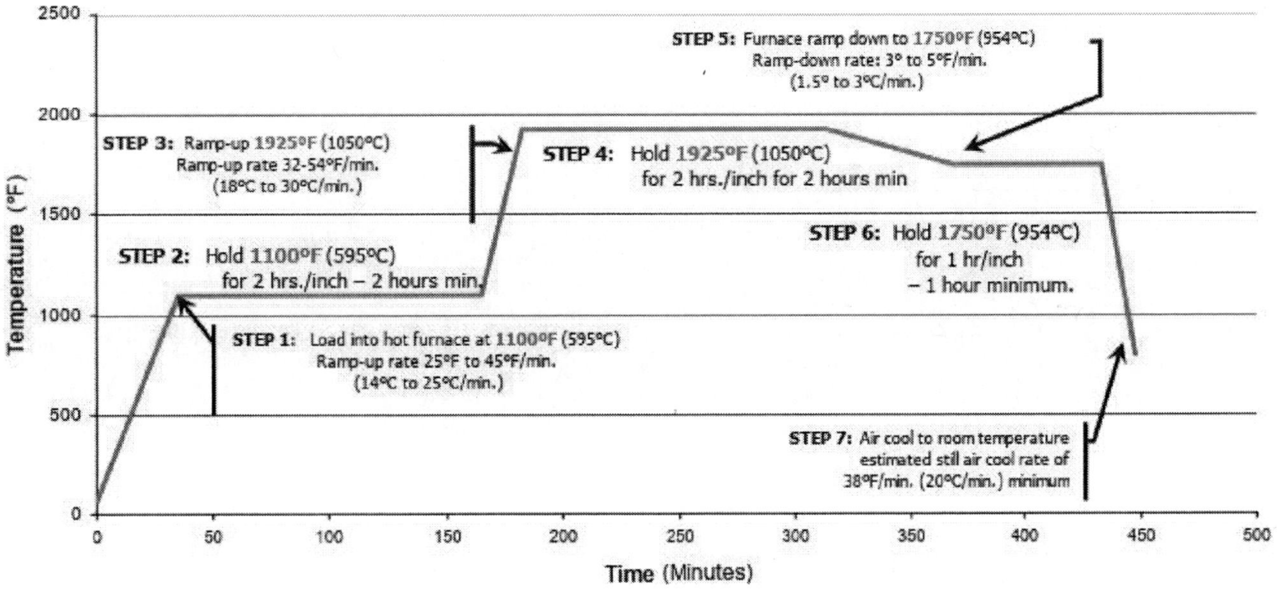

Figure 2. Optimized PWHT [18]

The Haldor Topsøe approach

Based on the study performed, Haldor Topsøe has the following current recommendations to minimize the risk for SRC:

<u>Type of material</u>

Depending on all other factors, Topsøe can recommend seamless pipes of 304H, 321H, 347H, 800H/800HT, 602CA. For Alloy 800H, the Ti+Al content shall be below 0.7%.

<u>Grain size</u>

To prevent SRC ASTM E-112 grain size shall be between 4 and 7 for 304H, 321H, and 347H. For alloy 800H/800HT, ASME requires grain size coarser than 5. Topsøe recommends grain size finer than 3.

<u>High residual stresses</u>

In order to reduce the stress levels of the weld, the weld geometry shown in Figure 3 shall be used. Flanges are designed with long lips to allow separation between weld and notch. Furthermore, because SRC is often initiated at stress risers or notches, sharp edges in cross section are avoided and fatigue resistant contour weld finish are recommended.

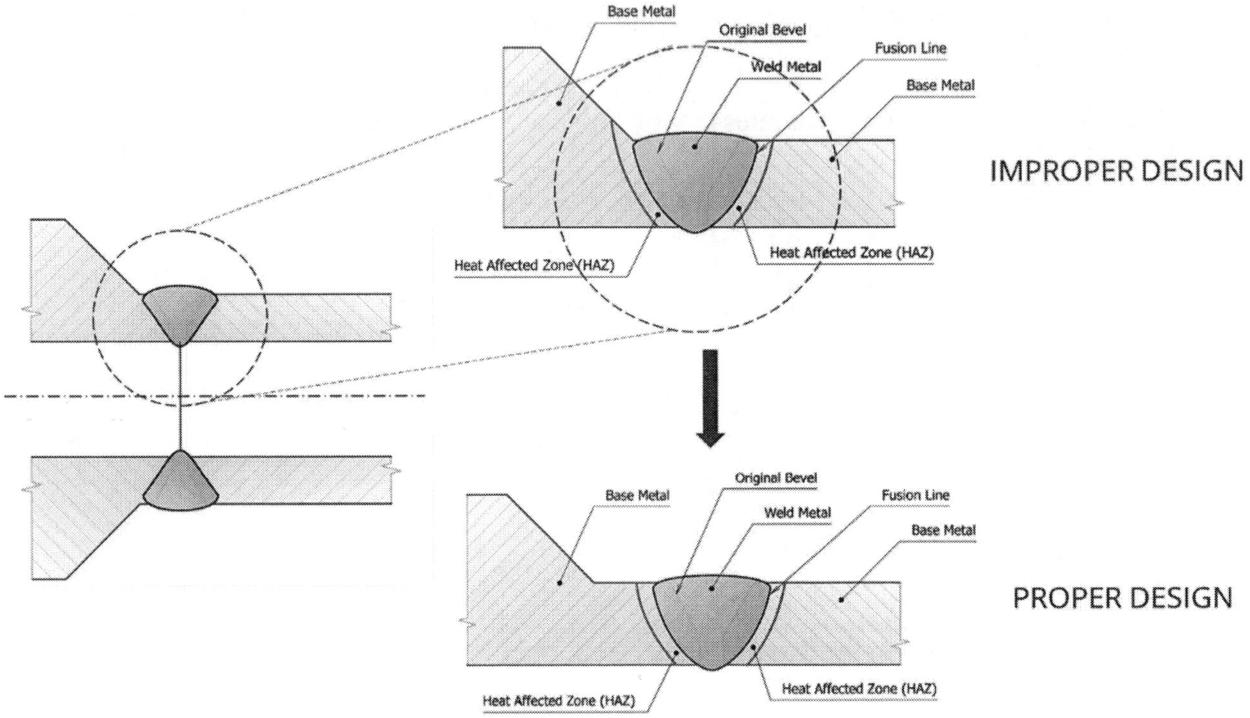

Figure 3: Weld design

Heat-treatment

Depends on thickness. Topsøe recommends a minimum of 3 hours at 890±10°C (1634±18°F).

Pipe sizes

Pipe thickness to withstand given service conditions depends on the creep strength of the alloy selected. The higher the creep strength, the lower the pipe thickness. I.e. for a given set of conditions, the thickness calculated for SS347H would be lower than the thickness calculated for SS321H.

Topsøe's approach is to determine the tube thickness based on the process conditions and select the most optimal material: either SS321H or SS347H. If the calculated thickness for SS347H is above the 35 mm (1.4 in), then Topsøe recommends re-studying the design.

Although SS304H is not Topsøe's preferred alloy if there is risk of SRC, it has been used with thickness up to 30 mm (1.2 in) with success. In Topsøe's experience, only one documented failure has taken place with thickness of 48 mm (1.9 in).

Summary

SRC is a phenomenon affecting many welds of austenitic stainless steels and Ni-base alloys during service in the temperature range 500°C – 700°C (932°F – 1292°F).

In Haldor Topsøe's opinion, it is difficult to rank the different alloys in terms of susceptibility, as SRC depends not only on the type of material, but also on the grain size, residual stresses from fabrication, presence of stress concentrators, welding method, filler metal and heat-treating conditions.

References

[1] Osmsbee C., Ammonia Technical Manual (2011), 89-93.

[2] Thomas C.W., Smillie M.J., Ammonia Technical Manual (2011), 101-111.

[3] Shoemaker L.E., Smith G.D., Baker B.A., CORROSION NACE Conference (2007), Paper No. 07421.

[4] van Wortel H., CORROSION NACE Conference (2007), Paper No. 07423.

[5] van der Westhuizen C.E., CORROSION NACE Conference (2008), Paper No. 08454.

[6] Dejaeger J., Guns L., Korkhaus J., Ammonia Technical Manual (1994),134-144.

[7] Stahl H., Smith G., Wastiaux S., Ammonia Technical Manual (2000), 21-24.

[8] Grubb J.F., Fritz J.D., CORROSION NACE Conference (1997), Paper No 97185.

[9] Shargay C., Anil Singh A., CORROSION NACE Conference (2002) paper no. 02478.

[10] Auzoux Q., Allais L., Caës C., Monnet I., Gourgues A.F., Pineau A., Journal of Nuclear Materials 400 (2010) 127–137.

[11] Hester M.W., Benac D.J., Ammonia Technical manual (2013), 199-211.

[12] Kant R., DuPont J., Supplement to the Welding Journal (2019), 29-49.

[13] API 942B (2017).

[14] van Wortel J.C., Stainless Steel World Conference (1999), 281-290.

[15] Lee H., Jung J., Kim D., Yoo K., Engineering Failure Analysis 56 (2015), 413-422.

[16] Pan J., Fan Z., Zong N., International Journal of Pressure Vessels and Piping 148 (2016) 1-8.

[17] Korkhaus J., Schwarz M., Luyckx G., Lesicki R., Verbruggen L., Ammonia Technical Manual (2000), 270-279.

[18] Messer B., Oprea V., Phillips T., CORROSION NACE Conference (2004), Paper No 04640.

[19] Dhooge A., Welding in the World 41 (1998), 206-219.

[20] Chabaud-Reytier M., Allais L., Caes C., Dubuisson P., Pineau A., Journal of Nuclear Materials 323 (2003), 123-137.

[21] Gendron T.S., Cleland R.D., Newman R.C., CORROSION NACE Conference (1995), Paper No 181.

[22] van Wortel J.C., Conference Baltica IV (1998), 637-647.

[23] van Wortel J.C., Stainless Steel Europe (1995), 47-49.

[24] Korkhaus J., Stainless Steel World Conference (1999), 27-32.

Lessons Learned in Nickel Catalyst Handling

The paper discusses the hazards associated with nickel dust along with methods of controlling exposure to nickel dust when charging steam reforming catalysts. The UNIDENSE® and UNILOADER® charging methods are also described along with some lessons learned when handling such catalysts

J. Brightling, J. Pach, A. Richardson
Johnson Matthey plc

C. Döring, S. Struck
Unidense Technology GmbH

Introduction

Nickel (Ni) containing catalysts can be found in the hydrodesulphurization (HDS, 1), pre-reforming (2), primary reforming (3), secondary reforming (4), and methanation (5) reactors of an ammonia plant (Figure 1).

Although Cobalt Molybdenum can be used in preference to Nickel Molybdenum to catalyse the HDS reaction, there are no readily substitutable and economically viable alternatives for the reforming and methanation catalysts.

With all nickel containing catalysts there is a risk of exposure from dust which may be generated during catalyst loading and unloading activities; and appropriate precautions are necessary.

The HDS, pre-reforming, secondary reforming and methanation reactors can be loaded from bulk containers thus helping limit the risk of exposure to dust.

The primary reformer (3) is different in that there are, in effect, several hundred parallel reactors each of which needs to be loaded individually, with compounded risk of exposure.

The effective operation of a primary reformer is critically dependent on both the correct choice and optimal installation of the catalyst to ensure even flow through the reformer tubes. In addition to minimising the methane slip, this also helps to maximise reformer tube life.

This paper discusses the hazards associated with nickel and nickel catalysts and then, with specific reference to the primary reforming catalyst, discusses methods of controlling exposure to nickel dust.

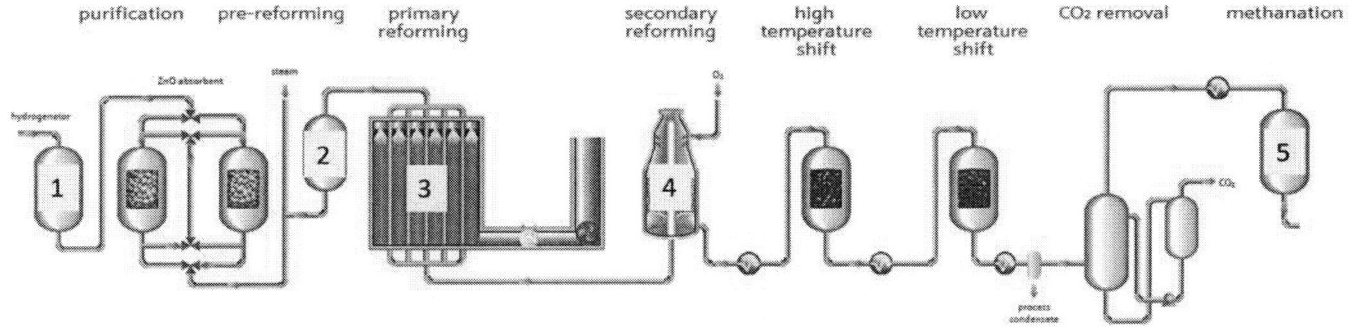

Figure 1. Nickel containing catalysts in an ammonia plant.

Nickel Dust Hazard

Over the last three decades, the health hazards of Nickel compounds have been studied extensively, by scientists and industry alike.

Human data and animal toxicity studies (using physiologically relevant routes of exposure) have confirmed the carcinogenic potential and respiratory carcinogenesis of certain nickel compounds. This data shows that the inhalation is the most devastating source of exposure for nickel and that the respiratory tract is the target organ for cancer. They also suggest that the different nickel species have different carcinogenic potentials.

The classification of nickel compounds according to the European Union Classification, Labelling and Packaging regulation (EU CLP) considers these differing potentials. Many soluble and insoluble nickel compounds are classified as Carcinogenicity 1A under the EU CLP. Category 1A classification implies these nickel compounds are known to have carcinogenic potential for humans based largely on human evidence.

By contrast, nickel metal is classified as Carcinogenicity 2, suspected human carcinogen, based on insufficient or negative evidence from human studies with suggestive evidence from animal studies

The carcinogenicity of nickel compounds is mainly an occupational concern. As such workplace standards such as Occupational Exposure Limits (OEL) are set to protect workers from these effects. There are contrasting national OELs across EU and Non-EU countries (Table 1).

Recognising this inconsistency in Nickel OELs, regulatory bodies in Europe have been working towards the implementation of EU wide OELs.

The process of introducing an OEL has been ongoing for some years. In March 2018, the first step towards EU-wide nickel OELs was taken when the European Chemicals Agency's (ECHA) Risk Assessment Committee (RAC) provided an opinion on the scientific evaluation of OELs for nickel and its compounds. [1] This opinion has served as basis for other stakeholders (the EU Commission and associated expert health and safety committees) in the setting of binding OELs for nickel compounds under the Carcinogens and Mutagens Directive (which is expected in 2020/2021). The current values which have been discussed and put forward for consideration at EU level for Nickel compounds are: 0.01 mg Ni/m3 (8hr TWA) respirable fraction (that which enters the nose and/or mouth during breathing) and 0.05 mg Ni/m3 (8hr TWA) inhalable fraction (that which penetrates down to the alveolar region of the lung).

These values are in line with Johnson Matthey's internally adopted occupational exposure limits for our own workers.

Country	Metallic Nickel and Insoluble Nickel species
Austria	0.5 (insoluble compounds, metal & alloys)[b] [I]
Belgium	1.0 (Ni metal), 0.1 (Ni subsulphide), 0.2 (other insoluble compounds)
Denmark	0.05
Estonia	0.5
Finland	0.05 (Ni compounds) [I] 0.01 (Ni metal + compounds) [R]
France	1
Germany	0.03 (Total compounds) [I] 0.006 (Ni metal + Total compounds) [R]
Hungary	0.1 (Ni metal + insoluble compounds), 0.01 (Ni subsulphide)
Ireland	0.5 (Ni metal) 0.5 (insoluble compounds)
Italy	0.2 (insoluble compounds)
Latvia	0.05 (Ni metal)
Luxembourg	1.0
Netherlands	0.1
Norway	0.05
Poland	0.25 (metal and compounds)
Portugal	1.0 (metallic Ni) 0.2 (insoluble compounds)
Spain	1.0 (metallic Ni), 0.2 (insoluble compounds) 0.1 (Ni trioxide)
Sweden	0.5 (Ni metal) 0.1 (Ni oxide, Ni carbonate) 0.01 (Ni subsulphide)
United Kingdom	0.5
Switzerland	0.05 [I]
USA	1.0 (OSHA PEL) [I]
China	1

a. 8-hour TWA (Time-Weighted Average) unless otherwise noted. All values refer to 'total' nickel unless otherwise noted.
b. A value of 0.5 mg/m3 applies to inhalable nickel metal and alloys, nickel sulphide, sulphidic ores, oxidic nickel, nickel carbonate, and dusts from Ni compounds and alloys; a value of 0.05 mg/m3 applies to soluble nickel compound in the form of "inhalable droplets."
I = Inhalable dust; R = Respirable dust;

Table 1. Current worldwide Nickel OELs

Hierarchy of Control

The Control of Substances Hazardous to Health Regulations (COSHH) [2], are intended to protect people from hazards of substances used in the workplace. This is achieved by risk assessment, control of exposure, health surveillance and incident planning. Although not unique to COSHH, a hierarchy of control (Figure 2) is a key feature.

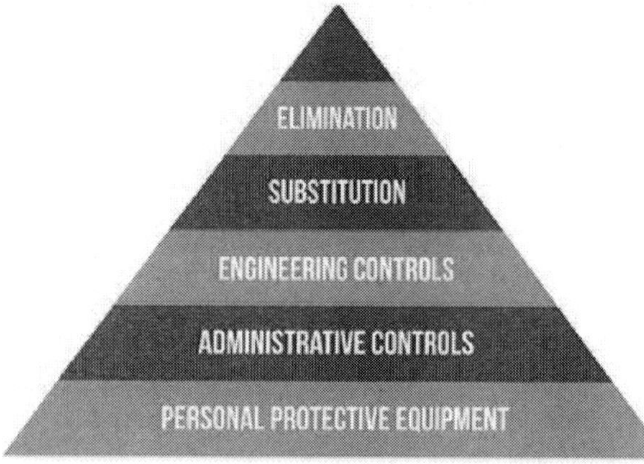

Figure 2. COSHH hierarchy of control.

Elimination – is best way to control exposure, but it is one that may be very difficult to achieve. Although not possible to eliminate the need for a catalyst, it is, however, possible to reduce the frequency at which the catalyst is replaced.

Substitution involves changing chemical substances and/or processes to less hazardous ones. With the exception of the HDS catalyst, there are no readily substitutable and economically viable alternatives

Engineering controls are steps to reduce emissions, such as containment, automation and additional practices described in this paper.

Administrative Controls are worker-based controls which may reduce the operators' exposure. This includes training, supervision and housekeeping.

Personal Protective Equipment (PPE) is the last line of defence and should not be relied on as a single control. PPE can however be very effective when combined with other controls.

Elimination – Why is the Catalyst Changed?

Typically, the primary reforming catalyst is replaced for one or more of the following reasons:

(a) catalyst breakdown;

(b) tube blockage;

(c) overheating of the catalyst;

(d) poisoning of the catalyst;

(e) thermal aging.

Catalyst breakage and blockage of the primary reformer tubes causes an increased pressure drop across the steam reformer, and if the effect is random it results in an uneven appearance of the tubes in the furnace. More fundamentally, it can lead to overheating of the catalyst and the tubes, loss of activity and a reduction in output. Overheating of the catalyst can also be caused by mal operation of the reformer. Loss of activity through poisoning by contaminants in the process gas can lead to carbon deposition. This can, in turn, lead to overheating, catalyst breakage and, in extreme cases, partial blockage of the tubes. Gradual loss of activity or thermal aging caused by progressive loss of nickel surface area through sintering places a limit on the life of a catalyst charge and, for any particular catalyst, this depends on the actual operating conditions. In practice the most important effects are catalyst poisons and carbon formation.

Common causes of increased pressure drop include:

- thermal cycles which cause breakage
- carbon laydown
- wetting
- loading problems

The impact of the above can be mitigated by selecting a catalyst with high intrinsic strength and favourable breakage characteristics. Decades of focus on these as well as other aspects has led to significant improvements in the capability of Johnson Matthey's (JM) catalysts, as illustrated in Figure 3.

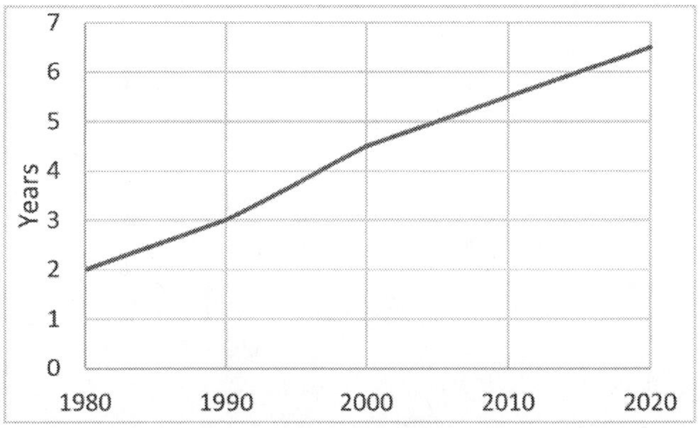

Figure 3. Average lifetime of primary reformer catalysts vs time

In addition to the economic benefits of improved reliability and lower operating costs, the reduction in catalyst replacement frequency leads to a significant reduction in the number of opportunities for exposure to nickel containing dust.

The mechanical properties of the catalyst arise from a combination of the inherent material properties of the support and its porosity which, in turn, depend on catalyst formulation and manufacturing techniques.

The mechanical strength of a catalyst pellet has been shown to be related to porosity and pore structure. The tensile strength decreases exponentially with the porosity, as shown by the Ryshkewitch-Duckworth equation.

$$\ln\left(\frac{\sigma_t}{\sigma_0}\right) = c_1(D-1)$$

Where σ_0 is the tensile strength of the material at full density ($D = 1$) and c_1 is a constant related to the bonding capacity.

Hence there is a direct trade-off between overall porosity and physical strength. Generally, ceramic catalysts will fail in tensile mode, where tensile stresses are developed within the catalyst structure (as shown in Figure 4). Whether or not crack propagation occurs through the porous substrate will be dependent on the porosity both with regards to size and connectivity.

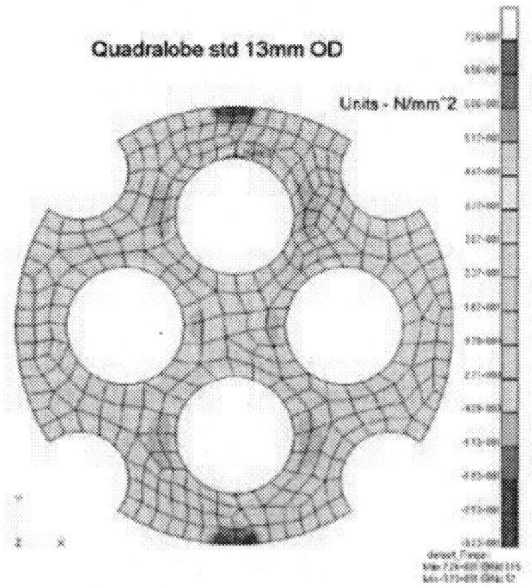

Figure 4. KATALCO QUADRALOBE™ Stress Map

It is well understood in the field of material science and, in particular, fracture mechanics that, 'blunt crack' tips have a higher threshold fracture toughness, when a crack meets a pore the crack will become blunt which will decrease the stress concentration, potentially stopping it from propagating further, preventing breakage. Cracks with sharp tips propagate easier than cracks with blunt tips (Figure 5).

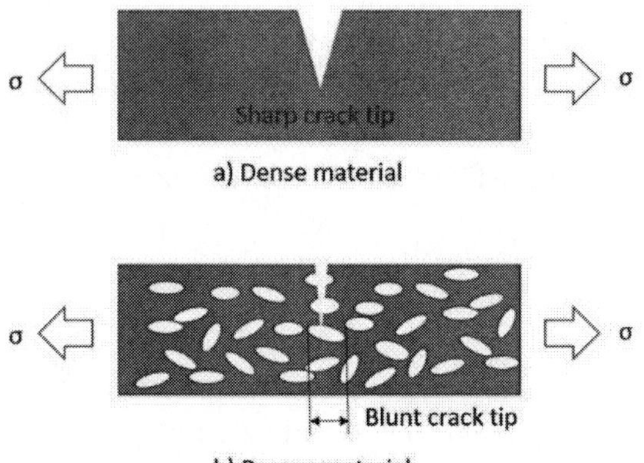

Figure 5. Crack propagation through (a) dense, and (b) porous material; showing how cracks blunt due to material porosity

The strength of the catalyst must be high enough to allow loading through conventional means without breakage but still have sufficient porosity to maximize activity.

The final consideration with regards to mechanical properties of the catalyst is how it will pack and break. In some circumstances the catalyst will need to fail, and it is beneficial from an operator's perspective that it does so in such a way as to minimize the consequent increase in pressure drop.

An example would be a temperature cycle during reformer start up and shut down. As the tubes initially heat up, the tubes expand more than the catalyst which leads to some settling of the catalyst. On subsequent cooling, the tubes contract more than the catalyst and stress chains can form, as illustrated in Figure 6.

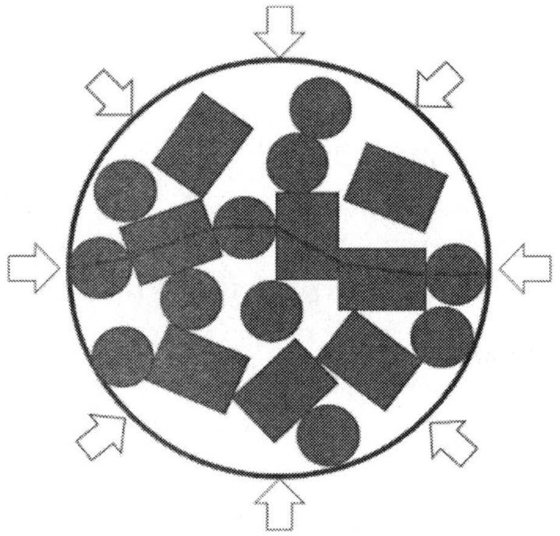

Figure 6. Stress chain (red line) during tube contraction

The size of the fragments created during breakage to release this stress chain are a function of catalyst shape as well as material properties.

KATALCO™ 57-series catalyst, a calcium aluminate supported catalyst, has been optimised taking all of these aspects into consideration. The calcium content increases the basicity of the surface to minimise carbon formation without introducing interactions that detrimentally affect the ease of reduction of the NiO. The porosity of the support has been optimised to provide a high surface area for dispersion on the active Ni, minimising diffusion resistance and imparting high mechanical strength. KATALCO 57-Q series catalysts has been designed to break into large pieces, as shown in Figure 7, to minimise any consequent increase in pressure drop.

Figure 7. Broken Q-shape supports showing resultant large pieces which have a small impact on pressure drop.

Alternative formulations with high strength and non-optimized porosity can lead to formation of many small fragments, which would lead to an unnecessarily high increase in pressure drop over the lifetime of the catalyst. This can also be the case if a suboptimal shape is used such as external features which can stack or interlock, facilitating build-up of stress chains. Intricate internal design can also lead to a greater number of smaller fragments being created when the pellets break, again leading to an unnecessarily high increase in pressure drop. The fact that such catalysts are less robust can also be seen during UNIDENSE® charging in that an increased number of fall arrest springs are necessary.

Engineering Controls

Catalysts Europe (of which JM is a long standing and active member) has created a Catalyst Handling Best Practice Guide. [3] This guide is intended for industrial catalyst users, handling companies and designers of catalyst systems. It covers catalyst specific guidance and is designed to supplement general safety practices.

Key engineering controls that should be followed during loading of catalyst into the reformer tubes include the charging method used, the use of nickel exposure controls and levels of ventilation during the loading.

Charging reformer tubes

There are several methods for charging catalyst in reformer tubes. Historically it was loaded from pre-packed catalyst in socks. In the early 1990s Norsk Hydro presented a paper on dense loading. [4]

The technique for primary reformer changing, which was subsequently given the brand name UNIDENSE®, involved catalyst being poured into the top of a tube which contained a rope fitted with "brushes" (metallic springs) which controlled the catalyst fall. The rope was pulled upwards as the catalyst level increased (Figure 8).

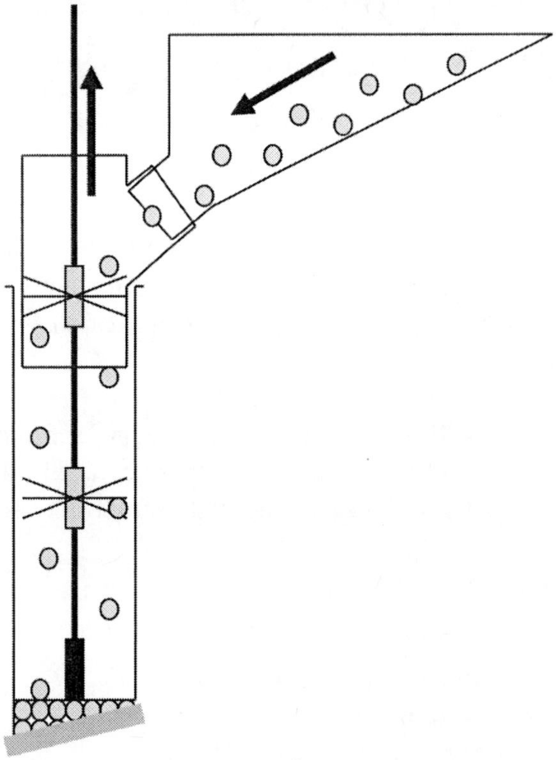

Figure 8. UNIDENSE® loading schematic.

The original method relied on manual extraction of the rope combined with manual pouring of catalysts into a funnel hopper at the correct rate, as shown in Figure 9. This required two people.

Figure 9. Manual UNIDENSE®

In more recent years UNILOADER® has been developed. This automates many of the activities which could impact the uniformity of the pressure drop.

UNILOADER® greatly reduces the risk of nickel exposure – there is one less operator at the tube top and it is enclosed with dust extraction.

The benefits are

- Faster and automated loading
- Eliminates human error
- Less labour
- Dust-free loading
- Greater consistency in loading

Figure 10. UNILOADER® with dust extraction

Nickel catalyst exposure controls

Catalysts should be handled as much as possible in well ventilated areas and in a way that avoids excessive formation of dust. Operators who handle catalyst must wear suitable protective body clothing, gloves and goggles.

Inhalation of dust must be avoided, and the appropriate occupational exposure limits should be strictly observed.

Ventilation and atmosphere monitoring

Periodic air monitoring can and should be carried out to determine personal exposure levels to airborne dust. These monitoring results can be used to assess the effectiveness of dust control measures and respiratory protection.

Table 2 gives the results of such monitoring, which was carried out in accordance with 29 CFR 1926.55, Appendix A and OSHA and NIOSH methods, whilst dense loading a primary reformer.

Employee samples for the bucket loader and the catalyst pourer were both above the American Conference of Govt Industrial Hygienists (ACGIH) and the OSHA permissible exposure limits (Table 3) confirming the need for appropriate respiratory protection when carrying out these activities as, in fact, was the case.

Sample Location	Ni 8 hour TWA (mg/m3)	Total dust (mg/m3)
Bucket loader	4.33	12.4
Catalyst pourer	0.483	1.5
Area	0.004 to 0.035	<0.22

Table 2. Summary of area monitoring during a loading

Limits	Ni (mg/m3)	Total dust (mg/m3)
ACGIH (TLV)	0.2	10
OSHA (PEL)	1.0	15.0
Area	0.004 to 0.035	<0.22

Table 3. ACGIH Threshold Limit Values (TLV) and OSHA Permissible Exposure Limits (PEL)

Personal Protective Equipment

Personal protective equipment (PPE) is employed as the final line of defence and should only be used in conjunction with other control methods higher in the COSHH hierarchy of control.

Many types of PPE can be utilised to reduce or prevent contact with catalytic material during the loading of reformer tubes.

Respiratory protection must be used throughout the loading of the catalyst. The type of respiratory protection to be worn depends on the dust levels generated, and the exposure limits for the materials involved.

Suitable approved half mask respirators (e.g. EN 140 or as required by local regulation) with high efficiency filters (e.g. EN 143 or as required by local regulation) should be worn by personnel involved in other activities such as the screening of catalysts, filling socks and cleaning up spillages.

All personnel close to catalyst handling activities should also wear approved respiratory protection. Similar considerations indicate that personnel involved with the maintenance of equipment which has contained catalyst, should wear suitable dust respirators. All personnel who are required to wear respiratory protection should be instructed in its proper use and limitations.

Good Practices

The use of a FFP3 or N99 dust mask during catalyst handling is required. The catalyst handling begins with the opening of the catalyst drum and ends with the final housekeeping of the work area. All plant personnel and visitors should wear PPE even if they are simply observing the work.

The catalysts for steam reformers are usually supplied in steel drums which must be transferred equally into smaller containers (such as buckets). All the tubes must be filled with the same amount of catalyst, which means each individual bucket must be weighed. Due to there being many internals on a reformer, such as the burner, an air supply and a gas supply, there is limited working space. Furthermore, systems and sensors for monitoring additionally restrict the working space. The transfer into buckets or containers must therefore take place in a suitable location near the reformer. Loading into storage containers has the advantage that the dust can be extracted and contained during the refilling process. However, this process can cause breakages in the catalyst.

The dust formation is at its greatest during the filling of the buckets. Hence personnel must wear appropriate PPE, dust coveralls as well as the mandatory half mask respirator. Once the weighing is complete, the buckets are transported to the appropriate tube, this is carried out at low speed to minimise dust formation.

The final possibility for dust formation is during the filling of the catalyst into a funnel, by either the manual loading method or into the pre-loading container of the Uniloader®. Due to the catalysts differing in their chemical and physical composition depending on the specifics of the installation, the loading rope is also adapted accordingly. Each loading rope is configured to the given requirements and modified, if necessary, during the corresponding test loading. Afterwards, the dust generated from loading the catalyst is extracted by the vacuum cleaner connected to the Uniloader®.

Pressure Measurements

When measuring pressure drop, dust release and exposure can be significantly reduced if the measurement is carried out slowly. The balloon should be inflated slowly and separately to seal the tube and thus no dust can escape upwards. The measurements can then be carried out safely. After the required air flow for the measurements is stopped and the sealing balloon has been depressurized, it can be released. The use of a dusk mask is absolutely mandatory.

Storage

When catalyst has been stored in sunlight for a long period of time, the physical properties of the catalyst may have changed. This can result in increased or decreased dust generation or in altered fragility compared to the freshly delivered catalyst.

Lessons Learned

The most effective way to avoid dust is to have pre-weighed bags filled by the manufacturer. Thus, the catalyst does not have to be decanted and weighed on site, meaning the only possibility of dust formation is when the catalyst is filled into the pre-loading container of the Uniloader®.

As a result it is highly recommended that the catalyst is packed into individual bags before being shipped to the customer alongside the use of the Uniloader® with an integrated dust extraction during the filling as this significantly reduces the development and release of the nickel dust. However, few people are currently taking advantage of these benefits.

Conclusions

Over the last few decades, the hazards of nickel have increasingly been recognised. Nickel based catalysts are however necessary for many chemical processes not least of which are steam reformers. The risk of exposure is greatest when catalysts are being charged and discharged. Risks can however be minimised by selecting catalysts and charging techniques which maximise the productive life of the catalyst and which minimise dust formation.

References

1. Committee for Risk Assessment RAC Opinion on scientific evaluation of occupational exposure limits for Nickel and its compounds ECHA/RAC/ A77-O-0000001412-86-189/F Adopted 9 March 2018
2. The Control of Substances Hazardous to Health Regulations 2002, UK Statutory Instrument 2002 No 2677, as amended
3. https://www.catalystseurope.eu/images/Documents/ECMA1004q_-_Catalyst_handling_best_practice_guide.pdf
4. Norsk Hydro experience with dense loading of catalysts in ammonia plants E Dessen, G Rynkveit, Y Haquet; 1992 AIChE Ammonia Symposium

Designing a Toxic Shelter, an Art and Science

BakerRisk has performed toxic shelter evaluations, air infiltration testing, and toxic risk mitigation effectiveness assessments for buildings for a range of clients and industry segments. Through these assessments, a number of common issues associated with toxic shelter designs and operations have been identified as opportunities for improvement. The purpose of this paper is to discuss well-designed toxic shelters, including timely, reliable gas detection and ventilation isolation, leak-tight boundaries, and establishment of an effective fallback plan.

An accidental release of anhydrous ammonia can occur at any facility that handles ammonia. The risks associated with an accidental release of ammonia need to be managed to protect personnel to a level consistent with the owner's risk management philosophy, which is typically accomplished through sheltering in place and/or evacuation. However, facilities that handle ammonia often have other hazards, such as blast and fire, which should be considered when designing toxic shelters. Most importantly, a hazard resistant shelter has to be fully functional for its intended use as well as providing protection to personnel from the applicable process hazards.

The paper discusses a case study of a highly effective toxic shelter building, the test results and method used to perform those tests, and insights gained from the evaluation. Blast and thermal hazards considerations, and their importance in a well-designed toxic shelter building, are also discussed.

Anthony G. Sarrack and Murtaza I. Gandhi
Baker Engineering and Risk Consultants, Inc.

Introduction

As part of the OSHA PSM requirements, companies are required to look at facility siting to determine potential impacts to onsite personnel. In the United States, most companies look to API 752/753 [1,2] for permanent and temporary structures, respectively, when conducting either consequence or risk-based facility siting. Historically, there has been a heavy focus on blast events as a result of incidents in industry which includes the 1980 Phillips Pasadena [3] 2005 BP Texas City explosion [4]; however, the upcoming revisions to API 752/753 [1,2] are expected to place an increased emphasis on toxic and thermal personnel exposure in permanent and temporary buildings. Based on the review of hundreds of buildings across a number of processing industries, a significant amount of work will be required to properly address toxic shelter-in-place under new requirements.

The authors have performed toxic shelter evaluations, air infiltration testing, and toxic risk mitigation effectiveness assessments for a range of clients and industry segments. Through these assessments, a number of common issues associated with toxic shelter designs and operations have been identified as opportunities for improvement. The purpose of this paper is to discuss well-designed toxic shelters, including timely, reliable gas detection and ventilation isolation, leak-tight boundaries, and establishment of an effective fallback plan. Blast and fire hazards should not be overlooked when making decisions regarding toxic shelter design and location.

Components of Toxic Shelter

The basic components of a toxic shelter are shown in Figure 1.

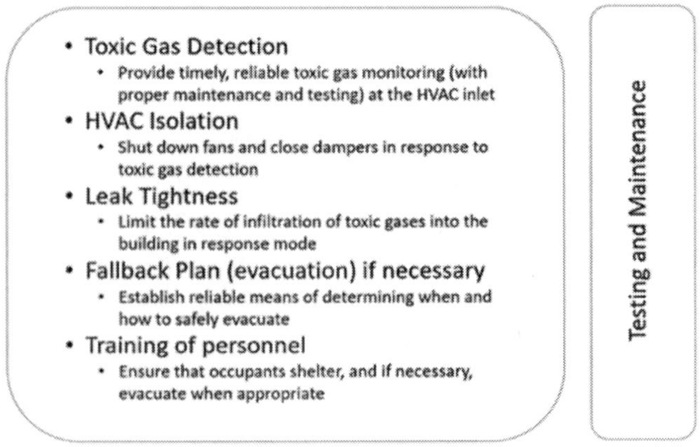

Figure 1. Elements of a Toxic Shelter

Additional protection (scrubbing, filtration, clean air supply, etc.) can be implemented to further reduce toxic risk for a given building, but this paper is limited to identification of common shortcomings found for these five basic components of toxic shelters and descriptions of how a well-designed shelter addresses each .

The paper will discuss a case study for what to expect in a highly effective toxic shelter. It will walk through a prolonged, severe toxic impact scenario and show the results of calculations that predict toxic impacts to various parts of the building as well as to the interior room where occupants would shelter for added protection in a highly effective toxic shelter.

Case Study Description

This case study features a toxic shelter located at an ammonia production facility with large anhydrous liquid ammonia storage tanks nearby. The accident scenario for the case study is a large leak that takes three hours for emergency response personnel to manually isolate and causes a continuous cloud of 5,000 ppm ammonia to engulf the shelter building (Figure 2).

Figure 2. Severe Toxic Cloud Impacting Shelter

Shelter Design and Operation

Toxic Gas Detection

The first component necessary to have an effective toxic shelter is toxic gas detection. A facility that produces and/or handles acute toxic hazards with potential for significant impacts to personnel (such as ammonia in the fertilizer industry or hydrogen fluoride / hydrogen sulfide in the refining industry) should provide effective toxic gas detection and alarms in areas where toxic releases are most likely so that personnel are reliably made aware of the toxic threat in a timely manner.

In addition to processing area toxic gas detection, each toxic shelter should provide detection at the heating ventilation and air conditioning (HVAC) fresh air inlet. A significant number of toxic shelters assessed by the authors were not equipped with toxic gas detectors at the HVAC inlet. Instead, many sites rely on process area detection and subsequent plant-wide alarms and emergency response announcements to notify occupants of toxic shelters that an event has occurred with an action to activate the toxic emergency response mode. As a result, the timeliness and reliability of response are both significantly degraded.

For sites that do have toxic gas detection at shelter locations, these are positioned at various locations, including near entrances, walls facing the hazard area, etc. While best practice is to have detection at multiple locations, most do not have detection for the air drawn in the building through HVAC systems where it matters the most (Figure 3).

Figure 3. Toxic Gas Detection on HVAC Air Inlet and Building Entrance

Furthermore, although providing toxic gas detection is a great first step, it is equally important to have the ability to view the concentration profile from the detectors at each location (i.e., at building entrance, at HVAC air intake, and within the building itself) in order to make informed decisions while located within the toxic shelter. For more information on this topic, see the fallback plan section of this case study.

It is important that the HVAC inlet toxic gas detector has a rapid response time because some scenarios may be so severe that even a short delay between toxic gas impact and HVAC isolation can result in the shelter environment becoming unsafe. This issue is particularly important for once-through HVAC systems and pressurized buildings. The detection system and interlocks should be maintained as systems important to safety, and be routinely tested to ensure high reliability.

HVAC Isolation

The second component of an effective toxic shelter is HVAC isolation. Ideally this activation will automatically occur immediately upon detection of toxic gas at the HVAC inlet. It is common for shelters to not have an automatic shutdown of the HVAC system based on toxic gas detection. The most common design for buildings assessed was to have an "Emergency Pushbutton" that placed the building in toxic emergency mode when activated. Best practice is to have detection at the HVAC inlet with an automatic interlock with an "Emergency Pushbutton" backup of the automatic interlock. Having the "Emergency Pushbutton" allows the shelter to be placed in toxic emergency response mode based on other indications, such as process area toxic gas alarms or other site-wide alarms. Note that depending only on manual actions (having no automatic interlock) can result in a degradation of HVAC isolation timeliness and reliability due to the possibility of human error.

Another common issue found with HVAC isolation is that the interlock addresses only the main HVAC system and does not thoroughly isolate all ventilation systems. It is common for buildings to have one or more exhaust fans that remain operational when the building is placed in toxic emergency response mode. The most common exhaust fans that should be isolated but are typically left in service are bathrooms, locker rooms, kitchens, and janitor's closets. Chemical hoods are also found within some toxic shelters, and it is not always safe to turn them off, so this challenge should be investigated in detail and be addressed in the toxic emergency response plan.

In the case study, HVAC is immediately shutdown upon toxic gas detection above the toxic emergency response concentration threshold at the HVAC fresh air intake. The interlock also shuts down all exhaust fans and demands all dampers closed. Alarms are initiated to notify occupants of the event, so they know to relocate to a shelter room and initiate toxic emergency response procedures.

Leak Tightness

Once a toxic shelter is placed in emergency response mode (HVAC is isolated), the amount of time that a shelter can protect its occupants depends on the concentration of the toxic gas that impacts the building, the lethality of the toxic material, the duration of the toxic cloud, and the infiltration rate (i.e., the rate at which air enters the building with HVAC isolated). For a given toxic gas, the leak tightness of the building determines how effective the shelter is at preventing conditions within the shelter from becoming dangerous.

Leak tightness can range dramatically from one shelter to another. The required leak tightness of a toxic shelter depends on the potential severity of toxic impacts to the building. Infiltration rates depend on the building size, shape, surface to volume ratio, and leak-tightness of gaskets, dampers, and penetration sealants. Infiltration rates also depend on wind speed. For the purpose of simplicity, all infiltration rates in this paper are based on a wind speed of 5 mph.

It is common for toxic shelters assessed to experience an infiltration rate greater than 1.0 ACH (air changes per hour). With an infiltration rate of 1.0 ACH, if the toxic gas concentration impacting the building is 30 times the dangerous concentration, conditions within the shelter would be calculated to become dangerous within 3 minutes. For example, this may be a 9,000 ppm cloud of ammonia impacting a building (immediately dangerous to life and health (IDLH) concentration = 300 ppm).

Most shelters assessed experienced an infiltration rate in a range of 0.3 – 1.0 ACH. With an infiltration rate of 0.3 ACH, if the ammonia concentration impacting the building is 9,000 ppm, conditions within the shelter would be calculated to exceed 300 ppm in 3 – 7 minutes.

A small number of shelters fall in the range of 0.1 – 0.3 ACH. At 0.1 ACH, if the ammonia concentration impacting the building is 9,000 ppm, conditions within the shelter would be calculated to exceed 300 ppm in about 7 – 21 minutes.

Very few buildings have an infiltration rate 0.03 - 0.1 ACH. Given an outdoor ammonia concentration of 9,000 ppm, the concentration within the building would be calculated to reach 300 ppm in about 21 – 70 minutes.

The examples above are based on an outdoor ammonia concentration of 9,000 ppm and reflect a dangerous concentration of 300 ppm impacting the shelter occupants. If the concentration of the material impacting the shelter were lower, the amount of time that occupants are safe would be extended. Conversely, if the outdoor concentration were higher, the times to lethality would be shorter. Figure 4 summarizes the configurations and conditions described above.

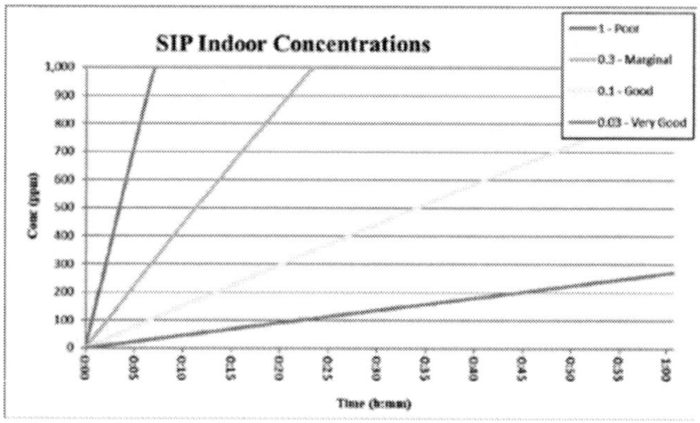

Figure 4. Concentration vs. Time for Example Leak Tightness Values

Having an interior room designated as the shelter-in-place (SIP) location dramatically improves the effectiveness of the toxic shelter by providing a secondary, independent barrier against toxic infiltration. Furthermore, the interior room is not subjected to wind forces, so it is possible to provide a highly effective barrier even under high wind conditions. Figure 5 shows that the concentration in the shelter building for an exceptionally leak-tight shelter (0.02 ACH – blue curve) is similar after one hour as the concentration in an interior room of a building that is only marginally leak tight (0.3 ACH – green curve) with a room leak tightness of 0.15 ACH (purple curve).

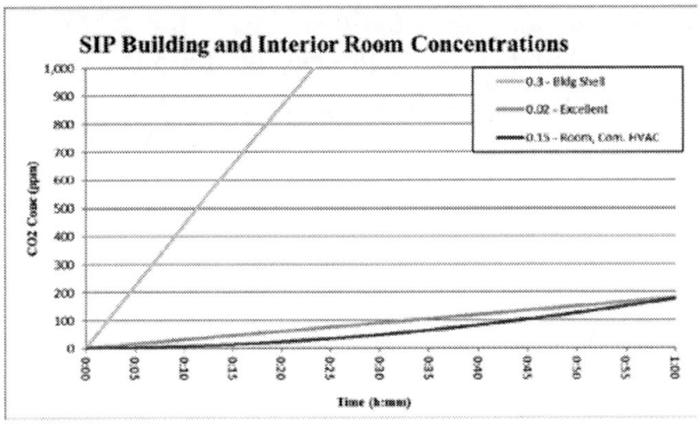

Figure 5. Concentration vs. Time for Shelter with Interior Room

It is difficult to design and maintain a building that is very leak tight, but designing and maintaining a building that is only marginally leak tight, combined with an interior room shelter that is moderately leak-tight is feasible in most cases. For existing buildings where improvements need to be made to the toxic shelter protection, the proper design of an interior sheltering option may be a viable and cost effective solution.

A well-designed interior room shelter has no low-resistance leak paths between it and the rest of the building. The most common low-resistance leak path between an interior shelter room and the rest of the building is a common HVAC system without any dampers. Most interior shelter rooms have a supply duct and a return duct without dampers, which is equivalent to having a large hole in the door of the room and degrades the level of protection provided by the internal shelter location. The next most common low-resistance path is having one or more unfinished walls (walls that terminate short above a suspended ceiling). The suspended ceiling affords little resistance to air exchange, and above the suspended ceiling, the room walls terminate, allowing air to freely flow between the shelter room and other parts of the building. Finally, many interior rooms have a low resistant leak point at doors without any gaskets.

Figure 6 shows the toxic shelter considered in this case study as a best practice approach. It shows the designated shelter location is an interior room with no windows, no exterior doors, and no HVAC ducts connecting it to the rest of the building or to outside air. It also shows vestibules provided at entrances to the building to reduce the impact of a door opening during a toxic emergency (late comer arrival or occupant needs to leave the shelter, for example).

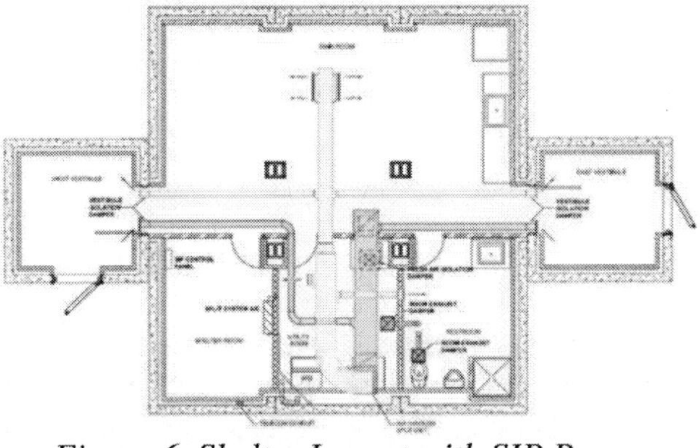
Figure 6. Shelter Layout with SIP Room

Figure 7 [5] shows an aerial view of this same building. Temperature is regulated in the shelter room by a split AC unit, so it draws no air into or out of the shelter room.

Figure 7. Shelter with vestibule

Despite the significant advantages gained by having an interior room serve as the toxic shelter for a building, this approach is not normally implemented. If an interior room is used, the HVAC system is often common to the rest of the building with no dampers to isolate the room. Interior room shelters often have one or more doors without gaskets. These shortcomings can be mitigated by providing duct tape, plastic, and training to occupants to ensure they seal up low-resistance leak paths in the event of a toxic emergency. Mitigating walls that terminate just above drop ceilings is much more difficult and cannot be effectively addressed by the occupants of the building at the time of a toxic emergency.

In addition to the toxic gas infiltration, shelters may also be designed to protect personnel against flammable gas and/or blast impacts. It is imperative that consideration be given to the vulnerability of a building to blast impacts from a vapor cloud explosion (VCE) before deciding to declare it a toxic shelter. If the response to an accidental release of hazardous material is to shelter, and people are told to shelter in a building that is vulnerable to collapse from a VCE, it is possible that they are an ignition source away from life threatening injuries.

Fallback Plan

The focus of this paper is toxic shelter design, but evacuation can also be an effective method of mitigating toxic risk. Designing an effective evacuation plan is simple for buildings that do not serve as toxic shelters. Designing and implementing an effective evacuation plan as a fallback plan for a toxic shelter is more complicated. An effective fallback plan for a toxic shelter includes the following components:

1. Toxic gas concentration monitoring within the toxic shelter.
2. Establishment of evacuation criteria (probably a concentration threshold for each applicable toxin).
3. Procedures and training.
4. Communication capabilities between shelter and emergency response personnel.
5. Adequate quantity of proper protective gear to allow safe evacuation.

If conditions within a toxic shelter degrade and approach a dangerous toxic gas concentration, it is appropriate to evacuate the shelter. Most shelters include a fallback plan. Few shelters have toxic gas monitoring and/or a logical criterion for transitioning from sheltering to evacuation. Without both of these critical components, the fallback plan cannot reliably be effective.

A well-planned fallback plan has evacuation criteria chosen to optimize the safety benefit afforded by sheltering, but also to avoid unnecessary discomfort or injury. A facility that handles acute toxic materials should have information on those chemicals, and the relevant information should be provided to shelter occupants so they are able to make informed decisions regarding whether to shelter or evacuate.

By nature, people will do whatever they believe optimizes their chance of survival. To optimize toxic risk mitigation through sheltering and evacuation, it is important that building occupants implement the plans. This is only likely to occur if they have a good understanding of the plan, the logic behind it, and truly believe it is their best option in the event of a toxic release. In addition, clear communication with a centralized emergency command center is helpful to ensure that the decision to evacuate or continue sheltering is well-informed, and that the evacuation route and muster point are optimized.

To create a level of understanding that results in proper action in a tense situation, it is important to provide shelter occupants proper training on the fallback plan, or the plan will probably not be followed. It is imperative that occupants understand the answers to the following critical questions regarding the fallback plan:

1. At what concentration do we evacuate, and why was that threshold selected?
2. What effects should I expect to encounter as the toxic gas concentration approaches the evacuation threshold?
3. What concentration constitutes a dangerous condition for the applicable toxins?
4. How severe might the toxic gas concentration be outside of the shelter, and what would be the effects of breathing gases at that concentration?
5. Where is the protective gear, and how is it used?
6. How effective is it, and how might it fail?
7. What precautions should be taken when using the equipment?
8. How will we know where it is safe to evacuate to, and what route should be taken to get there?

Figure 8 shows the decision tree for the shelter and evacuation plan as described above.

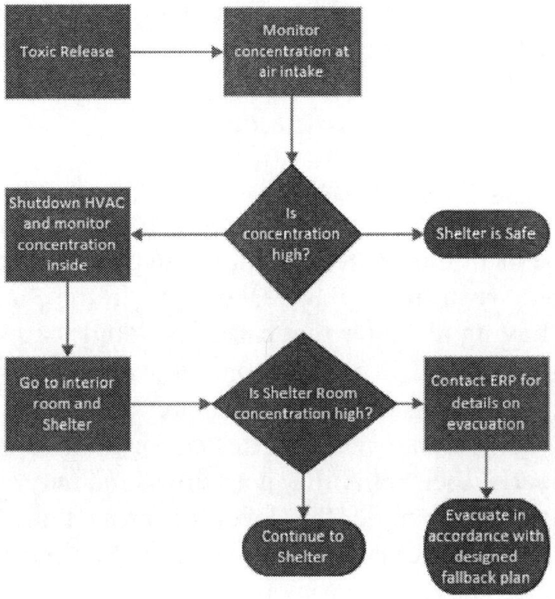

Figure 8. Shelter and Fall back Plan

Training

A toxic shelter's effectiveness depends largely on the understanding by the occupants of the building as to how the shelter works. Ideally, all occupants of buildings that are designated as toxic shelters are aware that the building is a toxic shelter, where they should muster in response to a toxic emergency alarm, and what actions are required to verify that the shelter is properly aligned for the toxic emergency. While occupants of most shelters are normally trained on the essential steps of sheltering, they are often unaware of all aspects to provide the most effective shelter possible.

Two main shortcomings include:

1. Occupants may not know how to turn off ventilation if interlocks fail to automatically isolate it, but they know what alarms indicate that a toxic emergency is in progress and where to muster in response to the alarm.

2. If fallback plans are not normally established, the level of training for them cannot be assessed. Occupants typically do not know when they should transition from sheltering to evacuation.

Maintenance and Inspection

The reliability of the protection depends on the effectiveness in testing and maintaining the components of the shelter. The strength of the protection is as strong as the weakest link in the chain. By having a shelter that can provide all the protections listed in the sections above is the first step in having an effective shelter. However, if the components are not tested or maintained properly, their reliability goes down and may significantly degrade the level of protection afforded in an actual release condition. Shelters are likely to have numerous issues like door gaskets not being replaced, detectors not calibrated on time, HVAC shutdown not tested regularly, and doors left open, among others. Hence, as shown in Figure 1, maintenance and inspection is necessary for the effectiveness of every component and in effect, for the overall successful operation of toxic shelter.

Testing Effectiveness of a Toxic Shelter

It is important that all toxic shelters be tested to quantify their effectiveness at mitigating potential toxic impacts. Knowing how effective a toxic shelter is at mitigating risk allows resources to be optimized in minimizing risk to the extent practical. For example, the importance of an effective fallback plan depends on how reliable and timely HVAC isolation is for the shelter, how tight it is once it is sealed, how likely a toxic plume is to impact the shelter, how intense the toxic plume is that may impact the shelter, and how long the toxic plume may be present. By evaluating these parameters, the toxic risk can be quantified, and the safety benefit of potential mitigation strategies can be determined.

Overall Shelter Protection Summary

As shown in Figure 9, there are many inter dependencies between the different protections afforded by a toxic shelter. The figure shows a simple event tree summarizing the potential outcomes, given the specifics of the accident scenario, design of the shelter, and functionality of the toxic mitigation systems.

1. Safe in shelter for duration of event.
2. Safe in shelter for long time, then safely evacuated.
3. Safe in shelter for long time, but eventually evacuated and perished.
4. Safely evacuated shelter shortly after impact.
5. Perished during evacuation shortly after impact.
6. Toxic impact.

Conclusion

Occupants of buildings can be effectively protected against toxic hazards if the building is properly designed and operated as a toxic shelter. Having an effective fallback plan and providing building occupants an appropriate level of training to ensure that they understand the fallback plan further reduces the associated risk. However, similar to an overall emergency response plan, one weak link in shelter design and philosophy may result in catastrophic impacts to building occupants and, therefore, the shelter must be designed in such a way to minimize weaknesses in the system. Unfortunately, a poorly designed shelter with inappropriate training and a poorly designed fallback plan may actually increase risk to its occupants by giving them a false sense of safety and preventing them from evacuating in a timely manner when that may be the best approach.

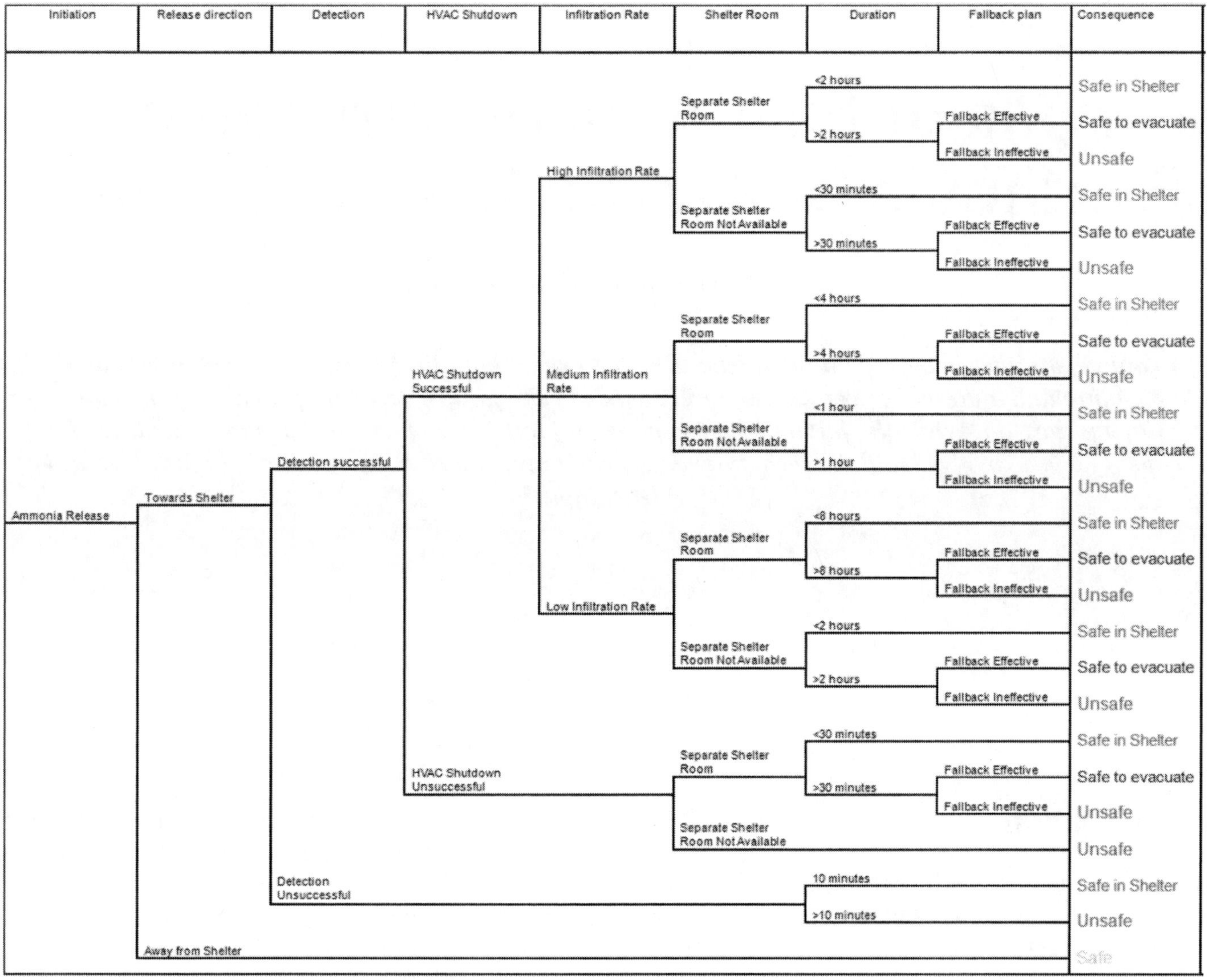

Figure 9. Event Tree for Toxic Shelter

References

1. API RP 752, "Management of Hazards Associated with Location of Process Plant Permanent Buildings", 3rd Edition, December 2009
2. API RP 753, "Management of Hazards Associated with Location of Process Plant Portable Buildings", 1st Edition, June 2007
3. OSHA, 1990. *The Phillips 66 Company Houston Chemical Complex Explosion and Fire, a Report to the President,* U. S. Department of Labor, OSHA.
4. CSB Report BP America Refinery Explosion, https://www.csb.gov/bp-america-refinery-explosion/
5. Fortress Protective Buildings (http://www.fortressprotectivebuildings.com)

Major Failure of an Ammonia Synthesis Converter Internals Due to Overpressure

A complete failure occurred on the inlet and outlet expansion joints for the first bed interchanger in an Ammonia Synthesis Converter. The direct cause of the failure was over pressure of the converter during start-up. While the failure of the expansion joints is the focus of the paper, the identified causal factors involved and the recommendations to prevent a re-occurrence are highlighted as key learnings.

Gary L. Acosta, Ousher Bhagwandeen, Nizam Abdool
Proman

Introduction

The inlet and outlet expansion joints for the first bed interchanger for the Ammonia Synthesis Converter on an Ammonia plant failed due to an excessive differential pressure during a plant startup on November 5, 2013. The failure occurred in an 1,850 MTPD (2,040 STPD) ammonia plant located in Trinidad and Tobago. This paper describes the failure that occurred, the investigations, the lessons learned, and follow-up actions necessary to prevent a re-occurrence.

System Configuration

The ammonia synthesis converter is a vertical "hot wall" design which incorporates three intercoolers and four radial flow catalyst beds. The internal gas flow pattern is arranged such that all the synthesis gas passes through all the catalyst beds.

The first catalyst bed contains a conventional promoted iron metal catalyst. The remaining three beds contain transition metal promoted graphite catalyst. All four catalyst beds use screens to form internal catalyst bed walls, which allows even gas distribution through the catalyst.

The graphite catalyst in Beds #2, #3 and #4 is not as robust as iron catalyst, and is highly susceptible to damage by moisture. Moisture will typically cause a reduction in catalyst activity if introduced to Beds #2, #3 & #4.

As seen in Figure 1, the catalyst bed interchangers #1, #2, and #3 are located within Beds #1, #2, and #3 respectively. Expansion joints are located on the inlet and outlet piping of each interchangers.

Synthesis gas feed is heated in the converter in a split stream of parallel flow to the interchangers #2 (between Bed #2 and Bed #3) and #3 (between Bed #3 and Bed #4), through manual valves

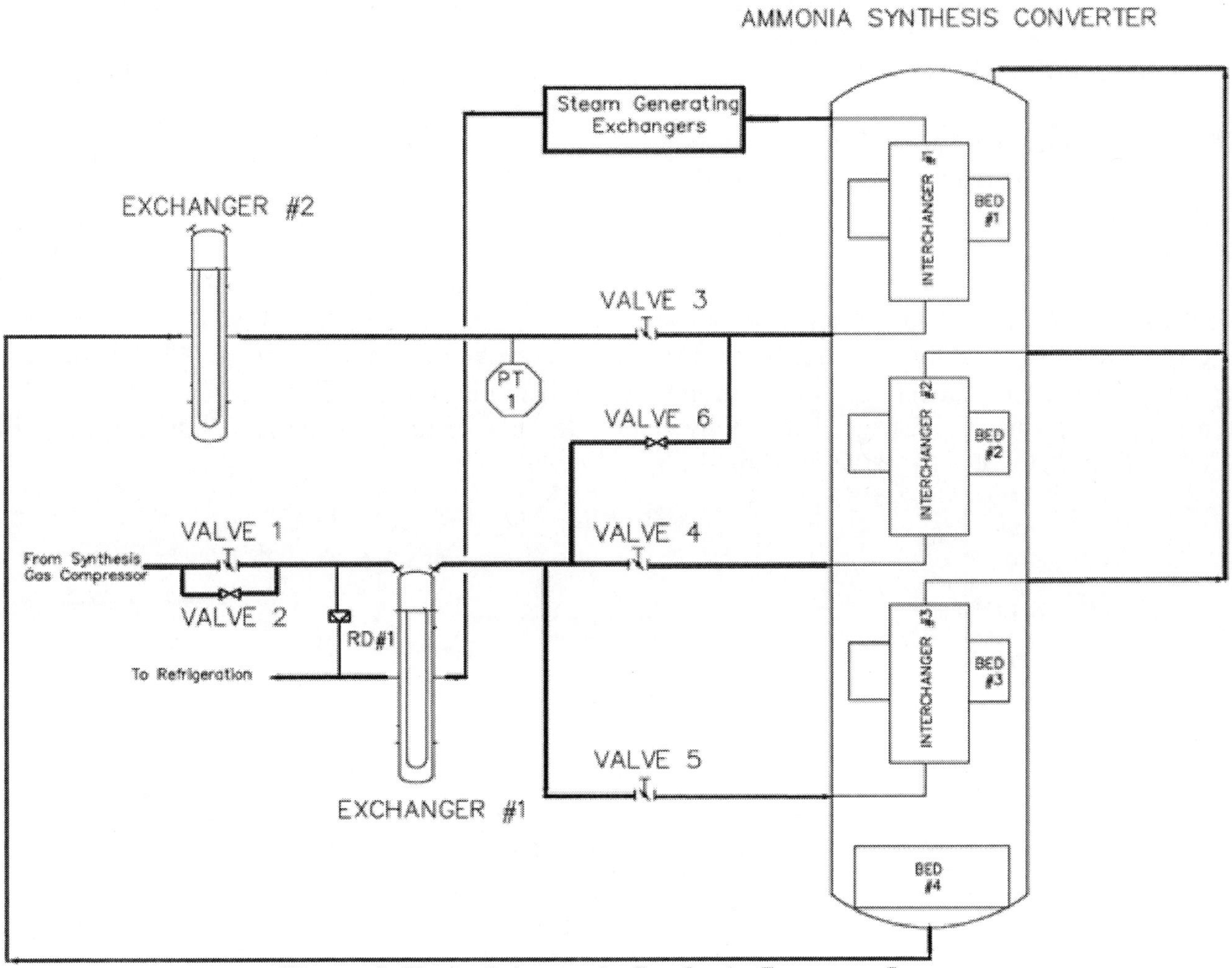

Figure 1. Typical Ammonia Synthesis Converter Loop

Valve #4 and #5, before entering the top of the converter where conversion to ammonia gas begins, traveling downwards through Beds #1 through #4, with cooling between each bed The effluent gas, that is ammonia gas together with unreacted nitrogen and hydrogen, then leaves the converter and is used to preheat boiler feed water in Exchanger #2. Passing through Valve #3, the cooled gas then enters Interchanger #1, absorbing heat from the reaction in Bed #1 before exiting the converter.

The effluent gas exiting Interchanger #1 is passed through steam generating heat exchangers before it is used to preheat the feed synthesis gas via Exchanger #1.

Incident Description

On November 5, 2013, the ammonia plant was being started. Synthesis gas supplied by the synthesis gas compressor provides synthesis feed gas at a discharge pressure of approximately 1300 psig (90 barg).

During startup, the Ammonia Synthesis Converter is slowly pressured up using a four-inch bypass valve (Valve #2), around the main twenty-inch inlet valve (Valve #1). This by-pass is controlled by manual gate valve.

Moisture reduces the activity of synthesis catalyst, especially the graphite based catalyst in Beds #2, #3, and #4. Because of this, the startup procedure requires the synthesis gas to be sampled for its moisture content prior to being introduced into the converter.

The only available standardized sample point at the time was downstream the shell side of Exchanger #1 as the final gas goes to the Refrigeration process.

A recently approved procedure to allow for sampling of synthesis gas before it is introduced to the converter via the parallel split flow through Interchangers #2 and #3, instructed that Valves #3, #4 & #5 be closed and Valve #6 should be opened. As seen in Figure 2, this would have allowed for flow from the synthesis gas compressor through Valve #6 and through the top bed interchanger for sampling of moisture at the shell side outlet of Exchanger #1.

It must be noted that this procedure was developed at a neighboring plant operated by the same service company in conjunction with the plant licensor. The neighboring plant utilized the new procedure on three occasions prior without incident and it was transferred unmodified, without review of the existing installation at the subject plant.

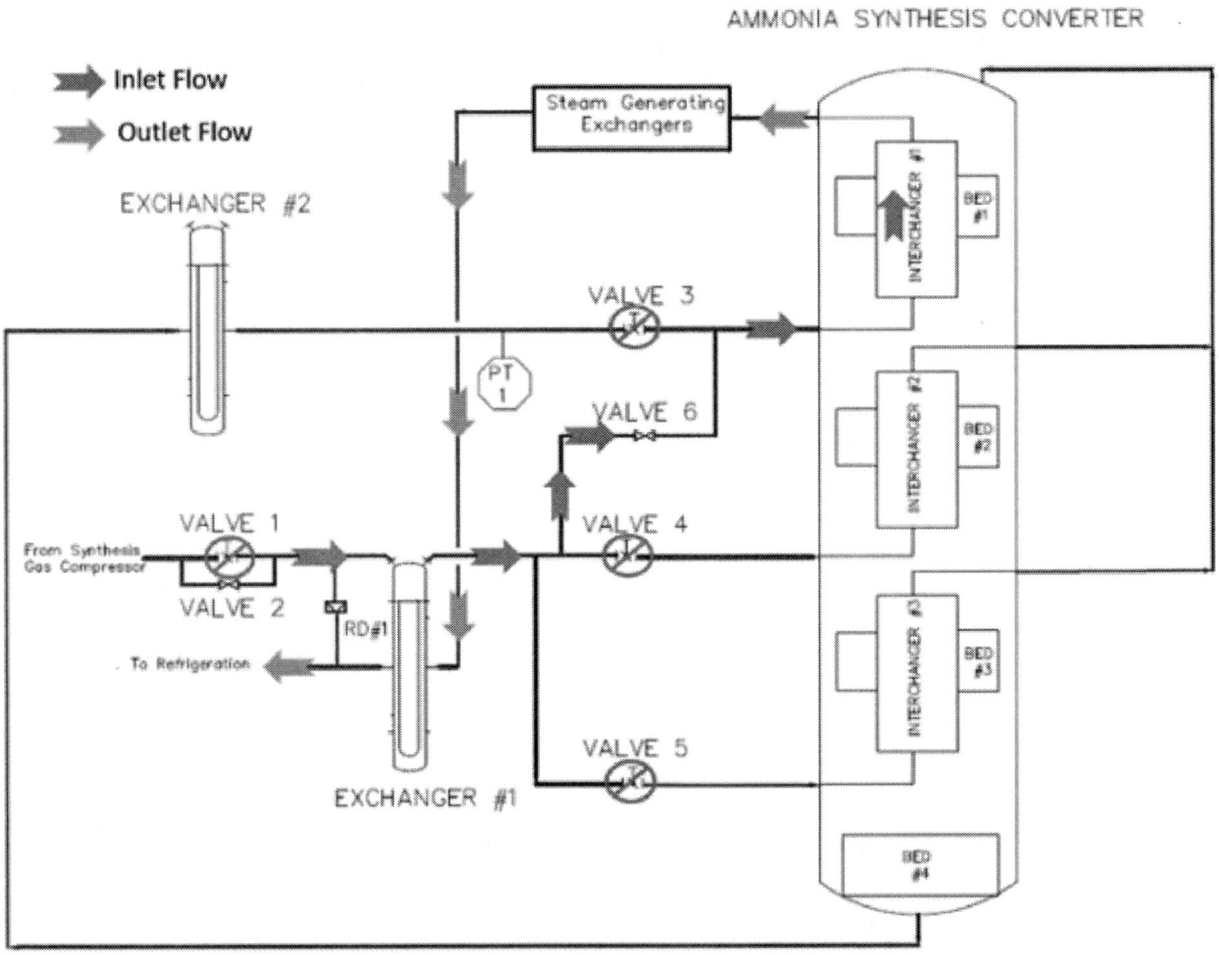

Figure 2. Typical flow path through Ammonia Converter during Gas Sampling

The subject plant underwent a major stabilization and upgrade project six months prior, which included the installation and commissioning of a new converter. In this new installation, a new four-inch manual by-pass valve (Valve #2) was installed around pneumatic Valve #1 in place of the previously existing one-inch by-pass valve. This four-inch by-pass installed to reduce the nitrogen consumption during the reduction of catalyst and to aid in the opening of Valve #1 by equalizing the high differential pressure across the valve during start-up. Gas was introduced across Valve #1, using the four-inch manual by-pass (Valve #2) with a field operator making minute adjustments, at 1:22pm on November 5, 2013. At 1:26pm, the downstream pressure measured in the Refrigeration area increased from 0 psig to 486 psig (0 to 33.5 barg). The field operator was required to listen for a flow across the valve and liaise with the Distributed Control System (DCS) Operator based on the system pressure. With Valves #3, #4, & #5 closed, the flow admitted through Valve #2 was uncontrolled due to the inability to make minute changes to the four-inch valve, causing the rupture disc RD#1 to rupture. RD#1 is rated at 290 psi (20 bar), and designed to protect the tubes of the Exchanger #1. When it ruptures flow is directed through the shell of Exchanger #1 and in the reverse direction of flow through the steam generating exchangers and into Interchanger #1 as seen in Figure 3.

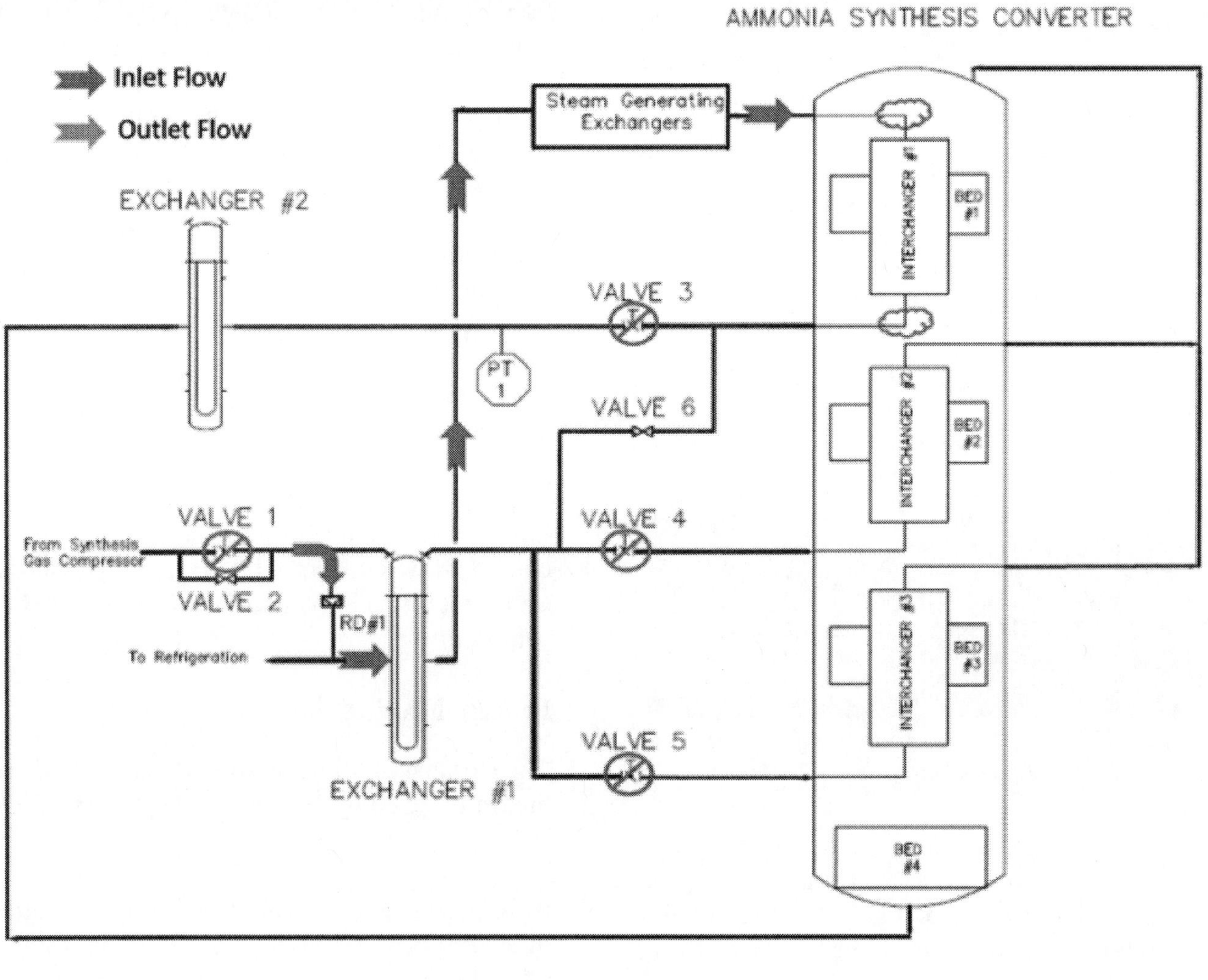

Figure 3. Flow path to illustrate reverse flow

The increase in differential pressure across the inlet and outlet expansion joints for the Interchanger #1 caused major failure of several components. The failure also promoted the by-passing of synthesis gas away from the catalyst beds once the converter was attempted to be placed into service.

Findings

Inspection Findings

Internal inspections of the Ammonia Synthesis Converter were performed to investigate the extent of damage and effect the necessary repairs to return the plant to production

Inspection of the top of Bed #1:
The top nozzle expansion joint was ruptured along the convolutions (Figure 4), while the other expansion joints seem to be deformed. The entire interchanger piping/elbow/expansion joint assembly appeared to be shifted in the direction away from the ruptured expansion joints and vessel shell and slightly upward. A small amount of catalyst was seen at the base of the annular space.

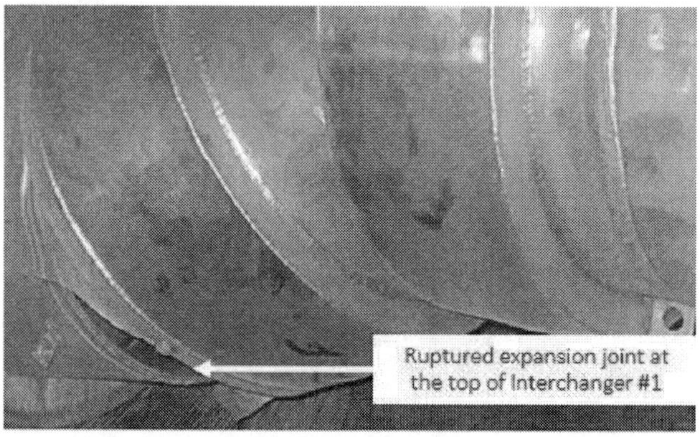

Figure 4. Ruptured expansion joint on Interchanger #1

A leak test of Interchanger #1 did not reveal any leaks after it was pressured up to 10 psig (0.7 barg) with nitrogen and held for 20 minutes. The top tube sheet of the exchanger was also visually inspected using remote video equipment through the outlet piping, but no issues were noted.

Inspection below Bed #1:
Both expansion joints and cover plates on the second nozzle expansion joint were deformed. Piping elbows of the Interchanger #1 bottom head and Interchanger #2 top head appeared to be very close to one another as a result of disfigurement of the second nozzle expansion joint.

A bulge was also detected on the straight piping entering the Interchanger #1 bottom head just below the downcomer can. Upon further inspection, it was discovered that there was also a bulge in the same section of the piping just above the downcomer can. In between the bulge, a reduction in pipe outer diameter (OD) of approximately 2.4 inch (60 mm) was seen.

Similar to the top of Bed #1, small amounts of catalyst were seen at the base of the annular space.

Inspection below Bed #2:
All internals of manway #3 were seen to be in satisfactory condition. The annular space was clean and clear and the perforated plate was in good condition. Leak checks were done on the expansion joints. No leaks were detected.

Inspection below Bed #3:
Catalyst was seen around the internal manway cover and was confirmed to be the metal catalyst. Leak test of the expansion joint found it to be acceptable.

Process Findings

The expansion joints on the Interchanger #1 have maximum design differential pressures of 72 psi (5 bar) for the inlet and 96 psi (6.6 bar) for the outlet. Process trends indicates that Interchanger #1 experienced a differential pressure greater than 150 psi (10.3 bar) between the shell and tube during the pressuring up of the converter whilst attempting the adopted procedure.

The rupture disc, RD#1, is designed to protect the tubes of the Exchanger #1. If a differential pressure between the shell and tube exceeds 290 psi (20 bar), the disc will burst/rupture to allow process to bypass the tubes.

The new procedure noted the requirement to ensure that the pressurization rate of the converter did not exceed 100 psi/hr (6.9 bar/hr) and was to be monitored by DCS pressure indication PT-1.

The inherent differences created a scenario that the expansion joints on interchanger #1 to be exposed to a differential pressure larger than its design.

The difference and highlighted concerns were that the DCS pressure indication PT-1 was located upstream of the closed Valve #3 and outside of the flow path for the sampling process. Therefore, panel operators did not see any change in the pressure while following the procedure. It was only realized following the incident that the pressure indicator on the outlet of the converter measured large pressure increases within a four-minute period.

Another difference was that the manual opening of Valve #2 to control the pressurization created a situation where gas was introduced too rapidly.

Based on the findings within the converter, the expansion joints for the inlet and outlet of the Interchanger #1 had to be completely replaced. In addition, the missing cover plate seen at the top of Bed #1 was also replaced and all catalyst noticed in the annular spaces were vacuumed.

To facilitate the removal of the inlet expansion joint for Interchanger #1, a section of neighboring pipe had to be removed. This section was reinstalled following the replacement of the expansion joint. The bulged pipe was not replaced. However, additional clearance was installed by grinding the downcomer can to allow for unrestricted movement.

The subsequent inspections and repair caused an additional plant downtime of twenty-one (21) days and incurred significant financial losses.

Discussion

Procedure Development

The sharing of knowledge across Ammonia production plants, especially those maintained and operated by the same company, is a common practice. The learnings and upgrades may be applicable across different sites of similar design.

The fact that previous reduced activity in the Ammonia Synthesis Converters at the different Ammonia plants were attributed to the introduction of moisture unto the graphite catalyst, meant that the previous operations had to be re-examined and modified. The drying and sampling of moisture content of the synthesis gas before it is introduced into the converter therefore was a critical step that had to be introduced to protect the catalyst within the converter.

The plant team was therefore tasked with conceptualizing and implementing a procedure which could satisfy the requirements of sampling without introducing possibly “wet” synthesis gas into the converter. At the time, the only available sample point within the loop was located downstream the shell side outlet of Exchanger #1.

The procedure therefore was developed to restrict the gas flow through Valves #4 and #5 to prevent gas going through the beds of the converter and to divert gas only through Interchanger #1 in order to get to the sample point. Two samples with a moisture content less than 0.5 ppm were required before pressurization of the converter was allowed.

Procedure Oversight

A fundamental flaw in the procedure however required operators to trend the pressure seen on PT-1 to ensure that the rate of pressure increase of the loop pressure during the sampling exercise was not excessive. This procedure was successful on the neighboring plant only due to the fact that Valve #3 was previously leaking through, allowing PT-1 to measure the loop pressure.

Since this converter, along with its manual valves, were replaced during a major plant stabilization and upgrade (S&U) some six months prior, Valve #3 was fully functional and provided proper sealing. Therefore, PT-1 was unable to measure the loop pressure during the sampling exercise.

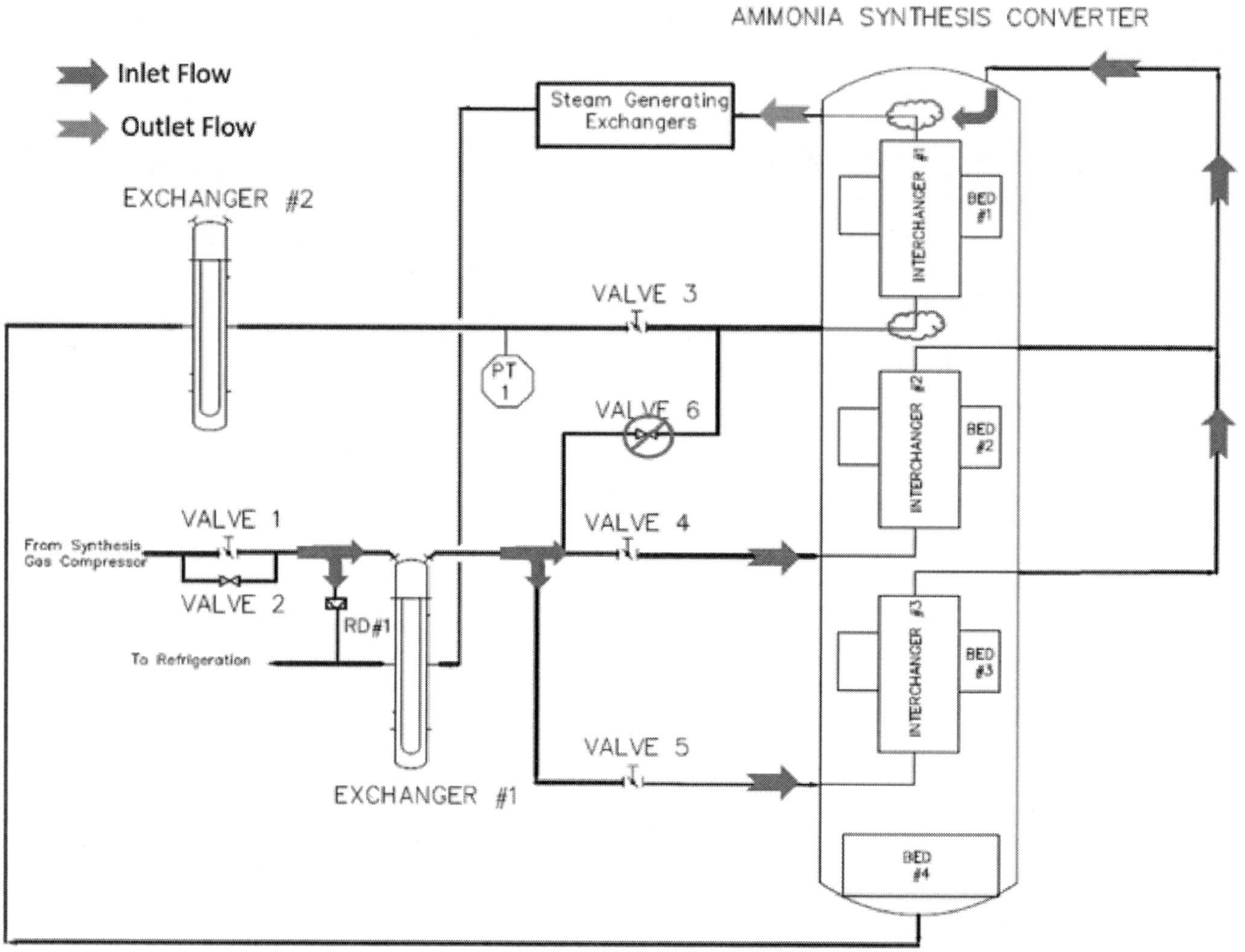

Figure 5. Synthesis Gas Flow Path following expansion joints failure

Another flaw in the procedure was that it did not take into consideration the size of the by-pass Valve #2 used to introduce the gas. During the S&U six (6) months prior, a four-inch valve was installed to facilitate the reduction of all four beds of catalyst since the one-inch valve would not have provided the required flow rates to reduce the catalyst. The replacement of this manual valve was approved and reviewed by a multi-disciplinary team prior to its installation.

The adopted procedure therefore did not consider the fact that operators were required to make minute adjustments on a four-inch valve instead of a one-inch valve to control the flowrates to the converter.

Misdirected/Reverse Flow

Rupture Disc (RD#1) is designed to protect the tubes of Exchanger #1 from a high differential pressure greater than 290 psi (20 bar). Pressure indications in the refrigeration section measured a pressure increases of 480 psi (33.1 bar) in a four-minute timespan. This therefore means that RD#1 would have been exposed to similar pressure differentials, causing it to activate and burst, directing flow from downstream Valve #2 to the shell side of Exchanger #1.

This new path allowed the flow of synthesis gas, in a reverse flow path, through the trio of steam generating exchangers and into the outlet of Interchanger #1.

The outlet and inlet expansion joints of Interchanger #1 is designed for a maximum differential pressure of 96 psi (6.6 bar). The introduction of synthesis gas feed at more than 480 psig (33.1 barg) would have caused damage to the expansion joints.

Note that the expansion joints on Interchangers #2 and #3 were undamaged due to the fact that the procedure called for the isolation of its inlet valves, Valve #4 and Valve #5.

Unfortunately, the overpressure and subsequent failure of the expansion joints were not recognized as gas sampling continued in the refrigeration area. Once the moisture content was acceptable, the start-up of the converter continued with the introduction of dry synthesis gas through Valves #4 and #5 and into the top of the converter.

The expected temperature increase across the catalyst beds after gas was introduced were immediately incorrect. This would have been due to the fact that the feed gas introduced through the top of the converter by-passed all of the beds as it took the path of least resistance through the now failed interchanger expansion joints and out of the converter.

The required reaction of the converter catalyst with the synthesis gas was by-passed preventing the production of ammonia from being completed.

Analysis and Cause

The activation of the rupture disc due to the increased differential pressure across it allowed for the reversal of flow into the interchanger. The rupture disc differential pressure increase was due to the inability to properly control the flow of synthesis gas using a manual four-inch valve.

The root cause of this incident was determined to be the uncontrolled introduction of synthesis gas to achieve the synthesis gas flow required for the sampling step. This compromised the ability to maintain the pressure differential parameters for the synthesis loop.

The design of the synthesis loop provided the panel operator with one point of pressure indication to determine the loop's pressure. This pressure indication however was isolated from the gas-sampling loop, effectively eliminating the operator's ability to properly measure and provide feedback to the field operator.

This therefore meant that it relied upon the field operator to listen to the valve to determine if the required flow rate was achieved. The management of change document which approved the replacement of the one-inch valve with the four-inch valve did not take into consideration routine plant start-up.

This change was made specifically for the reduction of catalyst following its replacement. Discussions with the plant designer was specifically related to the fact that a higher flow is required during the reduction of the catalyst and a minimum valve and line size of four inches was required.

The pressurizing of the synthesis loop is a complex task that heavily relies on the manual operation of valves and therefore opens the possibility of errors.

Another causal factor identified was the absence of a management system for document control or for changes to operating procedures.

While sharing knowledge across the plants was encouraged, a thorough and proper review of the procedures prior to adopting the procedure was required. If a physical walkdown of this system was performed on the facility, it would have identified the fundamental problems with the new procedure and may have prevented the failure.

Prior to the incident, there was no management of change documentation which adequately captured the change and thus encouraged a detailed review of the changes before it was implemented.

Recommendations

To prevent a reoccurrence, the following recommendations were made (Refer to Figure 6):

- Minimum stop of 15% on Valve #3:
- A physical minimum stop on Valve #3 will ensure that the PT-1 can monitor the pressure increase in the Converter as well as any possible misdirected flow, such as due to the rupture of RD#1, can be channeled out of the converter and hence does not create a high differential pressure scenario.
- Installation of DCS indications with alarms for the differential pressure between the inlet and outlet of Interchanger #1.
- Installation of automated converter pressurization scheme to provide a controlled and well measured system by which the converter is pressurized during a start-up.
- In order to properly sample for moisture prior to introducing gas into the converter, the start-up procedure will be altered to ensure sampling is performed on the discharge of the synthesis gas compressor.

Instrumentation Upgrades/Modifications

Valve #2 was replaced with two automated control valves that are in parallel with each other and the main control valve, Valve #1. Each valve is fitted with a digital valve controller that provides real-time position feedback and further analytics capabilities via the plant's asset management system:

- One one-inch automated control valve (Valve #2A) is used to pressurize the converter during routine start-up.
- One four-inch automated control valve (Valve #2B) which will be used during the reduction of catalyst

Three new pressure transmitters were installed to measure the loop pressure as indicated by PT-2, PT-3, PT-4. The median of these three pressure transmitters are then compared to PT-1 where the higher of the two will be selected as the loop pressure.

Wireless position feedback transmitters for manual Valves #3 through #6 and other valves on the outlet of the converter and to its start-up heater were installed. These wireless position feedback transmitters provide real-time and accurate valve position to the plant's DCS that has historical trending functions.

Pressurization Control Philosophy

The scheme was built such that the following permissive must be adhered to prior to allowing the pressurization of the ammonia synthesis converter. If any of these permissive are not adhered to at any time during the pressurization, the process is halted and isolated to prevent damage to the converter and its components:

- The synthesis gas compressor must be online

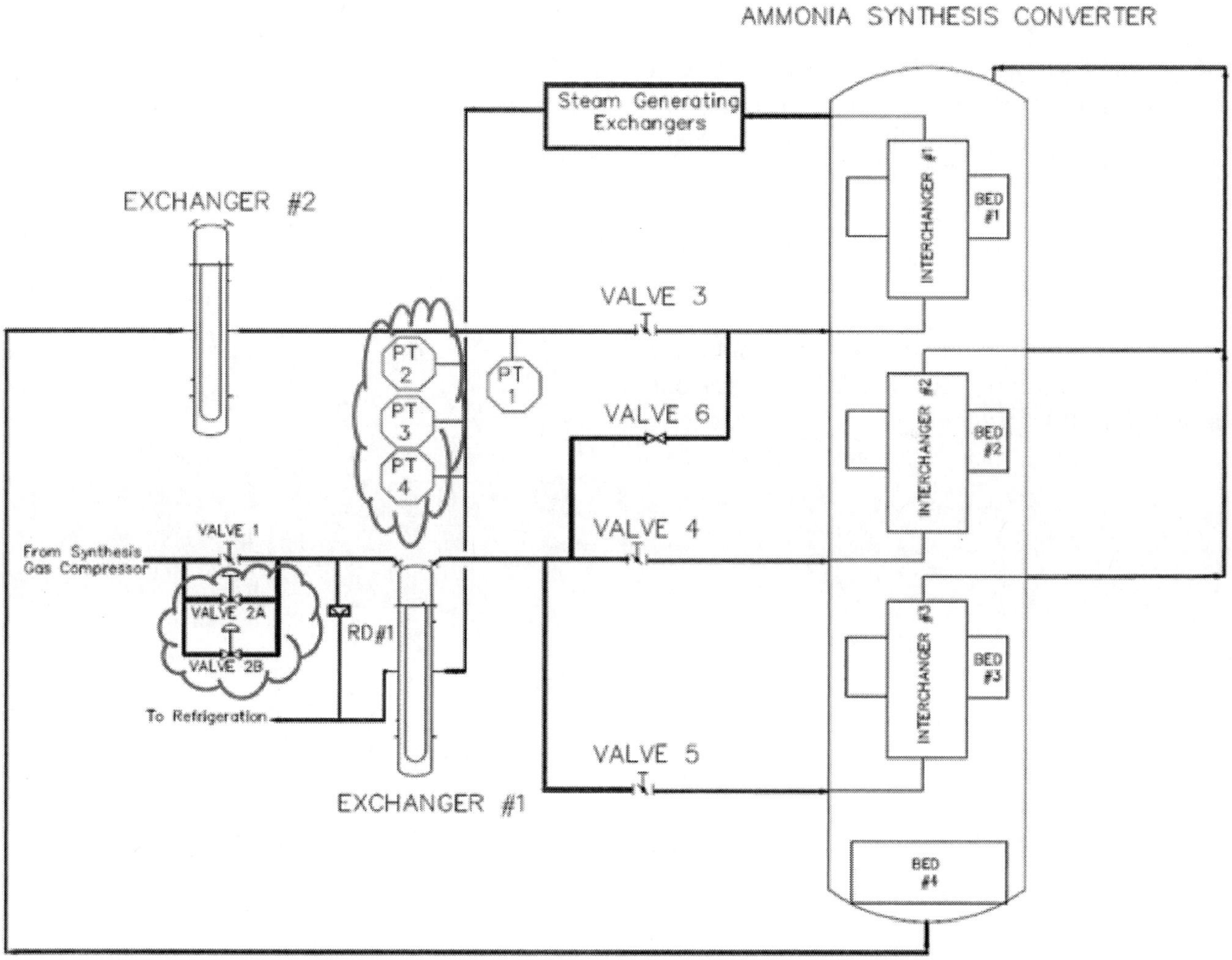

Figure 6. Modifications to Synthesis Loop instrumentation

- A series of valving arrangement with the converter and other valves relating to the system are required
- The rate of pressurization is no greater than 100 psi/hr (6.9 bar/hr)

A powered solenoid on the instrument air supply to the new automated control valves (2A & 2B) will only be energised and opened once the permissive above are met.

Once pressurization is ready to begin, the operator selects the target loop pressurization rate on the operator's screen. Using PID control logic, the DCS opens Valve 2A to achieve the required setpoint. Two alarms will alert Operators if the rate of pressurization increases past 95 psi/hr (6.6 bar/hr) and 100 psi/hr (6.9 bar/hr) respectively.

If any of the permissive mentioned above are not met, valves 2A and 2B are tripped to their close position.
An additional control also completed on Valve #1 which implements an interlock that prevents the opening of this valve unless the loop pressure measured by PT-1 through PT-4 is within 100 psi (6.9 bar) of the Synthesis Gas Compressor's discharge pressure.

Conclusion

Consequence of failure of expansion joints can be significant in certain situations. The exposure of personnel to unsafe conditions and the financial loss for repairs and downtime therefore enforces the need to have safeguards for existing systems and properly vetted procedures prior to implementation.

Modifications to systems such as these should be evaluated thoroughly considering the lessons learned from this incident.

Corrosion Failure of New CO_2 Stripper Reboiler in Benfield Service

The CO2 Stripper Reboiler was re-designed from a "U" tube to straight through flow to reduce tube side pressure drop and eliminate tube vibration. High heat transfer at one end of the exchanger led to excessive boiling, imbalanced liquid flow, higher temperatures, high velocity two phase flow, a vapor pocket, and loss of passivation. Localized corrosion of the carbon steel shell caused a leak after 8 months. Corrosion of carbon steel bolting caused another leak at 21 months. There was no corrosion to stainless steel components. Corrosion indicators were seen before the shell breach, but not fully understood.

Ryan E. Redd, Jordan Strunk
KOCH Fertilizer Enid LLC

Introduction

The Ammonia #1 unit at the KOCH Fertilizer plant in Enid, Oklahoma is a Kellogg designed, 1000 STPD (907 MTPD) nameplate plant, that was built in 1972. The capacity of the plant has been upgraded to 1650 STPD (1497 MTPD) through several upgrades.

The carbon dioxide (CO_2) removal system uses a potassium carbonate (K2CO3) solution with proprietary amine activator according to the UOP Benfield™ Process. This system directs hot saturated process gas through the tube side of the stripper reboiler exchanger where it gives its energy to the partial boiling of the Benfield™ solution.

In 2015 and 2017 the "U" tube bundle in the stripper reboiler suffered tube leaks due to vibration in the boiling zone of the exchanger. The decision was made to replace the heat exchanger and provide better support for the tubes to eliminate vibrations. This exchanger replacement was also seen as an opportunity to reduce the tube side pressure drop. Changing from a "U" tube bundle to a single pass tube arrangement would reduce pressure drop by 8-10 psi (0.6-0.7 bar). The re-design is illustrated in Figure 1.

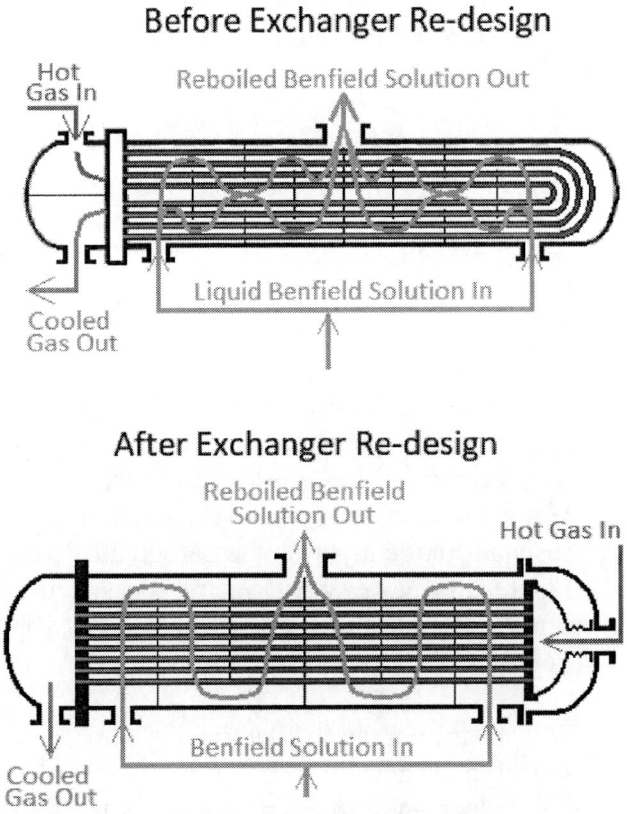

Figure 1. Heat Exchanger Flow Paths Before and After Re-design

We knew of similar design exchangers being used in reboiler service but did not know of any single pass floating head re-boilers in the UOP Benfield™ process. The UOP Benfield operating manual was consulted and nothing was found that indicated this exchanger style would not work in this application. The existing reboiler had 304 stainless tubes and a carbon steel shell, so the new reboiler was built using the same materials.

The replacement of the stripper reboiler was completed in July 2018 with results that seemed satisfactory at the time. The reduction in tube side pressure drop helped to achieve the plant capacity increase from 1550 to 1650 STPD (1406 to 1497 MTPD).

Early Warning Signs

Between the July 2018 startup and March 2019 three different groups within the plant made observations indicating corrosion was occuring:

1. The operations group found that the side-stream filters on the Benfield solution system were getting plugged with iron contamination more often than before.

2. Laboratory analyses indicated that the levels of Vanadium V_2O_5 (+5 oxidative state, active) were decreasing, and the VO_2 (+4 oxidative state, spent) levels were increasing. The function of V_2O_5 is to contribute an oxygen ion to iron which helps create a corrosion resistant oxide layer on the carbon steel. The lab results never went out of the specified range, so addition of chemical was not triggered.

3. The Instrumentation/Electrical group found multiple instances of iron fouling on magnetic float level instrumentation in the CO2 removal system.

All of these signs indicated iron/steel corrosion. Corrosion was suspected, but it was expected that any corrosion would be slow. Also, it was not possible to determine the location of the corrosion without a shutdown and internal examination of the equipment.

Incident Summary

On the morning of March 28, 2019 operators discovered Benfield solution dripping from the insulation jacket on the Stripper Reboiler exchanger. Within a short time the decision was made to shut down the plant.

Once the hot process gas flow to the exchanger was stopped, the leak rate increased significantly. The liquid leak was contained within the trench and sump system of the plant, and there was no off-site environmental impact from the leak.

The insulation was removed from the exchanger shell, and multiple holes were found on both sides of the shell approximately 10 ft (3 m) from the hot gas inlet tube sheet.

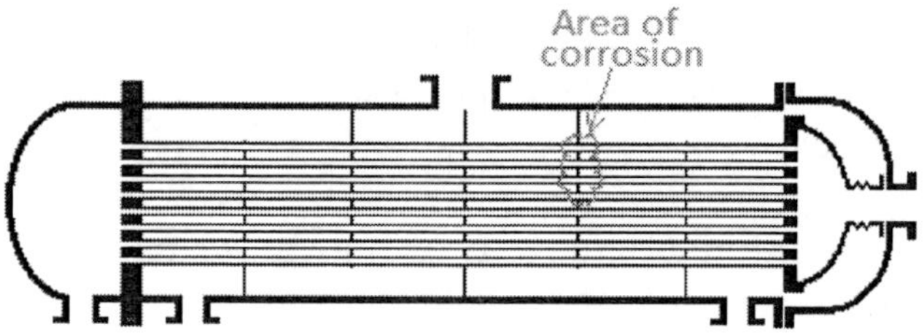

Figure 2. Location of Failure

Some of the holes coincided with one of the tube bundle baffles so that the baffle could be seen through the hole in the shell. It was interesting to note that the 304L SS baffle had apparently not suffered any corrosion, whereas the carbon steel shell had corroded or eroded through ¾" (19 mm) of material.

Finding Information

The heat exchanger was removed from the plant and sent to a shop for dis-assembly and repair. Once the tube bundle was removed from the shell, the full extent of the corrosion could be seen. It was clear that the worst of the corrosion occurred on the carbon steel shell next to the second baffle. There was also corrosion next to the intermediate vertical tube supports and at the horizontal seal bars. It looked as if the steel had been scooped out along several rough lines. An example of damage to the interior of the shell is shown in Figure 3, where the corrosion pattern can be easily seen. Figure 4 shows the 304 stainless tube bundle with the baffle, tube supports, and horizontal seal bars that correspond to the corrosion pattern in the shell.

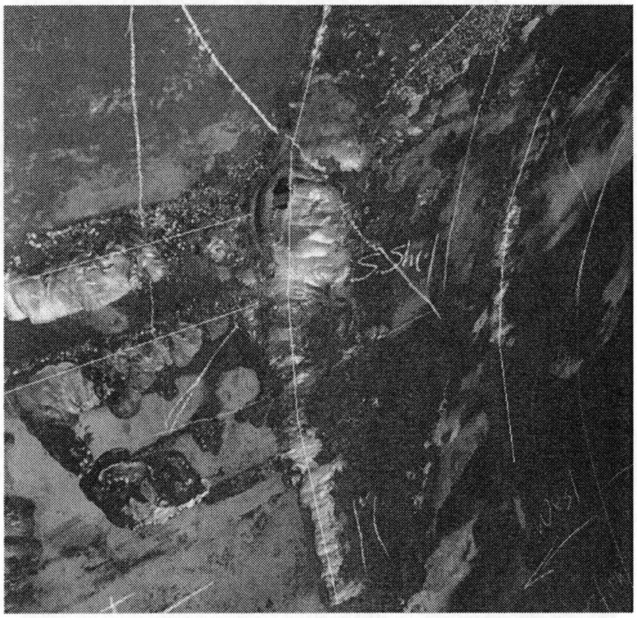

Figure 3. Corrosion of Shell corresponds with baffle, tube support, and horizontal seal bars

Figure 4. 304 Stainless Tube Bundle with baffle, vertical tube supports, and horizontal seal bars

Corrosion Analysis

A section of the shell with the worst corrosion, shown in Figure 5, was removed and sent to a metallurgical lab for analysis. The lab used visual examination, chemical analysis, hardness testing, scanning electron microscopy, and energy dispersive spectroscopy to determine the cause of the corrosion. Their conclusion was:

> *"The accelerated wall loss of the heat exchanger shell was due to cavitation and flow-assisted corrosion. Cavitation can occur in areas of boiling or pressure drops as the bubbles collapse. The cavitation removed the protective vanadium-rich oxide layers, allowing the surfaces to corrode and be eroded away by the high flow rates near the baffles."*

Figure 5. Corrosion on Interior of Shell with arrows showing the apparent direction of erosion/corrosion and dashed lines showing the un-affected weld zone.

Why Did This Happen?

Once we understood the full extent of the corrosion damage, we asked our engineering firm to model the exchanger with HTRI (Heat Transfer Research Institute) software not as one single exchanger, but as two separate units with equal surface area. We discovered that 76% of the heat transfer was taking place in the first half of the shell. The high vapor fraction in the first part of the shell was causing pressure drop or flow restriction, so only about 33% of the liquid solution was flowing to the hot end of the reboiler. The results for the first half of the exchanger were: higher temperatures, lower liquid flow rates, high vapor fraction, and high vapor/liquid velocity. We believe the combination of in-sufficient passivation by the liquid, elevated temperature, and high gas/liquid velocity dramatically increased the corrosion/erosion of the carbon steel. At the same time, the cooler side of the exchanger received 24% of the heat transfer into 67% of the liquid, which created very little re-boiling.

Mitigation

The shell was repaired using a stainless clad steel section where the worst of the corrosion had occurred.

In order to put the heat exchanger back in service, we concluded that we needed to correct the flow imbalance. A butterfly style throttling valve was installed in the line that feeds the cooler side of the reboiler. This would allow control of the flow of Benfield solution to that side of the shell, forcing more liquid flow to the hot side of the exchanger. (Figure 6)

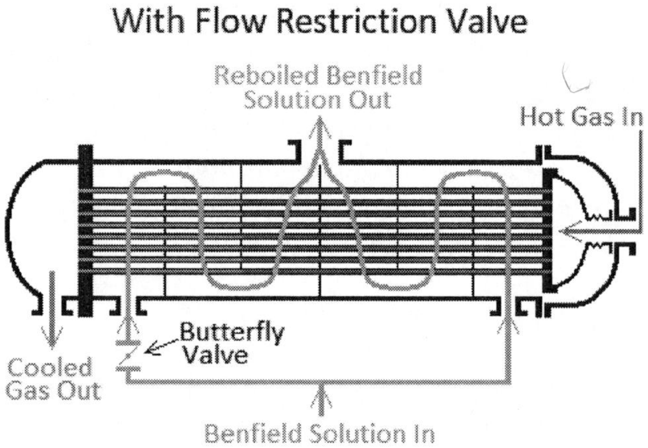

Figure 6. Arrangement Following Mitigation

Further Understanding

Once the exchanger was returned to service thermal imaging was used to assess the temperature profile.

Figure 7. Temperature Profile without Flow Adjustment

Figure 8. Temperature Profile with Flow Adjustment

Figure 7 is a thermal image showing the temperature profile without flow adjustment. In this view approximately 76% of the total heat transferred is applied to 33% of the liquid entering the exchanger on the hot side of the exchanger. The maximum temperature seen on the shell is 311°F (155°C). This temperature is well above the boiling point of the Benfield solution at this pressure and indicates that the vapor phase is superheated. It is also apparent that the area to the right of the second baffle is only half full of liquid, and half full of super-heated vapor. The apparent phase separation is forcing the vapors to bypass the second baffle instead of flowing under the baffle with the liquid.

The flow restriction valve was gradually closed until the best temperature distribution was reached. Figure 8 shows a more balanced temperature profile with a maximum temperature of 261°F (127°C), which is within the acceptable range for this process. The temperature difference across the exchanger cannot be overcome with a single pass tube bundle in this service.

The exchanger has run for approximately one year since the repair and mitigation were completed. Continued thickness monitoring since the failure has shown that the carbon steel shell is still experiencing corrosion, but at a slower rate.

Second Failure

On 29 April 2020 the stripper re-boiler suffered an internal failure, releasing hydrogen rich process gas into the potassium carbonate solution, and ultimately into the CO_2 stream. When the floating head cover was removed, a liquid level line clearly indicated that this area was only about ½ full of liquid during operation. This left the top half of the floating head bolts dry and unpassivated. Two of the grade B7 (carbon steel) studs on the upper half of the floating head were already broken off, and several more broke under torque from a hand wrench. In contrast, all the stainless steel, and the B7 studs on the lower half of the head that were covered by the Benfield™ solution were in near perfect condition. The failure of the studs surprised us because we thought high gas or liquid velocity was one of the driving factors in the corrosion/erosion of the shell in the first failure. This makes it clear that high vapor/liquid velocity is not required for carbon steel corrosion if the metal is hot and not fully wetted and passivated. Bolt stress could also have contributed to the corrosion.

The floating head was re-assembled with B8 (stainless steel) studs and the exchanger was returned to service. The B8 studs are not as strong as the B7 studs but do have sufficient strength to assure sealing of the floating head.

Key Learnings

1. The Benfield™ process includes corrosion inhibitors in the liquid solution. It is critical that any carbon steel be constantly covered and passivated by the liquid.

2. A "U" tube bundle distributes the heat across an exchanger better than a single pass bundle. Modeling should be used to understand the behavior of the exchanger before changing.

3. If there is a possibility of phase separation in the shell side fluid, the exchanger design should either eliminate or mitigate this.

4. Stainless Steel is corrosion resistant in the CO2 removal process, and should be used if constant passivation cannot be assured.

5. A rigorous management of change process must be used, and best knowledge sought out if re-design of equipment is considered.

Path Forward

Through two costly failures of this reboiler we have gained understanding of the failure modes. Our team has determined to replace the shell and floating head with stainless steel and install a vent line to eliminate the vapor pocket inside the floating head cover during our 2021 turnaround.

While we believe that this exchanger can be made reliable, we do not recommend a single pass in this service. We recommend staying with a U tube bundle exchanger for potassium carbonate solution reboiling.

ActiSafE - The Next Level of Safe Ammonia Synthesis Catalyst Reduction

Clariant has developed a safe, efficient and accurate method for measuring both water vapor formation and ammonia concentration during ammonia synthesis catalyst activation. Ammonia synthesis catalyst is delivered as iron oxide and the activation of the catalyst consists of a reduction of iron oxide to elementary iron which also forms water vapor. ActiSafE represents the latest technology for measuring water vapor formation at the outlet of the converter during ammonia catalyst activation, and provides several advantages such as continuous real-time measurements of the water vapor and ammonia concentration The real-time data allows plant operators to respond quickly to changes in water vapor formation and to optimize the reduction procedure, saving time, avoiding catalyst poisoning, obtaining a longer catalyst lifetime, and higher ammonia production.

In this paper, the fundamentals of ammonia reduction are described and an interesting avenue for the next level of ammonia activation is introduced. To illustrate the accuracy and advantages of ActiSafE, the case study of water vapor measurements during activation of Clariant's AmoMax® 10 ammonia synthesis catalyst at Yara's ammonia plants in Tertre and Yara Hull are demonstrated.

Dirk Sperling
Yara Technology and Projects

Scott Osborne, Ahmad Ghani, Stephan J. Reitmeier, Stefan Maier, Rene Eckert
Clariant Produkte (Deutschland) GmbH

Introduction

Ammonia synthesis catalyst is delivered as iron oxide and the activation of the catalyst consists of reduction of iron oxide to elementary iron, using hydrogen in the synthesis gas. During the reduction, oxygen adhering to the iron crystallites reacts with hydrogen and forms water vapor, which is separated by condensation in downstream processes.

The reduction does not produce nor consume heat in any noticeable amount. However, heat must be supplied to reach catalyst temperatures where the catalyst is reduced. It should be mentioned the faster the catalyst heating rate, the higher the formation of water vapor.

The water vapor formed is a poison for the synthesis catalyst, and has a direct impact on the formation of active sites; therefore, the formation of water vapor must be limited to certain concentration levels during the reduction. The main control criteria for the reduction of ammonia synthesis catalyst is a steady gas flow with a steady heat input, giving a well-controlled, moderate temperature increase in the catalyst bed(s) to keep exit water vapor levels within the desired limit.

Based on the information above, it should be clear, that controlling the water vapor produced via the catalyst heating rate is extremely important during the reduction. High heating rates can lead to excessive water generation, resulting in low catalyst activity, while too low a heating rate might cause a delay in production and lost revenue. Therefore, an accurate and continuous measurement of water vapor formation during the catalyst activation is crucial.

By using ActiSafE, a Non-Dispersive Infrared Technology (NDIR) based technology, the formation of water vapor can be monitored continuously, so the speed of reduction can be controlled and optimized to obtain the highest catalyst activity, fastest activation time, shortest time to production, and thereby better profitability of the plant.

Impact of Water Vapor Formation on Catalyst Activity

Proper activation of the ammonia synthesis catalyst is very important, mainly because the pore structure is developed during the reduction. The catalyst pore structure is a very important parameter for catalyst activity because it facilitates gas access to the active sites. A slow reduction provides a fine pore structure and a large surface area in addition to a maximum number of active sites, hence ensuring the highest possible catalyst activity.

The formation of water vapor as a product of reduction has a negative impact on the formation of the pore structure and it will poison the catalyst active sites. Furthermore, too high a level of water vapor will inhibit the ammonia reaction, which is exothermic, and the heat of this reaction is needed to assist in faster heat-up of the downstream catalyst beds since total heat input is usually limited by the capacity of the start-up heater. As a world-leading company within the field of catalysis, Clariant did multiple investigations to assess the impact of water vapor formation on catalyst activity during the reduction. Figure 1 illustrates that maximum catalyst activity (this is reflected by ammonia concentration at the outlet) is obtained at low water formation during the reduction. In this experiment, water formation was adjusted by varying the flow rates, while maintaining the heat input and pressure constant. The results also show AmoMax outperforms the other ammonia synthesis catalysts (magnetite type) over the range of relevant water vapor formation.

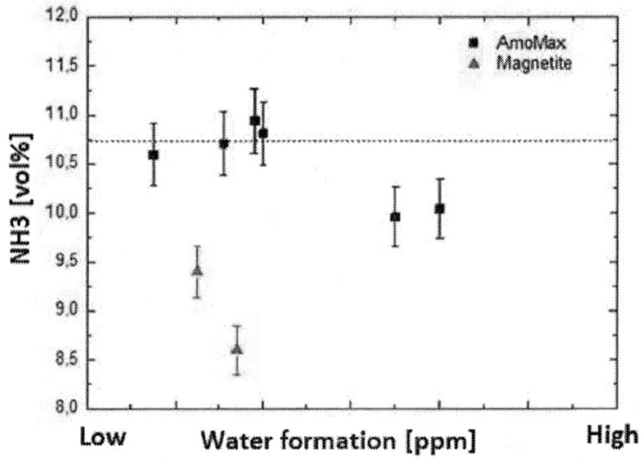

Figure 1. Impact of water concentration on catalyst performance

In this experiment, the maximum water concentration during reduction was adjusted by varying the flow rate of feed gas. A low flow rate results in a high-water concentration and vice versa.

Conventional Water Measuring Method vs. ActiSafE

Due to the adverse effects of water, it is important to be able to determine the water content at the outlet of the converter correctly. Currently, three analytical methods are widely used to determine the water vapor formation at the outlet of the converter during the reduction. These commonly used methods are known as Ascarite, Carbide method, and Karl Fischer titration method. All these methods have some operational challenges such as:

- Human errors, as the accuracy of the method requires highly experienced laboratory technicians.
- Methods one and three are far too cumbersome to be used by ordinary laboratory technicians. These methods are very time-consuming and provide information about the water content of the sample only after a certain delay (the result of each reading may take 3 - 4 hrs).
- These same methods require a lot of chemicals and a high number of technicians.
- Condensation in the sampling line, may not be detected on time, giving an error reading.
- Due to the discontinuous nature of the measurements (time delay), the speed of catalyst activation is normally slow, this might cause a delay in production that may cost a lot of money. If the water formation exceeded the limit, ammonia reaction can be inhibited and will prolong the activation. For example, one-day production loss can cost 700,000 USD for 2,000 MTPD (2200 STPD) ammonia plants (assuming an ammonia price of 350 USD/MT).
- The time delay of the measurements will not allow plant operators to counteract with any sudden increase in water concentration (e.g. by reducing the temperature) before permanent damage is caused to the catalyst.

In contrast, the ActiSafE method developed by Clariant provides several advantages, which include:

✓ **Continuous real-time measurement of H_2O and NH_3**

Continuous real-time measurements of the water concentration in a range of 100 – 6,000 ppm while also measuring the ammonia concentration at the same time.

✓ **DCS system connectivity**

There is no requirement of manual operation as the instrument has an analog output and can be connected to the DCS system of the plant for monitoring the real-time values in the control room.

✓ **Quick reaction time**

The real-time data allows plant operators to react quickly to changes in water content and to optimize the reduction procedure, thus saving time, avoiding catalyst poisoning, which will result in longer catalyst lifetime and higher ammonia production.

✓ **Faster and safer reduction**

By using ActiSafE, the reduction itself can be sped up within a safe range and obtain the highest possible activity without any risk of production delay.

✓ **High activity and optimized operation conditions**

Measuring ammonia concentration at the converter outlet, enables to optimize the performance of the catalyst by adjusting the bed temperature, hence ActiSafE will not only ensure high catalyst activity but also optimize catalyst operating conditions and thereby the highest possible profitability of the plant.

ActiSafE Demonstration

ActiSafE has delivered on its promises when it has been tested in the field. During the field test, the accurancy, reliability and practicality of Clariant ActiSafE have been evaluated.

ActiSafE equipment consists of a single benchtop device, providing easy installation and use. It only needs to be supplied with electricity and lined up to the sample line, where care should be taken to avoid condensation. For optimum utilization, the device should be connected to the DCS via the analog interface, providing real time data in the control room.

The real-time measuring of both water vapor formation and ammonia concentration at the outlet of the converter was measured, the accuracy of water vapor measurements was compared to Karl Fisher titration method, and the results reveal very high accuracy. The accuracy of the ActiSafE system is presented in Figure 2.

The Karl Fisher titration method requires some time to collect samples, meaning that an average value is obtained corresponding to a time interval, this means a peak of water vapor if any, would not be detected as highlighted in red in Figure 2. Figure 3 below represents the accuracy of the Clariant ActiSafE ammonia measurements when compared to the Gas Chromatography (GC) measurements.

As illustrated in Figures 2 and 3, the accurancy of ActiSafE measruments has been verified during the catalyst activation. Furthermore, the figures underline as well the reliability of the ActiSafE measurements.

Yara

As a world leader in production of nitrogen-based fertilizer, Yara is always adapting and pushing forward for new technology conversion leadership and away from traditional methods. For this reason, Yara selected Clariant ActiSafE to control the speed of their ammonia synthesis catalyst activation at their plants, which are namely Tertre and Hull.

ActiSafE showed reliable results during synthesis catalyst activation at Yara and was in principle able to optimize and speed up the activation process without stressing the catalyst with too high-water vapor formation. The information was available also during process disturbances, at any time of the day, without the requirement of additional laboratory staff or a resultant time delay. In addition, instantaneous concentrations and not averaged values are measured giving a more realistic picture of the situation when compared with other techniques such as the Karl Fischer titration method. A more realistic picture will increases the confidence the operator has with respect to full control over the reduction procedure. This is crucial as ammonia synthesis catalyst reductions are quite rare events for the production site and the majority of the operators will experience them just once in their work life.

Conclusions

Ensuring longer ammonia synthesis catalyst lifetimes with maximum ammonia production is now possible, thanks to Clariant's state of the art ActiSafE technology. During the reduction of AmoMax® 10 at Yara Tertre and Yara Hull, ActiSafE was successfully applied to monitor the water vapor formation and ammonia concentration with high accuracy. The reduction speed was optimized and increased without stressing the catalyst with too high-water vapor formation.

As a conclusion from Yara, ActiSafE service provided additional benefits, with less effort compared to the traditional methods and increases the process safety due to improved monitoring of the catalyst reduction. Finally, a well performed and monitored reduction was a key contributor to the good ammonia synthesis catalyst performance at Yara.

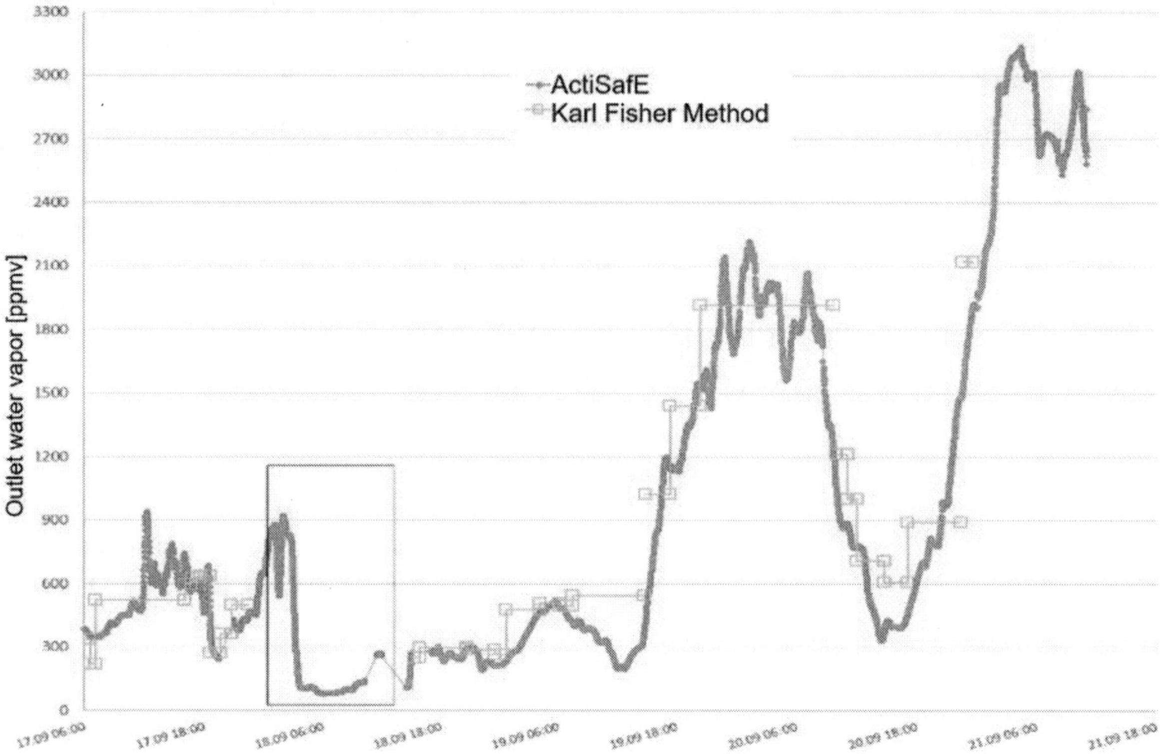

Figure 2. Karl Fischer method vs. ActiSafE during ammonia activation

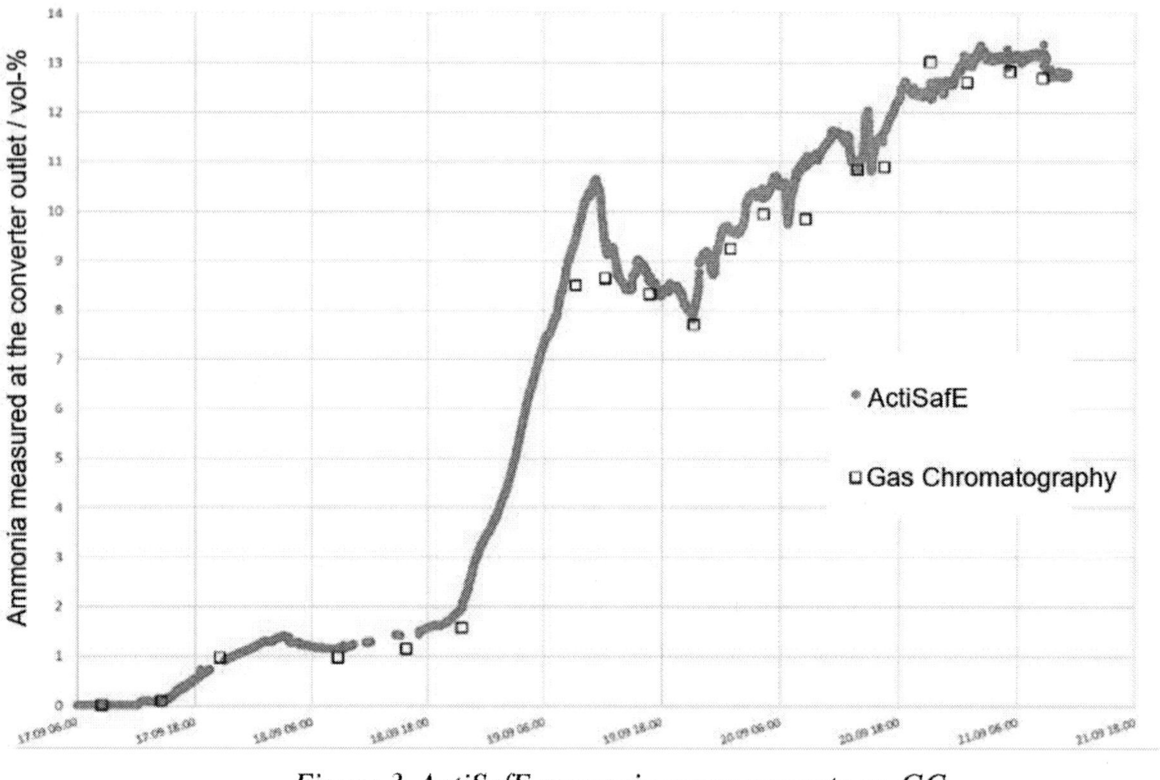

Figure 3. ActiSafE ammonia measurements vs. GC

Failure of a 316H Stainless Steel High Pressure Steam Superheater Header

The outlet header of a high pressure steam superheater located in the convection section of a primary reformer failed catastrophically leading to significant damage to surrounding equipment and extended plant downtime. This paper describes the header design, fabrication, plant operation, and metallurgical investigation that led to the identification of stress relaxation cracking as the principal damage mechanism. The failure occurred despite compliance with the ASME code.

Charles Thomas
Quest Integrity NZL Ltd

Maher Hussein, Mohamed Medhat
Egyptian Methanex Methanol Company S.A.E.

Introduction

The Egyptian Methanex Methanol Company S.A.E. (EMethanex) operates, in a joint venture with Egyptian partners, a 3500 MTPD (3850 STPD) methanol facility at Damietta on the Mediterranean coast. The plant has operated since 2011. In that time, the plant has operated continuously and satisfactorily other than ongoing issues with gas supply which has resulted in a number of outages. In 2019, the plant suffered a major shutdown caused by the catastrophic failure of the outlet header of the steam superheater in the primary reformer convection section. This failure was dramatic, caused damage to surrounding infrastructure, was heard over a large part of the local town, and led to an extended outage for repair but thankfully resulted in no injuries to personnel or in the neighbouring community.

Background

The high pressure (HP) steam superheater is located in the convection section of the primary reformer. It is the first coil to be exposed to the flue gas exiting the reformer radiant section. Figure 1 shows the location of the HP superheater within the convection section and the layout of the tubes and header. The inlet and outlet headers for the steam superheater (1E-0301) are identical in design. It was the outlet header, which operates at a higher temperature, which failed.

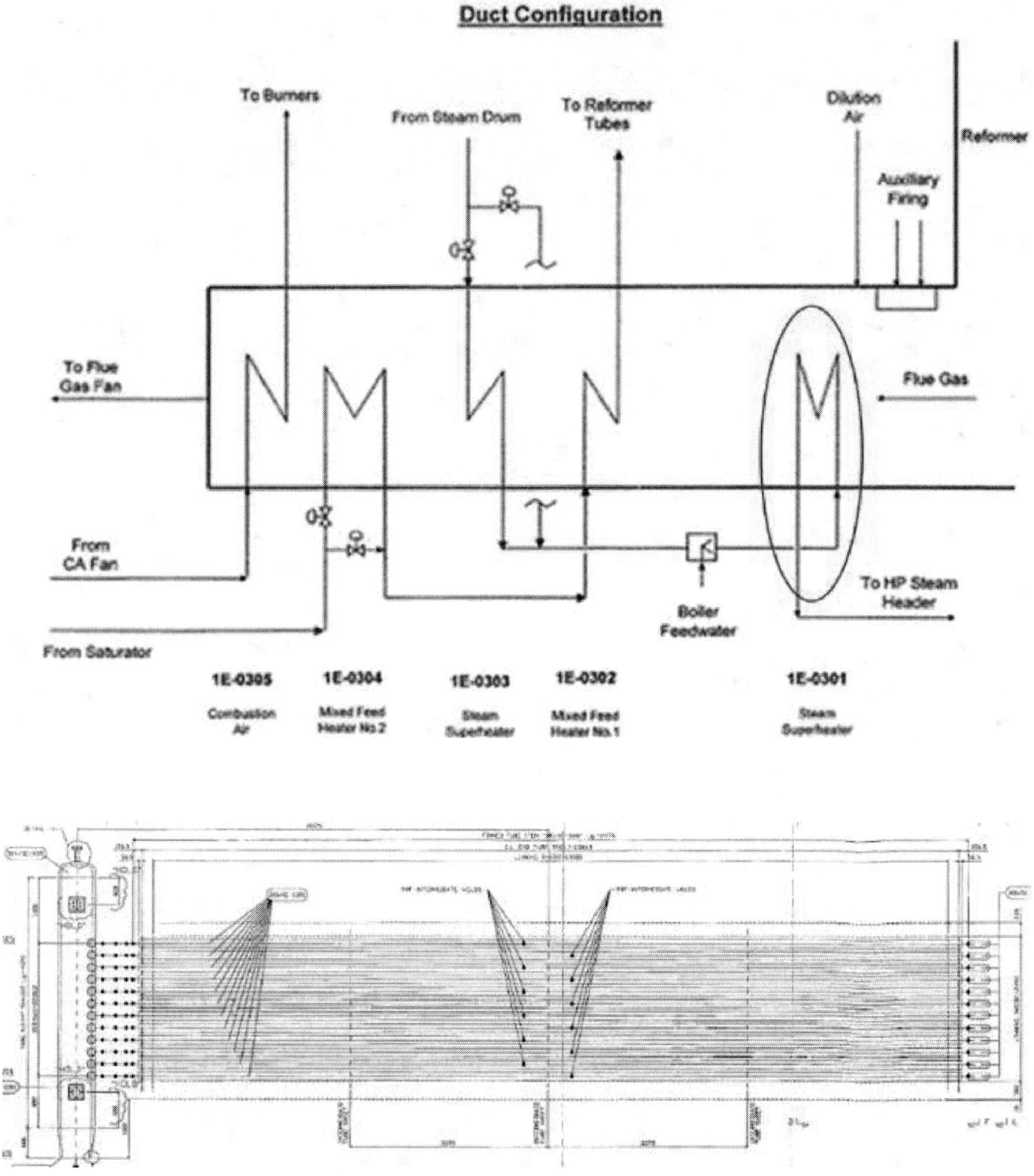

Figure 1. Duct configuration and superheater design

Operating and design data for the HP superheater coil and outlet header are shown in Table 1.

	Operating Data	Design Data
Steam Pressure	110-115 barg (1595–1670 psig)	125 barg (1815 psig)
Outlet Steam Temperature	510-520°C (950-968°F)	545°C
Flue Gas Temperature	960-1020°C (1760-1868°F)	
Outlet Header Material	SA-240 Gr316H Stainless Steel	
Outlet Header Dimensions	Diameter 609.6 mm (24 in) Wall Thickness 46 mm (1.8 in)	

Table 1. Operating and Design Data.

Operating data in the time up to the failure was reviewed and no unexpected or abnormal operating conditions were identified. This data revealed that at the time of the incident, pressure fell extremely rapidly from normal operating pressure to 10 barg (145 psig) in 5 seconds.

The headers were vertical in orientation and located in an insulated box on the side of the flue duct. Although not subject to direct contact with flue gas, their external environment may have been hotter than the plant design basis. Both inlet and outlet headers were identical in design and were manufactured from 316H stainless steel plate, and cold bent to a circular profile and then seam welded. Although designated in pipe terminology as 20NB schedule 120 they were in fact designed and manufactured to meet the requirements of the American Society of Mechanical Engineers (ASME) Section I – Rules for Construction of Power Boilers standard. The remains of welds used to locate temporary clamps (so called "dogs") to hold the cold bent cylinder closed while welding took place could still be seen on the external surface. The seam weld did not interact with any of the tube weldolets.

Figure 2 shows the failed header in the immediate aftermath of the incident. The header had ruptured, and a large flap of metal opened up after two independent cracks had formed along both sides of the seam weld. These cracks propagated along the seam weld approaching each other until they merged by jumping across the weld, at which time the failure occurred. The upper crack originated on the internal surface while the lower crack originated at the external surface.

Figure 2. The failed header and the full length of the crack

The crack surfaces had a rough but uniform topography with no indications of arrest marks that might indicate fatigue damage. Similarly, no obvious defect was found to indicate the origins of the cracks. The header surfaces were smooth with no indication of corrosion. Indeed, markings from the time of manufacture could still be seen.

The Damage Mechanism

Laboratory examination determined that the material conformed to the requirements of alloy 316H in all respects. A number of samples were removed for laboratory examination. The material was found to conform to the compositional requirements of alloy 316H. Figure 3 shows a section of the seam weld containing the crack. The crack in this sample ran from the weld toe on the external surface. The origin was in the heat affected zone but deviated from the weld towards the inner surface and is shown in more detail in Figure 4.

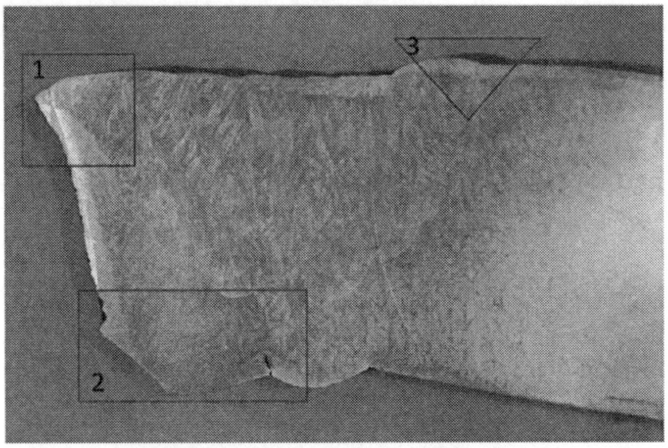

Figure 3. Section through weld and the crack. The fracture surface is to the left

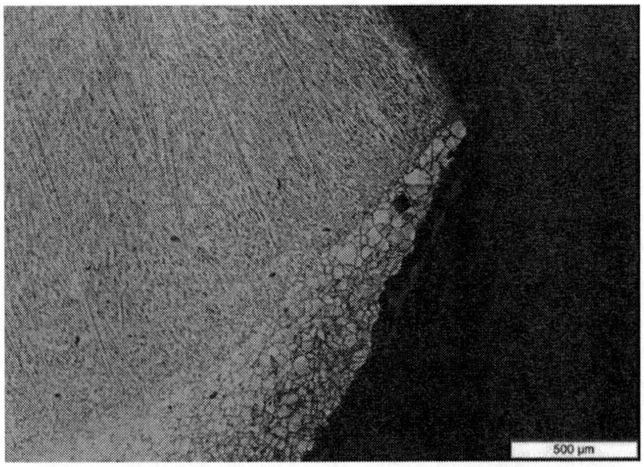

Figure 4. The fracture origin. The diamond shape is a hardness test indent

The material was heavily cold worked and the crack ran parallel and close to the boundary between the cold worked material and the material that was effectively annealed by the heat of welding at the time of manufacture. The hardness of the cold worked material was up to 300HV compared with expected hardness of solution annealed material in the order of 180HV.

Creep could be readily dismissed as a damage mechanism due to the operating temperature and absence of features associated with creep. Similarly, the fracture surface showed fatigue had not been active, nor was there any evidence of corrosion or stress corrosion cracking related mechanisms.

The metallurgical observations identified the failure mechanism as stress relaxation cracking. Stress relaxation cracking occurs when relaxation by conventional plastic deformation processes are inhibited. Relaxation is a small amount of creep or plastic deformation, typically driven by secondary or residual stresses. The deformation acts to eliminate that stress and this is what occurs in a stress relief heat treatment. In some circumstances, however, conventional plastic creep is prevented and the relaxation of imposed stresses occurs by intergranular cracking i.e. the material has very poor creep ductility. Van Wortel [1] has summarized the characteristics of stress relaxation cracking as being:

- Cracking is intergranular and isolated creep like cavities are present ahead of the crack.
- Cracks are in the heat affected zone, weld metal or cold worked material.
- Frequently, a metallic filament is present. In austenitic materials such as 316H, the filament is typically nickel rich.
- Cracking occurs in area where material hardness is elevated typically exceeding 200 HV.
- The temperature range is 550-750°C (1022-1382°F), the temperature of peak susceptibility being material specific.
- Many failures occur early in life typically within the first year of service.

These factors were met but not in total. In this case the temperature of operation was below the stated range. Other sources [2, 3] have reported that damage by this mechanism may occur at temperatures as low as 500°C (932°F). In either case, the operating temperature of the super-heater header was at the extreme low end of the commonly accepted range for stress relaxation cracking.

The header is heated by radiation from the surrounding hot flue-gas that can leak into the header box from the flue gas duct and insulated header box walls and cooled by the steam flow through it. Convection within the header box and the lower steam flow will make the metal temperature at the top of the header higher than the steam temperature. Calculating this temperature is difficult so design codes require an allowance of at least 25-30°C (45-54°F) above the maximum steam temperature. In the absence of a measured tube metal temperature, it is expected that the temperature in the failed section was at least 545°C (1013°F) and possibly as high as 560°C (1040°F), which would be in the lower end of the normal temperature range for stress relaxation cracking and just within the lower bound for creep.

This may explain the absence of other typical characteristics in this case. No voiding was associated with the crack and the failure occurred after eight years of service. The presence of nickel rich filament, commonly associated with stress relaxation cracking, was observed (Figure 5).

Figure 5. Detail of the fracture surface showing the filament characteristic of stress relaxation cracking and evidence of cold work

The poor creep ductility that leads to cracking is frequently caused by a hardening mechanism. In alloy 800H, for example, ductility loss may be due to gamma prime precipitation. In 316H material there is no obvious hardening mechanism other than grain size effects. In this case, there were high levels of cold work due to the original manufacturing method. Cold worked material is more susceptible to stress relaxation cracking [1]. In addition, thinning of the material and flattening of the surface adjacent to the weld were indicative of cold working during manufacture. The header did however conform to the ASME code which allows cold strain of up to 20% for design temperatures below 580°C (1076°F) [4]. The failed header satisfied both of these criteria, having a calculated cold strain of only 8% and a design temperature of 545°C (1013°F).

Regarding stress levels, a finite element analysis revealed the cold bending used to manufacture the header had generated residual stresses of approximately 400 MPa (58 ksi) at the external surface. Unlike welding residual stress, this would act as a primary stress around the full circumference of the header and would be sufficient to drive stress relaxation cracking. The location of cracking is likely to be that of maximum stress, being a combination of manufacturing and welding residual stresses.

The Replacement Header

The need to source material quickly in order to minimize plant downtime required the use of alloy 347H, rather than 316H, for the header replacement. Like all austenitic stainless steels, this material is not immune to stress relaxation cracking. [1], but considerable effort was expended to mitigate this risk. The manufacturing procedure was modified with the objective of minimizing residual stress, eliminating cold work, and avoiding precipitation of strengthening carbides and embrittling sigma phases.

The header barrel was solution annealed after the seam weld was produced. It was manufactured according to ASME SA-312 piping code using an autogenous weld and annealed as required by this code. This would eliminate the residual stress associated with the main seam weld and the cold work generated by bending. After the tube weldolets were attached and other welding completed, the principles developed by Messer et al. [5] were used to optimise heat treatments to minimise the risks of future cracking. This involved a combination of stress relief and solution anneal heat treatments in a carefully controlled sequence, with the intent of avoiding similar damage in the future by applying heat treatments that were not part of the original manufacturing process.

Closing Remarks

The failure of the HP steam superheater was a major event that resulted in an unplanned and extended plant outage, and could have been much more serious. Stress relaxation cracking of austenitic alloys has been recognised for some time. The use of these alloys appears to have become more common since the turn of the century due to the relative reduction in price compared to ferritic materials of similar strength. This has led to an increase in the number of reported stress relaxation failures. The ASME code has recognised this problem for alloy 800H pressure vessels built to ASME Div. VIII, and introduced a post weld heat treatment for design temperatures above 1000°F (538°C) [6]. Such heat treatments are not required for other austenitic alloys unless cold work strain and design temperature limits are exceeded. [4]

The experience from this event suggests this is inadequate to avoid in-service stress relaxation limits. It is considered that requirements for post weld heat treatments of new builds using austenitic alloys at elevated temperature should be considered.

For existing applications, the need for inspection to ensure freedom from cracking is important. The plant had a risk-based inspection system in place. Stress relaxation cracking however, had not been identified as a possible damage mechanism due in part to compliance with the ASME code and the relatively low operating temperature. This experience has raised the awareness of stress relaxation cracking as a potential damage mechanism and inspection test plans have been introduced to monitor susceptible plant equipment. The lessons learned from this incident have resulted in significant changes to the risk management of stainless-steel components operating at elevated temperatures in all Methanex facilities. [7]

This has introduced additional problems related to non-destructive inspection. The sound portion of the seam weld in the HP superheater outlet header and the equivalent weld in the inlet header were inspected for damage using ultrasonic methods. These inspections reported cracking was also occurring in these welds. Subsequent investigation found minimal damage. It is thought that beam bending of the ultrasonic wave due to microstructural variations was responsible for false interpretation that weld root cracking had occurred. These false positive results have led to ongoing investigations into ultrasonic testing procedures with the aim of improving sensitivity and reliability of these methods.

References

[1] H. van Wortel, Control of Relaxation Cracking in Austenitic High Temperature Components. Corrosion 2007 Paper 07423, NACE 2007.

[2] L.E. Shoemaker et al., Fabricating Nickel Alloys to Avoid Stress Relaxation Cracking, Corrosion 2007 Paper 07421, NACE 2007.

[3] ANSI/API RP 571 Damage Mechanisms Affecting Fixed Equipment in the Refining Industry, Third Edition, March 2020.

[4] ASME Boiler and Pressure Vessel Code Section 1, 2017 clause PG-19.

[5] B. Messer et al., Optimised Heat Treatment of 347 Type Stainless Steel Alloys for Elevated Temperature Service to Minimise Cracking, Corrosion 2004, Paper 04640, NACE 2004.

[6] ASME Boiler and Pressure Vessel Code Section VIII, 2017 clause UNF-56.

[7] P. Tait, Lessons Learned from the Rupture of a Steam Superheater Header, Inspectioneering Journal, Jan/Feb 2021.

Eliminating the Safety Risks of a Fired Start-up Heater

The ammonia industry has experienced a number of incidents related to fired start-up heaters, compromising safety, and in some instances causing severe operational impact.

When we consider these experiences, it becomes clear that causes of the incidents are not only related to the design and operation of the fired heaters, but also to the intermittent nature of their operation. The paper will point towards potential causes and points of concern.

An alternative to the fired start-up heater is the electrical heater. Topsoe is capable of supplying electrical start-up heaters which are located internally in the ammonia converter as well as external heaters that are located in a separate pressure shell, and some of our experiences will be presented.

Michael Hultqvist, Svend Erik Nielsen, Avneesh Sharma
Haldor Topsoe A/S

Introduction

There are several aspects of the operation of a fired heater, which cause major safety concerns within a process facility. As an operator, it is important to be aware of the phases of the operation which are most critical with respect to safety. In the case of fired heaters, it is clear that in particular the light-up of the heater from cold conditions is critical. In fact, according to a literature study by J. Dugué, [1] the light-off sequence led to 70% of the explosions and 90% of the casualties. The main causes were the following:

- Improper purge as a result of too short purge time or too low flow rate.
- Inadequate or incorrectly followed procedures.
- Delayed ignition causing build-up of a fuel cloud.
- Improper fuel-air ratio causing unstable burner flame.
- Liquid in the fuel gas system.
- Inerts in the fuel gas system.
- Tube failures causing release and ignition of combustibles.
- Insufficient maintenance leading to difficulties in ignition of burners or pilot burners.

In general, the industry has responded to these risks, and the resulting accidents, by improving the design of heaters and upgrading the materials of construction. The design of control and trip systems were also improved. In later years focus shifted toward operating procedures and advanced protection layers such as burner management systems. Despite these improvements, the human factor and intermittent operating nature of the fired start-up heater are still major concerns to many operators, and whatever means to mitigate the risk and concern should be considered by the plant owners.

In this paper we will present an overview of concerns and considerations with focus on the specific application of the fired start-up heater in an ammonia plant.

Published Incidents

To start out, we have made a survey of incidents related to fired start-up heaters reported in the Proceedings of the last 60 years from the AIChE Ammonia Safety Symposium. A brief account of the reported incidents is given below.

In 1973 an explosion occurred in the start-up heater of Shellstar Ltd.'s ammonia plant at Ince, England. It was identified that tube rupture in the start-up heater was the cause. The following investigations led to the conclusion that the underlying reason was a combination of temperature variations within the start-up heater and the fact that the heater had been in continuous operation during a two-year period, ultimately leading to creep failure of the heater tube. [2]

In 1979 a fire occurred in the start-up heater in Monsanto Company's 545 MTPD (600 STPD) ammonia plant in Louisiana. While there were fortunately no reported injuries, it was reported that the temporary repair of the fired heater led to a delay in the start-up of four days. Later the top coil section was replaced, and the material was upgraded from low alloy steel to stainless steel. A number of other preventive actions were also taken, for instance installation of a flow orifice on the syngas outlet flow from the heater with an interlock to ensure trip of the heater on low flow. [3]

A fire occurred in 1983 in the start-up heater of a 1250 MTPD (1378 STPD) ammonia plant in Ludwigshafen, Germany. In this case the fire was also initiated by coil rupture, here at a circumferential weld. The subsequent analysis of the event focused on metallurgy of the coil and led to upgrading of the coil material. [4]

A start-up heater tube also failed in 1985 in Standard Oil's ammonia plant in Lima, Ohio. It was concluded that nitriding contributed to the failure of the tube. [5]

Another incident occurred in 2005 in a 600 MTPD (661 STPD) ammonia plant in Australia, in which a rupture of a coil in the fired start-up heater led to subsequent fire/explosion in the heater. No injuries were reported, but the fired heater had to be temporarily repaired and later complete replacement of the heater coils were conducted. The root cause analysis showed the following.

- The coil rupture was a result of overheating of the coils since there was no syngas flow through the coil due to the inlet valve being blocked in.
- A flow signal had been forced to defeat the low syngas flow trip.
- This forced signal was given to allow ignition of pilot burners prior to introducing syngas to the start-up heater.
- Pilot burners were ignited prior to syngas introduction, since they had previously been unstable due to poor reliability of the flame scanners.

While the poor reliability of the flame scanner contributed to the incident, it was concluded that the root cause was in this case the practice of circumventing the trip system. [6]

In 2008 a coil rupture in the start-up heater occurred in an ammonia plant in Belgium. External corrosion of the outlet piece of the coil was found to be the cause. [7]

The examples above illustrate quite well that these learnings have led to design improvements, but despite efforts over the last decades in order to mitigate the risks the human factor still plays a significant role. Even though advanced safety layers have been introduced, it has not completely led to elimination of severe incidents.

Publicly available data show that coil rupture due to metallurgy related issues or over-firing in relation to low or no flow is the predominant failure mode of the fired start-up heaters.

Apart from the above referred cases, we have at Topsoe knowledge of several other unpublished cases where events in other areas of the plant led to the failure of the fired start-up heater. In the following section, we will provide some insight into these cases.

Haldor Topsoe's experience through the years

We have knowledge of several cases where the observed performance during the initial start-up of the ammonia plant of the fired heater does not match the required performance as per design. Time is very critical during a start-up, and in many cases the plant has been shut down for an extended period. Consequently, the operators compromise critical operating parameters such as draft, excess air and stack temperature to optimize the transferred duty and expedite the plant startup. This has in some cases even resulted in safety functions being bypassed.

In one case a flange leak in the ammonia loop led to a cloud of synthesis gas forming and when reaching the fired start-up heater, it was ignited, and an explosion occurred. This explosion was caused by a side effect, and if operational procedures about proper purging, check of combustibles gases etc. were adhered to such, an explosion would not have happened. The fact is that operational procedures are not always followed, and in particular related to fired start-up heaters this may have catastrophic outcome.

In a second case, the trip system was bypassed, and the pilot burner was ignited without proper procedures for checking for explosive atmospheres being adhered to, and since an explosive atmosphere was actually present, a severe explosion occurred tearing the start-up heater apart in two pieces.

A third case shows the same picture. It turned out that there was a coil leak in the start-up heater, but prior to ignition of the start-up heater proper explosive meter measurements were not conducted, and the result was an explosion.

For subsequent start-ups the requirement for the start-up heater is lower than the initial reduction of the ammonia synthesis catalyst, as the catalyst in the ammonia converter has already been activated and will ignite at lower temperatures. What we have observed in this connection is that operators sometimes forget and neglect the performance deficiency and associated risk, and that means also that proper root-cause analysis of the seemingly low performance of the start-up heater is most often not prioritized at a later date. Since the fired start-up heater is operated only occasionally, actual experience shows that the overall focus with respect to maintenance procedures, and improving operational procedures is not given a high priority. This is where things suddenly become dangerous and risky.

Recommendations for Risk Mitigation

What can be done with respect to design?

As process licensors today we generally recommend installation of a burner management systems, although we have no knowledge of severe incidents in the fired start-up heaters being caused by delayed ignition or unstable burner flame.

However, based on the cases described in this paper it is clear that special attention must be given to the design and metallurgy of the heater coil. During start-up the coil material temperatures are high and both nitriding and hydrogen embrittlement should be considered. Evidence of these phenomena were found in the investigations following some of the start-up heater failures presented above, and it was concluded that they were contributing factors. The heater coil metal temperatures during start-up are in the range where low alloy steels lose strength relatively quickly with increased temperature and it is important to consider the risk of creep failure due to overheating of the coil.

Accordingly, based on our experience, it is our recommendation to use stainless steel material for all coils. To our knowledge, no incidents with failure of coils has been reported where coil material is ferritic stainless steel.

What can be done with respect to operation?

While design of the fired heater and the safety system have been a major contributing factor in the recorded incidents, it seems more and more clear that the actual incidents in recent years have only been allowed to occur when the established procedures were not followed or when safety functions were bypassed.

We consider that the occurrences are related to the fact that the heater is in most plants being operated infrequently which increases the risk of mistakes or wrong decisions being made. Summarizing, it seems that most incidents could have been prevented if the following conditions had been satisfied:

- Maintenance procedures for the fired start-up heater were in place and being followed.
- Known operational issues related to the start-up heater were addressed by conducting proper root-cause analysis and conducting the necessary subsequent actions, despite its intermittent operation.
- Operational procedures were always updated and followed.

Most, if not all, plant operators would state that this is indeed the case for their plants, but as mentioned above reality seems to be different.

Consequently, to avoid taking any unnecessary risks, and to eliminate the human factor, Topsoe proposes to use an alternative solution to the fired start-up heater and thereby mitigating the risk associated with the open flame of the start-up heater, namely eliminating the ignition point entirely by introducing an electrical start-up heater in the plant design.

Electrical Start-up Heaters

The electrical start-up heater has in the past been applied in numerous ammonia plants in various configurations. For more than 60 years Topsoe has supplied electrical heaters, installed either internally within the ammonia converter or externally in a separate pressure shell. In both cases the heater element is a cable consisting of a copper core surrounded coaxially by a ceramic powder and a stainless tube. The electrical insulation is provided by the ceramic powder, while mechanical strength is provided by the outer stainless-steel tube.

Operation of electrical start-up heaters

Electrical start-up heaters are easy to operate and do not rely on complicated procedures, as is the case for a fired heater. In fact, the electrical start-up heater can be activated by the push of a button in the control room. The electrical system is based on step less thyristor control, which allows fast and simple control of the heater duty. An example of the control panel screen is shown in Figure 1.

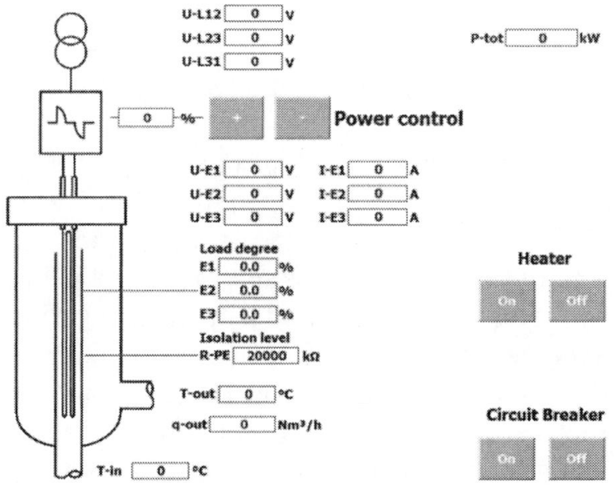

Figure 1. Main Control panel screen of an electrical start-up heater.

If an upset occurs in the plant causing loss of reaction in the first catalyst bed of the converter, then the electrical heater can be engaged in short time and thereby preventing cooldown of the converter.

Safety of electrical heaters

The electrical heater eliminates the risk of fire or explosion of the fired heater due to absence of open flames, fuel systems and hot pressure bearing surfaces, and as a side effect eliminates emission of carbon dioxide.

Therefore, in a risk assessment context, we identified the main risk as being low or no flow of synthesis gas while having an engaged heater leading to risk of overheating of the heater elements, a similar case to that of the coil rupture scenario of a fired heater. However, there is a large difference between the potential consequences of these scenarios. The electrical heater has no hot pressure bearing parts and there is therefore no risk of fire or explosion. Therefore, the risk scenario is purely an economical scenario, resulting from damage to the electrical heater elements themselves.

The potential safeguards against overheating of the heater elements include pressure drop measurements with trip actions like those applied in the case of a fired heater, but the electrical heater design proposed by Topsoe has an additional inherent safeguard.

Since the resistance of the copper core is temperature dependent, the temperature of the heater is indirectly measured by measuring the resistance. There is a temperature gradient across the heater, so therefore the measured temperature will be an average temperature. This is accounted for in the design by simulation of the heater during various scenarios, and then finally applying the most critical scenario for the heater design. In the control systems alarms are then provided at 90% and load 95% and a heater trip is provided if maximum load is exceeded.

The general mechanical integrity of the heater is monitored through the isolation resistance relay, and a low alarm is provided.

External heaters and internal heaters installed in the main converter of world scale ammonia plants will typically consist of two or three sections, each with their own electrical system. This provides these with an internal spare, since the start-up may continue based on the remaining sections if one section loses its functionality.

It should be mentioned that we have seen no catastrophic failures of electrical heaters. The few incidents observed have been related exclusively to performance and have been in electrical heaters in operation for more than 20 years. We also have no knowledge of cases where the start-up of the ammonia plant was delayed significantly due to the electric heater.

Another potential risk to be considered is undetected gas leakage to the surroundings through an electrical feedthrough into the pressure shell. An external electrical heater is typically installed at a certain distance from the converter itself, and in a less critical area. In comparison, we have never experienced a leaking feedthrough in an internal heater, but we still consider this a potential risk scenario in the design phase, which needs to be included in the project specific HAZOP. If the outcome of the evaluation for a specific case is that this scenario is unacceptable, then Topsoe proposes to install a gas detector as an additional safeguard.

Design capacity

The required design capacity of the start-up heater is in all cases determined by the heat input required to heat the reducing gas to a temperature which allows the catalyst to reach a temperature of well above 400°C (752°F), while maintain a space velocity which allows the water concentration in the reducing gas to be kept sufficiently low during the reduction. Naturally this causes the capacity of the start-up heater to be closely related to the installed catalyst type and volume in the ammonia converter.

The actual required design capacity will depend on the feed gas flow distribution for the given converter and start-up heater and will be discussed in the following paragraphs.

External electrical start-up heater

External electrical start-up heaters can be installed without modification of the converter itself and will still provide the benefit of improved safety. The external electrical start-up heater is the solenoid shaped heater type, which is shown in Figure 2.

Figure 2. Solenoid shaped heater element.

Typically, an external heater will consist of three elements located in parallel to the converter cold shot piping. During start-up the gas is routed to the start-up heater, as is the case for a fired start-up heater. The flow distribution of the feed gas to the converter will therefore be the same for the two types of external heaters. Accordingly, the requirements with respect to transferred duty and outlet temperature will also be the same. The concept of an external electrical start-up heater in connection with an S-300 converter is illustrated in Figure 3. The flow scheme of this concept is shown in Figure 4.

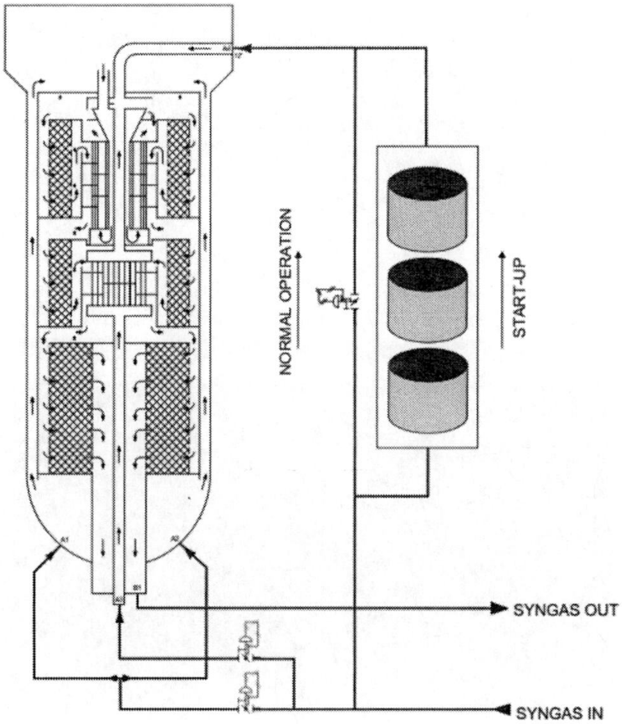

Figure 3. External start-up heater configuration.

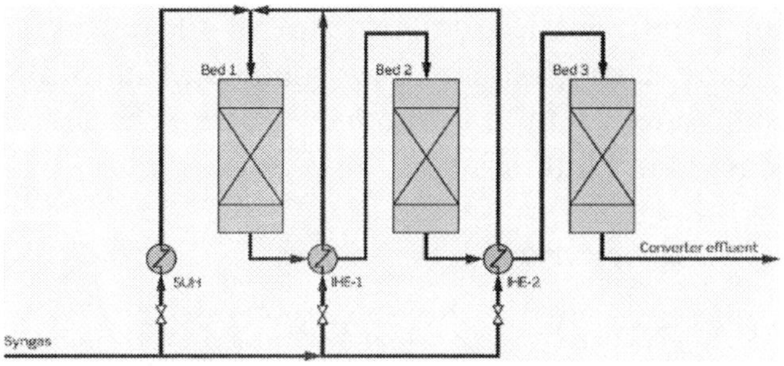

Figure 4. Flow scheme of an external start-up heater in combination with an S-300 converter.

Internal electrical start-up heater

The latest version of the internal start-up heater proposed by Topsoe today is of the hairpin type, and is located in the center pipe of the converter. A hairpin type heater is shown in Figure 5.

Figure 5. Hair pin type heater element.

An internal start-up heater is from a process point of view installed downstream of the inter-bed heat exchangers. The flow scheme of an S-300 converter with an internal start-up heater is shown in Figure 6.

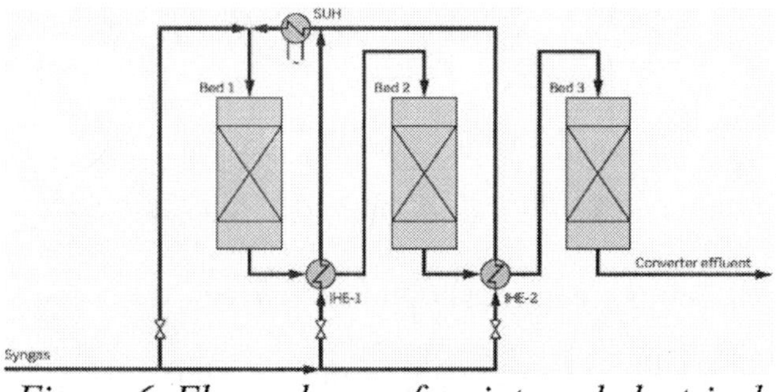

Figure 6. Flow scheme of an internal electrical start-up heater installed in an S-300 basket.

The internal heater provides the same improvement in safety as the external electrical start-up heater, but has several additional advantages when compared to an external start-up heater.

- Its location downstream of the interbed heat exchanger(s) allows the design duty to be lowered by more than 50% compared to the external heaters, since heat recovery from the internal heat exchanger is improved greatly.
- There is no requirement for a separate pressure shell as the heater is located inside the converter pressure shell.
- For a grass root plant there is significant saving in cost to piping upstream of the converter.
- Since there is flow through the internal heater during normal operation and it is located upstream the first catalyst bed, it allows the operator to respond even faster to a loss of reaction in the first catalyst bed.

The only drawback of the internal heater is the marginally larger converter reactor vessel. All the above advantages are relevant for grass root plants, and our evaluations have shown that for such cases the overall investment will be lower for an internal heater than for an external heater, whether electrical or fired.

Conclusion

In summary, we conclude although the design and operation of fired start-up heaters has improved over the last decades, these units continue to be a safety concern in the industry, since operational procedures are not followed carefully enough.

This risk can be eliminated by introducing an electrical start-up heater. Such heaters have been supplied and safely operated for several decades, and the new and improved design of an internal electric heater for the S-300 converter is our latest development.

Additionally, the electrical start-up heater provides a number of benefits related to investment cost and improved operation. Improvements are also with respect to emissions, and for some plant owners the absence of carbon dioxide emissions from an electrical heater is an added benefit in relation to new installation, since the carbon dioxide emissions from the fired start-up must otherwise be addressed when obtaining the required environmental permits.

References

[1] Dugué, J. "Fired equipment safety in the oil & gas industry". Energy Procedia, Vol 120 (2017).

[2] Mitcalf, J. "Failure in Start-Up Heater Tube". AIChE Symposium, Vol 17 (1974).

[3] Kokemor, F.K. "Synthesis Start-Up Heater Failure". AIChE Symposium, Vol 22 (1979).

[4] Rall, W. and Spaehn, H. "Rupture of the Start-Up Heater Coil in the Synthesis Loop of the New Ammonia Plant". AIChE Symposium, Vol 25 (1984).

[5] Mack, R.J. and Schultz, F.J. "Failure of Ammonia Converter Start-Up Heater Tube". AIChE Symposium, Vol 27 (1986).

[6] Pattabathula, V. and Jordan. L. "Synthesis Converter Start Up Heater Failure". AIChE Symposium, Vol 47 (2006).

[7] Warzée, Michel. "Fire in an Ammonia Synthesis Start-up Heater Synthesis Start-Up Heater Failure". AIChE Symposium, Vol 50 (2009).

[8] Ramsussen, C.-V. "Advances in Electric Heaters for Safe Startup of Ammonia Plants", AIChE Symposium, Vol 41 (2000).

Failure Analysis and Field Repair of Urea Reactor Shell Rupture

The outer shell of a Multi-layered Urea Reactor ruptured while the plant was in operation. The plant made an emergency shutdown, samples were collected, and an investigation carried out to establish the root cause of the failure. The paper explains the detailed analysis performed, along with corrective action implemented and a health assessment for safe working of the equipment. The paper also includes references to similar failures in old multi-layered Urea Reactors subsequently and the mitigation plans to avoid such failures.

Ravinder Singh, Dr. Krishnan Sivaraman
Larsen & Toubro, India

Tushar Bhagwat
Rashtriya Chemicals & Fertilizers, India

Introduction

The Urea Reactor is a critical component in any Fertilizer plant. Urea is formed in the Urea reactor by the reaction of ammonia and carbon dioxide at elevated pressure and temperature. During normal operation of the Urea plant, there was a sudden sound and a rupture was found in the outer shell layer of the Urea Reactor. Figure 1 shows the rupture of the outer shell, with a tearing of 1.5 m (4.9 ft) length. This multi-layer Urea reactor was supplied in 1982 and in operation for more than 35 years, with regular maintenance work carried out every 2-3 years, which included inner liner inspections and repairs.

This paper describes the investigation in terms of physical and metallurgical findings for the multi-layered shell, detailed analysis, root cause of the failure, lessons learned, and the health assessment results. Furthermore, the paper covers an innovative approach implemented to perform an in-situ replacement of the defective liner segment during the emergency shutdown. Some modifications and the preventive actions which were implemented to avoid reoccurrence will also be discussed.

Figure 1. Tearing of Urea Reactor Outer Shell

General Technical Data

Urea reactors are usually vertical pressure vessels, with either mono-wall, multi-layer, or multi-wall construction. Because of the corrosive nature of the medium, the interior surface of the vessel is lined with corrosion resistant materials.

The design and construction features of the subject Urea Reactor is shown in Table 1 and a diagram of features in a multi-layer urea reactor is shown in Figure 2.

No	Parameter	Value	
1	Construction	Multi-layered (7 layers)	
2	Total shell thickness	115.8 mm	
3	Layers	No. & thk	MOC
	Inner Liner	1 x 7 mm	SA240 TP 316LM
	Dummy layer	1 x 4.5 mm	JIS SS41
	Outer layers	7 x 15 mm	SA724 Gr A
4	Code of construction	ASME BPVC Sec VIII Div-2, Ed-1980	
5	Design Pressure	169 kgf/ cm2(g)	
6	Design Temperature	200 °C	
7	Operating Pressure	159 Kgf/cm^2(g)	
8	Operating Temp	190 °C (Top) 170 °C (Bottom)	

Table 1. Design and Construction details

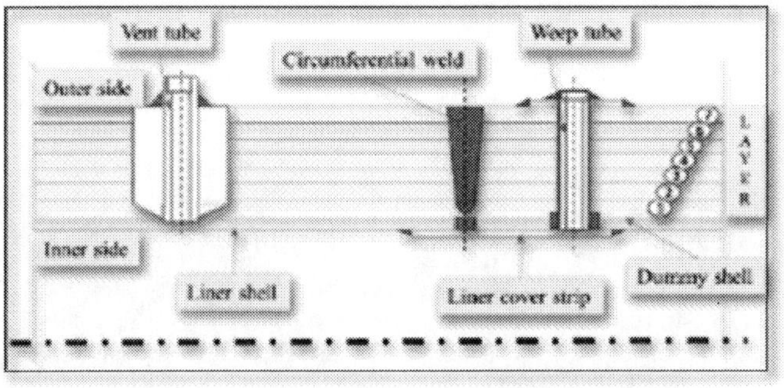

Figure 2: Construction features of multi-layer pressure vessel

Initial inspection at Site

- Visual inspection of the Urea Reactor showed tearing of the outer shell layer resulting in a 1.5 m (4.9 ft) wide opening near the davit arm location, as shown in Figure 3a. Deflection of the davit arm and bulging near the ruptured zone was observed.
- Heavy scale formation was observed on the outer shell as shown in Figure 3b. Thickness of the shell layer was reduced (~13.5 mm (0.53 in) against original 14.9 mm (0.59 in)), due to corrosion under insulation.

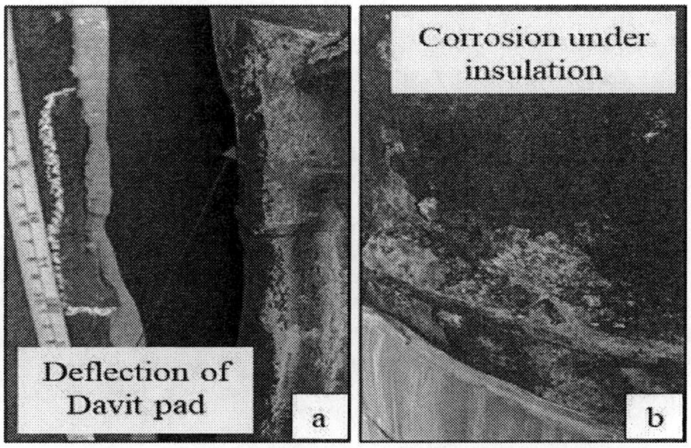

Figure 3. Tearing of outer shell and scale formation beneath insulation

- The fracture surface is shown in Figure 4, which shows pre-existing crack face and fresh rupture face. This pre-existing crack seems to have initiated near the edge of fillet weld between the manway davit arm pad plate and the outer shell layer.

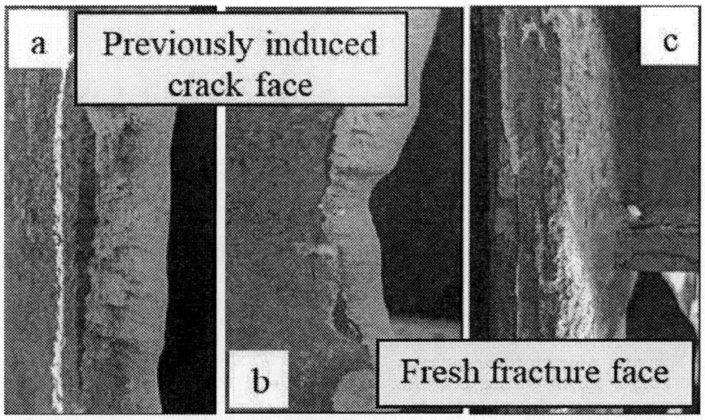

Figure 4. Fracture surface on (a-b) Shell layer side (c) Davit arm pad side

- The reduction in thickness of the outer shell layer near the ruptured zone was found to be ~2-3 mm (0.08-0.12 in).

- No traces of carbamate/residue were observed in the ruptured portion of the shell/immediate inner shell surface in the vicinity of failure.

- An air test was performed to find the leakage in the liner at the top portion of reactor (Top head + Shell). The test was not successful, due to weep holes blockage. Further steam purging was carried out to clear the weep holes. Deposits were observed at the weep hole opening, and collected for analysis (Figure 5).

Figure 5. Deposit collected from weep hole

- Visual inspection and dye penetrant test (DPT) of the inside liner surface showed signs of corrosion on the longitudinal and circumferential weld joints, as shown in Figure 6. Corrosion was also observed on the weld built-up area buttering (old cleat removal area). Localized dents and bulges on liner surface were also observed.

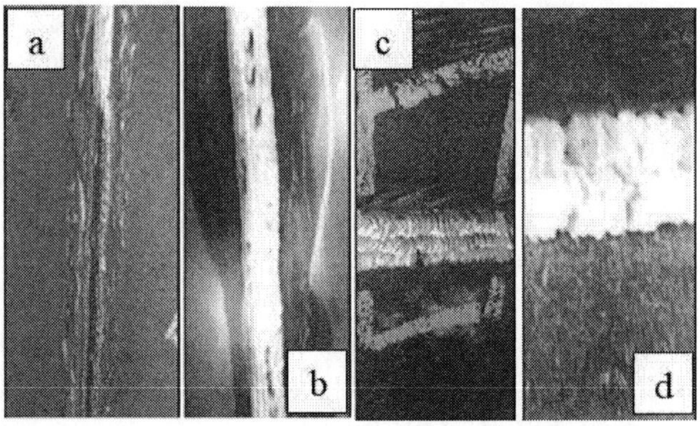

Figure 6. Corrosion on liner Longitudinal seam (a-b) Circumferential seam (c-d)

- After cutting and removal of the outer shell layer, the inside surface of the layer revealed discoloration towards weep hole-W9, as shown in Figure 7.

Figure 7. Inside surface of outer shell layer

- Air testing was conducted after the blockage in weep holes was removed. This revealed leakage in the liner long seam, in the area between the 1st and 2nd trays, as shown in Figure 8a. The location of leakage was then verified by dye penetrant test (Figure 8b and 8c).

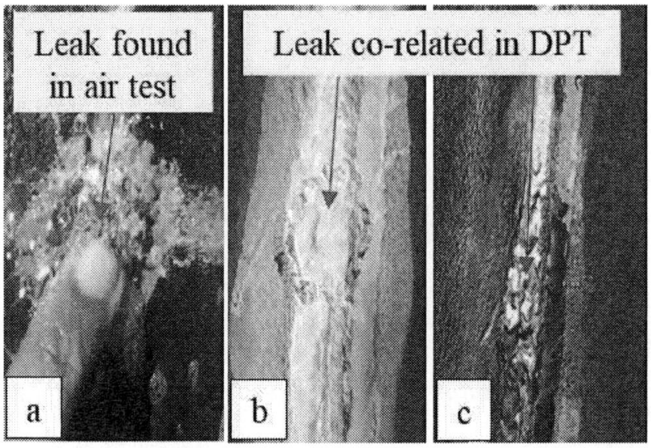

Figure 8. Leak found in air test and verified in DPT

- After cutting the liner plate at the leakage location, the inner dummy shell and 1st layer shell were found to be corroded through as shown in Figure 9.

Figure 9. Corrosion of dummy shell and 1st shell layer due to liner leakage

Investigation

To find out the cause of failure, a detailed investigation was carried out on extracted samples. The following tests were conducted:

- Chemical Analysis
- Mechanical Testing
- Visual examination
- Macro examination
- Scanning electron microscopy (SEM) & Energy dispersive spectroscopy (EDS)
- Microscopic examination

Chemical analysis

Chemical analysis by Optical emission spectroscopy was carried out to confirm the chemical composition of the shell layer. The results confirmed a match to the expected composition. These results are shown in Table 2

Element (%)	Result	Requirement of SA 724 Gr A
C	0.16	0.22 max
Si	0.39	0.60 max
Mn	1.38	0.92 - 1.72
P	0.013	0.025 max
S	0.006	0.025 max
Cr	0.025	0.29 max
Ni	0.003	0.28 max
Mo	0.002	0.090 max
Cu	0.007	0.38 max
V	0.043	0.090 max

Table 2: Chemical composition of samples

Mechanical testing

Tensile, impact, and hardness testing were carried out in the longitudinal and transverse directions. Tensile properties met the requirements of the material specification, as shown in Table 3.

Mechanical Properties	Longitudinal direction	Transverse direction	SA 724 Gr A ®
Yield Strength (MPa)	591	506	485 min
U.T.S. (MPa)	758	723	620-760
% Elongation	22.76	25.72	19 min
Impact at 0°C (Joules)	208, 232, 220	114, 94, 100	----
Hardness in (HBW)	215, 219, 219		----

Table 3. Mechanical properties

Visual and Macro examination

Samples collected from the failure location were thoroughly examined visually and using a stereoscope. The following observations were made:

- The outside surface of the shell layer showed presence of heavy scaling, as shown in Figure 10.

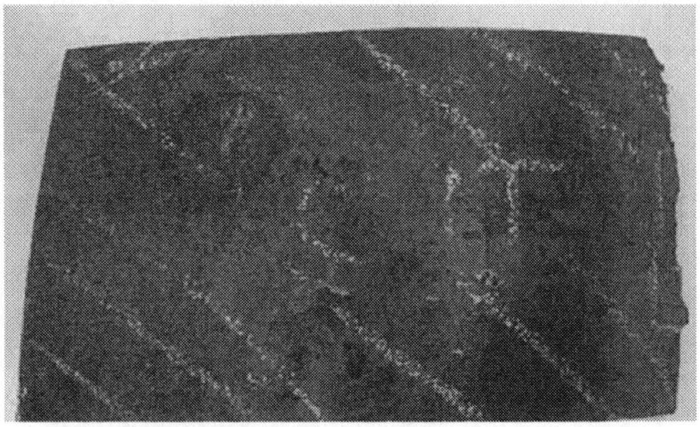

Figure 10. Corrosion under insulation on outside surface

- The inside surface shell layer revealed discoloration as shown in Figure 11. The residue was analysed to be iron and oxygen.

Figure 11. Brownish discoloration on inside surface

- A pre-existing crack (~3 mm (0.12 in) depth) was observed on the fracture face, and shown in Figures 12 and 13. This pre-existing crack appears to have initiated near the edge of the fillet weld between the manway davit arm bracket pad plate and the outer shell layer, as shown in Figure 14.

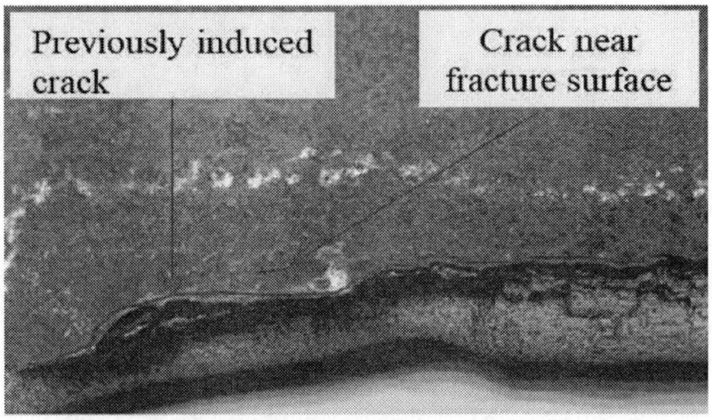

Figure 12. Failure location - Crack near fracture surface

Figure 13. Failure location - fracture surface pad plate side

Figure 14. Macro examination previously induced crack and fresh rupture – 10X

- Additionally, a small crack was observed near the fracture surface, which appeared to be generated due to mechanical force during fracture of shell layer (Figure 15).

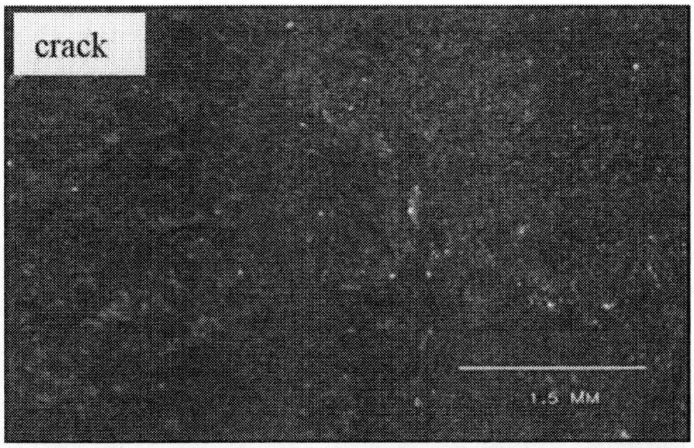

Figure 15. Crack beside fracture surface - 20X

- After visual and macro examination of the extracted samples, specimens containing the failure locations were cut for subsequent analysis. Specimen containing the fillet weld (Pad plate to outer shell layer weld) showed presence of lack of fusion with shell layer (Figure 16).

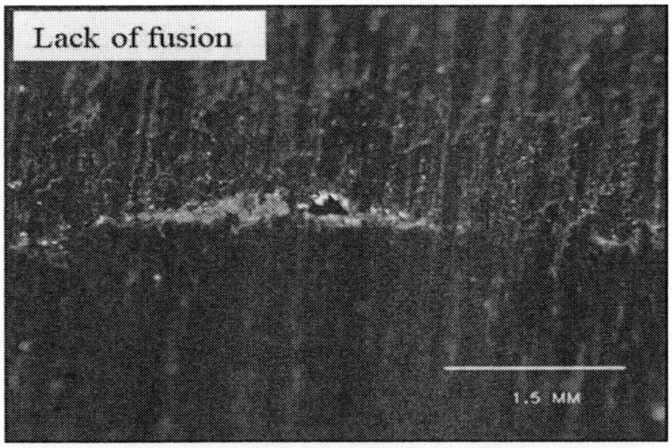

Figure 16. Discontinuity in fillet weld between pad plate to shell layer - 20X

SEM & EDS Analysis

SEM-EDS analysis was carried out on the specimens cut from extracted samples, scale formed on outside surface of shell layer under insulation, and on deposits collected from the weep hole.

- SEM of shell layer side part of the fracture surface clearly revealed a boundary between pre-existing crack and fresh rupture, as shown in Figure 17. Maximum depth of the crack was found to be ~3.7 mm (0.15 in).

Figure 17. SEM Image for boundary between previously induced crack and fresh rupture

- A specimen containing the pad side part of the fracture surface that showed lack of fusion in the weld revealed a micro crack in the base metal adjacent to weld edge, as indicated in Figure 18. The edge of the weld was clearly revealed, which appeared to be the initiation point of the pre-existing crack.

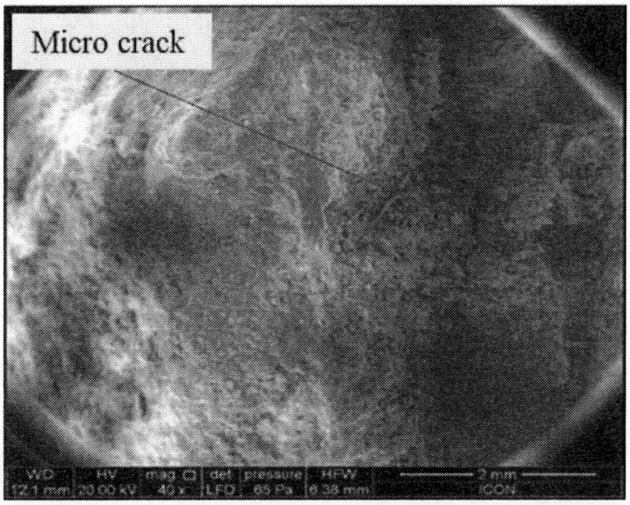

Figure 18. SEM Image for crack on fracture face near weld

- EDS revealed the presence of significant amount of oxygen and iron, along with other elements like sodium, calcium, and manganese on the pre-existing crack face and at the crack beside the fracture surface. (Figures 19 and 20)

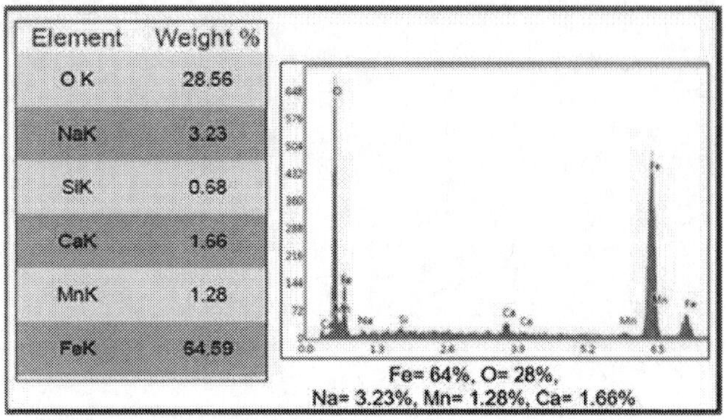

Figure 19. EDS on Previously induced crack face

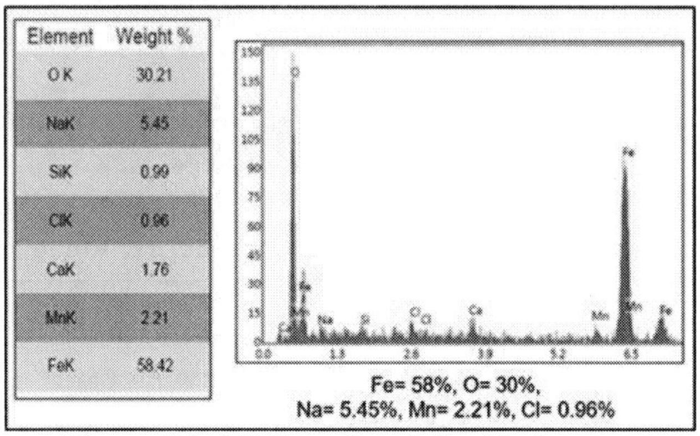

Figure 20. EDS on Crack beside fracture surface

- A significant presence of oxygen and iron, along with elements like sodium, manganese, calcium, and chlorine was observed on the crack face near the outside surface of shell layer. (Figure 21) A lower amount of oxygen was noticed on the rupture face as compared to the pre-existing crack face. This indicated low degree of oxidation on fresh rupture face.

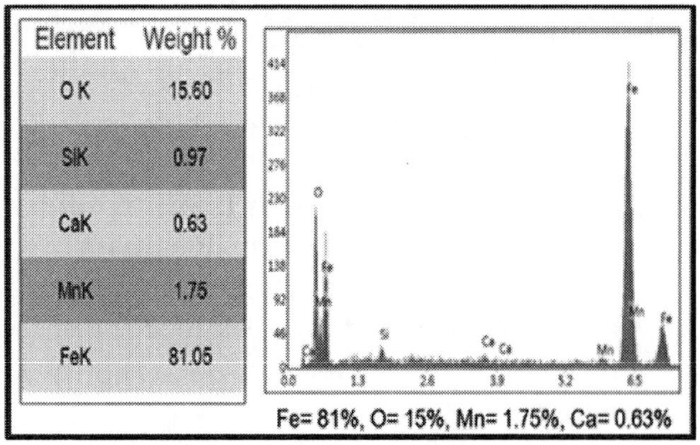

Figure 21. EDS on rupture face

- Scale formed on the outer layer of the shell consisted mainly of iron and oxygen. (Figure 22)

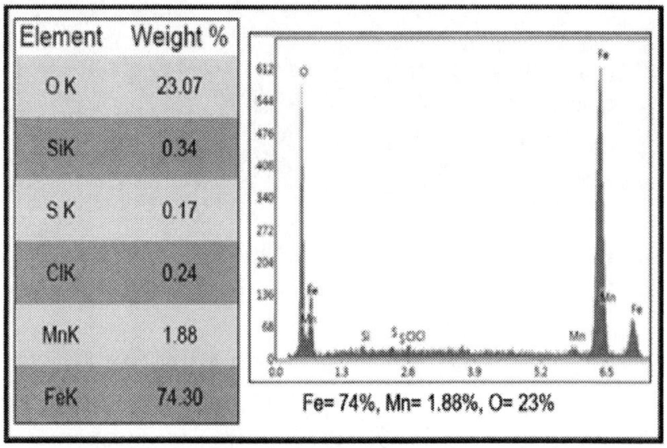

Figure 22. EDS of scale formed on shell layer

- The deposit collected from the weep hole showed significant presence of nitrogen, oxygen, and carbon. (Figure 23)

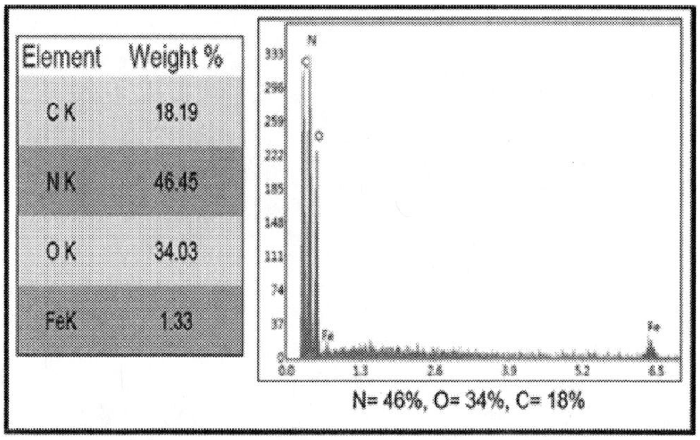

Figure 23. EDS of deposit collected from weep hole

Microscopic examination

After SEM-EDS analysis; the cross section surface of the specimen was polished and etched with 2% Nital and observed under a microscope. The following observations were made:

- Microstructure of shell layer showed tempered martensitic structure. (Figure 24)

Figure 24. Microstructure near shell layer from outside - 50X

- Pre-existing crack appears to be initiated from edge of the fillet weld and propagated in heat affected zone (HAZ) in the outer shell layer. (Figure 25)

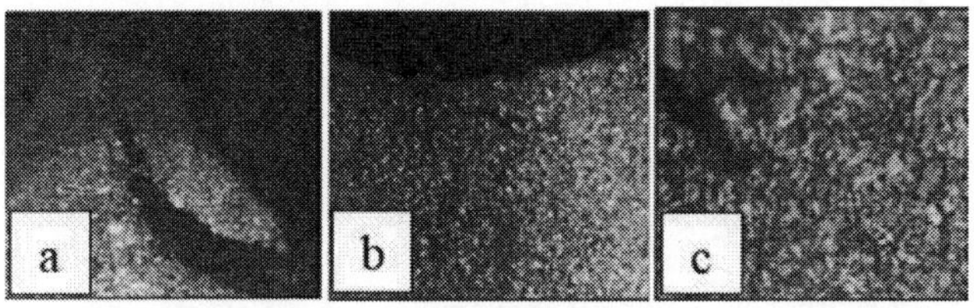

Figure 25. (a) Crack initiation point - 50X, (b-c) Crack propagating in HAZ - 50X and 100X

- Secondary cracks beside the fracture surface were also observed (Figure 26)

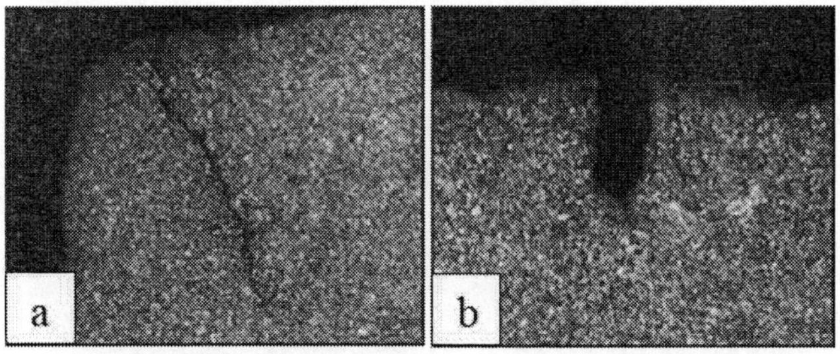

Figure 26. Crack beside fracture surface on shell layer - 50X and 100X

Weldability test

Weldability of the shell layer material was verified by welding butt joint and clearing the dye penetrant test and radiography test.

Summary of Observations

- The chemical composition and mechanical properties of outer shell layer conforms to SA 724 Gr A, as per drawing requirement. Microstructure of the material was found satisfactory. Weldability test carried out using shell layer material was also satisfactory. Based on this, it inferred that material had not degraded during service.
- During visual inspection at site, heavy scaling due to corrosion was observed on the outside surface of the outer shell layer. Thickness of the shell was found to be reduced by 1.5 mm (0.06 in) due to scaling. EDS of the scale formed on the outer layer under insulation was found to be mainly oxygen and iron. This indicated existence of moisture and general corrosion under insulation on outer shell layer over the time.
- The lack of fusion in the davit pad to outer shell layer fillet weld was observed in visual examination and SEM of the specimen containing the fracture surface. The presence of such a discontinuity at the edge of the weld increases the possibility of crack initiation and propagation over a period of time.
- Visual, macro-examination, and scanning electron microscopy of extracted samples showed presence of a pre-existing crack (~3 mm (0.12 in) depth) on the fracture surface, which is clearly differentiated from the fresh rupture face. It appeared to be initiated at the edge of the weld between the pad-plate and shell layer. This was also verified by micro-examination, which clearly showed crack initiation at the edge of the fillet weld and propagation through the outer shell layer.

- The inside surface of the reactor showed signs of corrosion on the liner long seam, circumferential joints, and buttering welds (old cleat removal area). During the air test, weep holes were found to be blocked due to solidification of carbamate in the leak path. The air test performed after the weep holes were cleared showed leakage in the liner long seam. Corrosion attack was observed on the dummy shell layer and 1st shell layer in the vicinity of the liner leakage.
- EDS testing of the deposit collected from the weep hole showed significant presence of nitrogen, oxygen, and carbon. This showed carbamate leaked through liner was solidified and deposited at the weep holes. The inside surface of the extracted outer shell layer revealed discoloration towards weep hole-W9 side, which was likely the path of flowing media generated due to weep hole blockages.
- EDS analysis showed the significant amount of oxygen and iron, along with other elements like sodium, calcium, magnesium, aluminium, and silicon, on the pre-existing crack face and additional crack beside the fracture surface. Presence of oxygen and iron indicates formation of iron oxide on the crack face. Other elements appeared to be contributing by the insulation material and water coming from external source.

Observations from recent repairs in Fertilizer plants

Besides this shell rupture, there have been similar instances of urea reactors leaking through liners and ending up in long shutdowns from 35-45 days globally [1-4]. Applying innovations, we have designed and developed a special dedicated machine used for in-situ rolling of liners inside vessel, which is patented and used for reducing shutdown durations. Typical causes of these failures and remedies to prevent them are summarized below:

- Vent holes are plugged from outside as these are supposed to be removed post erection. There is no path for the pressure to be released in case there is a pressure built-up in the shell inter-layers.
- Vent tubes are clogged / blocked and the leak detection system does not work, increasing the risk of pressure boundary failure by corrosion / rupture in case the liner leaks.
- Random dye penetrant testing of the liner welds, compared to 100% examination, likely results in defects being missed.
- Plant owners concentrate on the liners from inside and seldom check the pressure boundary from outside. Random non-destructive testing of pressure shell welds should be institutionalized in every shutdown.
- Attachments for liners are welded with partial penetrations. This increases the risk of corrosive medium attacking the crevices and resulting in liner failures.

Lessons Learned

✓ Liner fabrication and welding should be performed by approved contractors.

✓ Liner welding should be performed using fixtures to restrict the lifting of liner during welding and creating excessive air gaps.

✓ The optimum weld sequence should be developed to reduce the liner shrinkage during installation.

✓ The weld process parameters and filler wire selection should be optimized to restrict the heat input during pass partition plate welds.

✓ All attachments on liners shall be full penetration welds.

✓ Vent holes should be inspected and verified as open following the erection of lined vessels, and also during each turnaround.

✓ Periodic quality checks for weep tube leak detection system, as specified by the process licensor, is mandatory.

Conclusion

Based on the observations and analysis, the following conclusions can be drawn for the shell rupture:

- Material of the outer shell layer was not degraded. However, thickness of the outer shell layer was found to be reduced by 1.5 mm (0.06 in), possibly due to corrosion under the insulation of the outer shell layer over time.
- Visual examination revealed the presence of a pre-existing crack, which initiated near the edge of the fillet weld between the manway davit arm pad plate and the outer shell layer. This could be due to utilization of the davit arm beyond its design load for various handling activities during earlier shutdowns. Furthermore, discontinuity in the form of lack of fusion increased the possibility of crack initiation and propagation over a period.
- Leakage in liner coupled with blockages of weep tubes during operation resulted in process media to enter the shell inter-layers, which in turn exerted pressure on outer layer. Thickness reduction due to corrosion under insulation and localized reduction in thickness due to pre-existing crack made the outer layer unable to withstand the operating pressure. This led to bulging of outer shell layer and subsequent shell rupture.

Based on the experience of past urea reactor repairs globally; it is inferred that the liner failure for a urea reactor can result in a major catastrophe and lead to long shutdowns or the replacement of the vessel. The weep tube leak detection system is the lifeline of lined vessels. Quality controls must be adhered to with regular monitoring of these systems, as specified by the Process Licensors.

References

[1] Thielsch, H., and F. Cone (1993). Failures, Integrity and Repairs of Petrochemical Plant Reformers. Ammonia Technical Manual, Vol. 33, AIChE, New York.

[2] Thielsch, H, (1952). Stainless Steel Applied Liners. Welding / Research Supplement, 31,321s.

[3] Zheng, J. Y., Deng, Y. J., et al. (2005). Experimental investigation on dynamic response and fracture characteristics of discrete multi-layered thick-walled explosion containment vessels. Explosion and Shock Waves, 25(6), pp. 506-511.

[4] Jojima, T. (1979). Urea reactor failure. Ammonia Plant Safety (and Related Facilities), 21, pp. 111-119

Explosion Risks in High Pressure Scrubbers of CO_2 Stripping Urea Plants

Over the history of Stamicarbon's design of CO_2 Stripping Urea Plants there have been changes in the understanding of the risks presented by the mixture of flammable gas and oxygen in the synthesis process. This has led to modification to the design of high pressure scrubbers, as well as new requirements on the source of oxygen required for passivating the metals in the synthesis section and the acceptable amount of hydrogen allowed in the process.

When high pressure scrubbers are replaced the plant owner must balance the options to improve the safety of the plant with the practical challenges of implementing each option. A risk based assessment process should be used to properly identify the best option for a specific customer, based on cost, operating conditions, and regulatory requirements.

This paper presents Stamicarbon's findings over time of the risks presented by flammable gases in the high pressure urea scrubber, and design changes that are available to mitigate this risk in new and existing plant installations. The paper presents a risk based approach to optimize the selection of the risk mitigation design options.

Harald Jorritsma, HSE Engineer
Stamicarbon bv

Background

During the 2000 Stamicarbon symposium the paper titled "Explosion risks in urea plants: The Stamicarbon Y2K update" was presented. With respect to explosion risks, it was concluded that hydrogen is the dominant component contributing to an explosion in a urea plant. Concentrations of hydrogen higher than 5 mol%, in the presence of oxygen, should be avoided in process equipment and pipelines.

This paper also pointed out that customers and regulators continuously raise their expectations with respect to the safety performance of petrochemical plants. To achieve this, operators, manufacturers, and licensers must periodically update their safety standards and design specifications.

Among other sources of information, experimental research and incident case history provide new insights regarding the development of risk scenarios and subsequently the way to modify the design their process or equipment.

Based on the design philosophy in 2000, a standard Stamicarbon grassroots urea plant would incorporate a high pressure (HP) scrubber of the sphere design, with a design pressure of ~160 barg (2320 psig), in conjunction with catalytic removal of hydrogen in the CO_2 feed stream to the urea plant.

In this configuration, the hydrogen converter reduces the likelihood of the presence of an explosive atmosphere within the HP scrubber, whilst the sphere explosion dome volume mitigates the consequences for the plant environment in case of an internal deflagration reaction in the scrubber.

Recently, several operators of cylindrical high pressure scrubbers (often called cigar type) have asked Stamicarbon to supply new (identical) cigar type high pressure scrubbers in Safurex® infinity duplex stainless steel. Most of the early generation cigar type high pressure scrubber are characterized by a very limited explosion dome volume, and as such do not comply with the latest Stamicarbon design standards, as they have developed over time.

Incorporation of the safest option, the sphere high pressure scrubber with an adequate explosion dome volume, is not always technically feasible or may require disproportionate investments to adapt existing structural work (load limitations) and/or piping arrangements.

This paper elaborates on the exploration of design solution alternatives in the context of a risk based approach.

Process Safety Perspective

Physical properties of Gas Mixtures

For an explosion to occur oxygen and flammable gases, with a composition within flammability limits, must be present with an ignition source. In urea plants, the combustible components of concern are ammonia, hydrogen, and any other combustible components that may be present, such as methane.

The flammability limits of mixtures containing ammonia, hydrogen, and methane at elevated temperature and pressure are well known. Knowledge of these flammability limits is an important pre-requisite in safety studies. However, the flammability limits, as reported, do not account for the considerable difference in reactivity, with respect to combustion, that exists between ammonia/air mixtures on one hand and hydrogen/air mixtures on the other. It is also well known that the minimum ignition energy for hydrogen/air mixtures is much lower than ammonia/air mixtures.

In the late 1990s Stamicarbon conducted practical experiments to obtain more in depth knowledge of the behavior of NH_3/H_2/air mixtures. The project focused on the reactivity of oxygen containing process gas mixtures with respect to deflagration to detonation transitions in pipelines under various conditions (e.g. composition, pressure, and temperature) that are typical for a urea and melamine plants. The experimental apparatus consisted of a 20 m (66 ft) pipeline with an internal diameter of 59 mm (2.3 in). In this pipeline gas mixtures of ammonia, hydrogen, and air were introduced and the likelihood of detonation of these gas mixtures, as a function of their composition, was tested at an (initial) temperature of 150°C (302°F) and (initial) pressures up to 150 barg (2175 psig). The most important parameter measured was the flame speed.

The main results of the project can be summarized as follows:

- For ammonia/air mixtures over the entire concentration range, flame speeds were sufficiently low to prevent detonation: The maximum flame speed measured (near the stoichiometric ammonia/air ratio) was 1.8 m/s (5.9 ft/s). (Transition of flames to detonation develops once a flame accelerates to 800-1000 m/s (2625-3280 ft/s)!).
- The experiments showed that up to 5 mol% of hydrogen could be added to these ammonia/air mixtures and flame speeds remained far from detonation.

At higher hydrogen concentrations, especially in the presence of small amount of ammonia, distinct flame acceleration was observed. It cannot be ruled out that in actual process piping turbulence produced by internal fittings could accelerate the flame propagation sufficiently for detonation to occur.

The test rig was also used for explosion tests on gas mixtures containing ammonia/hydrogen as combustibles mixed with oxygen enriched air. These experiments resulted in an unpleasant surprise: during an experiment with oxygen enriched air (40 mol% O_2/60 mol% N_2) a detonation occurred that was so violent that the test rig was severely damaged. The pressure during this event was measured at 3700 barg (53,600 psig)!

The mixture that caused this extreme reaction contained 28 mol% ammonia and only 3 mol% hydrogen. Considering the findings using "normal air", this violent detonation was rather unexpected. Because of this incident, Stamicarbon stopped further safety research with oxygen enriched air systems. The present Stamicarbon Y2K guideline does not allow for use of oxygen enriched systems in a urea plant.

Effect of an explosion

In a urea plant we can distinguish two different effects of an explosion:

- The direct effects such as rupture and fragmentation of equipment. These direct effects create a threat to people in the vicinity of the equipment concerned.
- Rupture of equipment results in a release of the hazardous chemicals contained by the equipment. This effect often results in a much larger risk, as the urea synthesis section contains large amounts of ammonia at elevated temperature and pressure. Ammonia is a toxic substance and is corrosive to the eyes, skin, and respiratory tract. Inhalation of vapor/fumes can cause severe breathing difficulties (lung edema). In cases of high exposure there is the risk of death. These effects justify a differentiation in the safety approach, depending on the amount of ammonia that would be released in cases of loss of containment.

Grassroots Design Philosophy

In the late 1970s several hydrogen related fires and explosions occurred in China. These incidents led to an increased awareness of the safety aspects in this respect and triggered a revised safety philosophy. It initiated the following design changes for the licensed CO_2 stripping plants:

- Adding a hydrogen combustion reactor to remove hydrogen in the carbon dioxide feed.
- Changing the total condensation of inert vapor in urea synthesis to a partial condensation process, and including the required sufficiently large pressure relief volume.

In the 1980s it appeared that total condensation in a urea synthesis plants with CO_2 stripping provided process advantages, and the safety philosophy was updated again to provide an acceptably low residual risk for a scrubber total condensation concept in urea synthesis sections. This subsequently resulted in the sphere type of high pressure scrubber which still represents Stamicarbon's standard design. The sphere is designed such that it incorporates a sufficiently large explosion pressure relief volume (dome) compared to the volume in which a deflagration explosive mixture can be present. Next to this dome volume the hydraulic pressure of that dome is prescribed as a design figure for the dome (typically 150% of the maximum synthesis operating pressure).

If the vapors from urea synthesis are fully condensed in a high pressure scrubber the passivation air in combination with hydrogen and other flammable gases can form a mixture that may lead to deflagration. This total condensation is very possible in diverse operating conditions of the plant, since the inert vapor is contacted with a solvent (carbamate or water) in the scrubbing part of the high pressure scrubber. The maximum effect of a deflagration is a fast pressure increase inside the scrubber and ultimately the rupture and fragmentation of the packed bed in the scrubber, leading to an immediate threat to operators and release of large volumes of toxic ammonia. This event is categorized as an event of the highest consequence.

Given the low ignition energy of the flammable mixture at urea synthesis conditions, an ignition source is assumed to be present anytime. The probability of such an event can be reduced by introducing a hydrogen removal unit in which the combustible component hydrogen is catalytically removed from the CO_2 feedstock. However, the likelihood of the event is not always reduced enough to meet Stamicarbon's current safety design standards. For this reason, measures to reduce the consequence of an internal deflagration are to be considered.

Note: The pressure relief volume protects the process only against deflagration explosion events. The pressure relief volume is not adequate to protect the process against detonation explosion events and thus enriched oxygen containing combustible hydrogen mixtures are prohibited to be present.

In Stamicarbon plants equipped with a hydrogen combustion unit and feeding air (not oxygen enriched), there are no cases known of scrubber fragmentation caused by an internal deflagration.

The effect of these design changes on a typical risk matrix is shown in Figure 1.

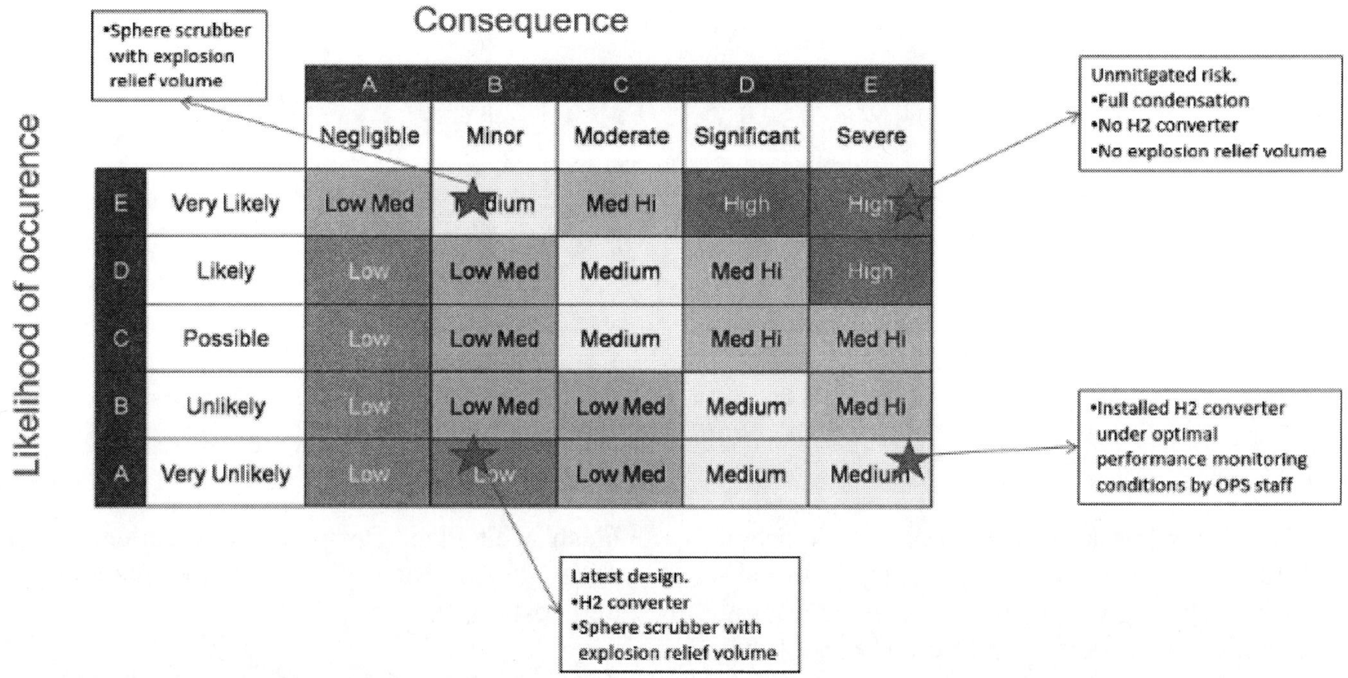

Figure 1. Combination of both the sphere type high pressure scrubber and a hydrogen converter results in the highest possible safety performance, a low risk category plotted in the risk matrix above.

In 2015 Stamicarbon began offering the Adiabatic Flash Melt Plant Design. In this design the high pressure scrubber becomes obsolete. A flash tank is placed in between the high pressure stripper and the low pressure recirculation section, and the urea solution leaving the high pressure stripper is flashed at a pressure of ~20 barg (290 psig) (medium pressure) before it is released into the existing recirculation section.

The vapor released by this flash, together with the vapor leaving the high pressure reactor in the synthesis section, are condensed in a newly installed shell and tube heat exchanger which is a pre-evaporator. In this pre-evaporator, heat is recovered from these vapors by condensing them against a urea solution from the low pressure recovery section. The recovered heat of condensation is used to pre-concentrate the urea solution leaving the recirculation section before it is collected in the urea solution tank. The principle of the proposed Adiabatic Flash Melt Plant Design is illustrated in Figure 2.

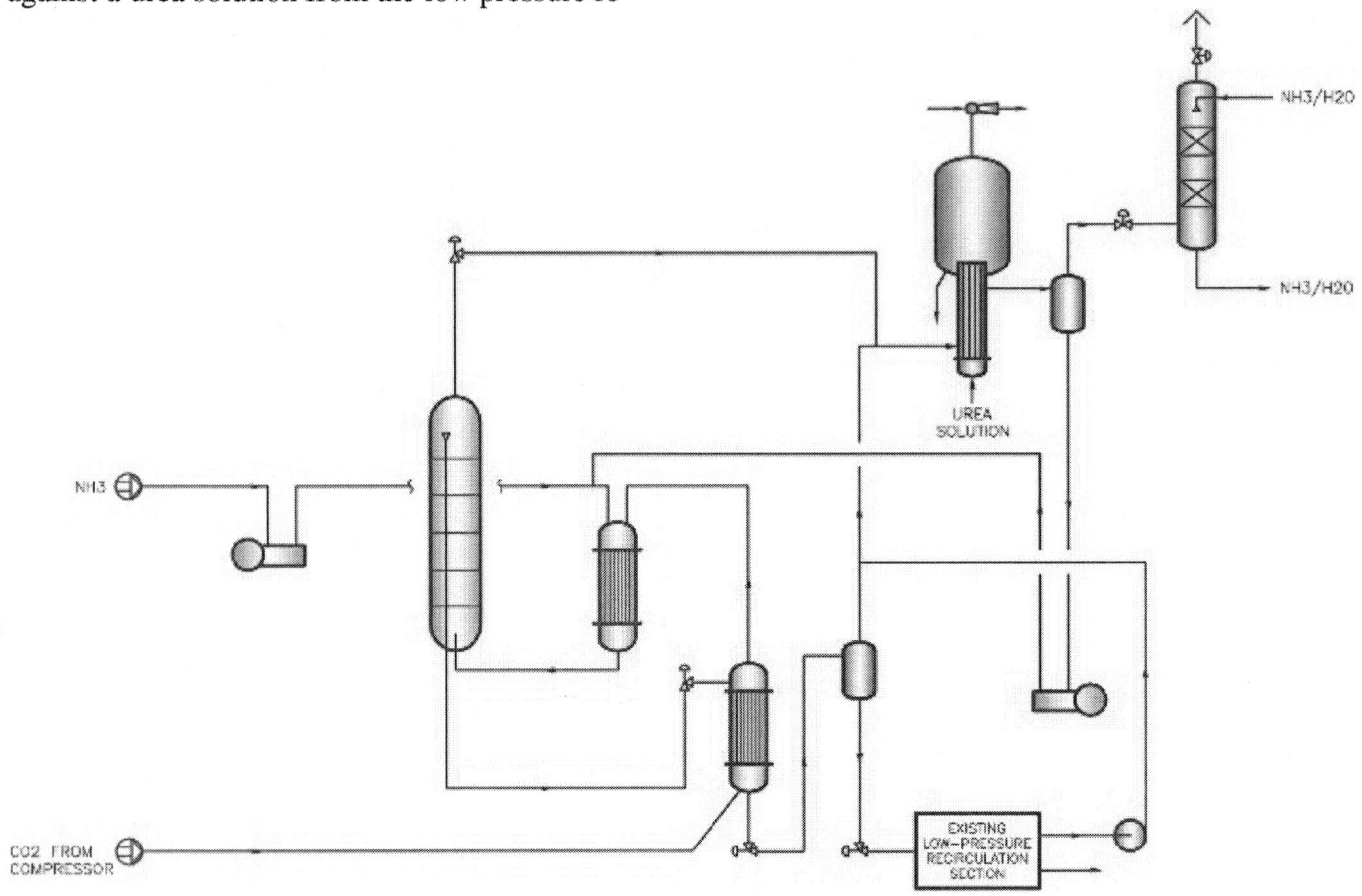

Figure 2. Flowsheet adiabatic flash melt plant design

The total condensation of ammonia and carbon dioxide in the inert vapor takes place in the absorber located downstream the pre-evaporator at a much lower pressure of ~4 barg (58 psig), before the inert gases are released into the atmosphere.

The absorber is designed to be deflagration proof, so in case of an undesired event loss of containment is not expected. Thus, the Adiabatic Flash Melt Plant Design is an excellent opportunity to safeguard an existing urea plant at limited costs.

Moreover, the Adiabatic Flash Melt Plant Design gives the opportunity to decrease the stripping efficiency and thus the steam requirements of the existing stripper. Consequently, the operating cost of the plant is decreased significantly by the application of this retrofit, compared to the original plant design.

Design philosophy existing facilities

Recently, several operators of cigar type high pressure scrubbers, of the total condensation type, have requested Stamicarbon to supply new (identical) cigar type high pressure scrubbers in Safurex® as a replacement.

Incorporation of the safest option, the sphere high pressure scrubber with adequate explosion dome volume, is not always technically or financially feasible due to the restrictions in weight or available space.

Mitigating the consequence of an internal deflagration by design

Several design options exist, and have been applied in order to minimize the effects of an explosion in the scrubber.

"Dome area" constructions

These are provisions that allow a deflagration in a section which may contain an explosive mixture (areas marked yellow in Figure 3) to expand in a much larger volume, which is designed to accommodate this event. After expansion the resulting total pressure will still be well below the rated design pressure of the enclosure. This construction has already been applied in several variations of high pressure urea scrubbers:

In case of an explosion in the protected part of the equipment, the internals will be damaged, and the production process must be stopped for repair. However, the mechanical integrity of the high pressure scrubber sphere will not be jeopardized, and a loss of containment scenario can be avoided.

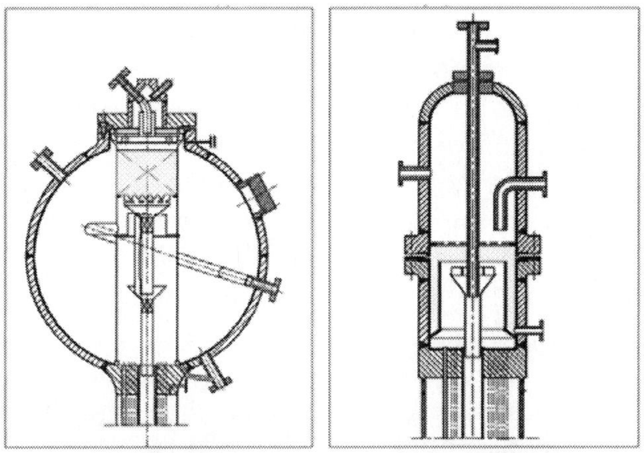

Total condensation sphere type HP scrubber — Partial condensation cigar type HP scrubber

Figure 3. Explosion relief volume options (yellow area indicates where a combustible mixture can exist).

Rupture discs

This provision can theoretically protect the equipment against a high pressure excursion in case of a deflagration by venting the contents to the environment. It is obvious that the rupture disc discharge should be vented to a safe location to avoid the same consequences as a failure of the enclosure. As the rupture disk must be sized to handle a deflagration, the release will be significant and instantaneous. As no secondary containment is foreseen, a single rupture disc scrubber design without a pressure relief volume construction is not acceptable.

Deflagration proof equipment

In case an enclosure would be designed for the pressure developed by an internal deflagration the enclosure would mechanically be able to withstand the internal deflagration. This mitigation by design is practically feasible for low pressure equipment. Designing medium or high pressure vessels to be deflagration proof would result in very high design pressures resulting in high equipment costs and for that reason is not practical.

Deflagration proof design is recommended for low pressure equipment in which explosive gas-mixtures with high hydrogen concentrations (>5 mol% at total condensation) can occur. In a plant with low hydrogen concentrations (<5 mol% at total condensation) and equipment containing an insignificant ammonia content a deflagration proof design is not deemed necessary. A threshold value of maximum 5 mol% hydrogen in the fully condensed gas is recommended to distinguish between "high" and "low" concentration in this respect. It should be noted that in pipelines, at higher hydrogen concentrations, flames could accelerate into detonations. Detonation proof design of pipeline systems is not possible.

Adiabatic flash revamp concept

In certain occasions, clients request the replacement of an existing cigar type, total condensation scrubber because the existing scrubber is at the end of life, and the location of the newly installed scrubber would be limited by weight and/or space restrictions.

The scrubber can be replaced with a scrubber of the partial condensation type. However, in this case the inert vapor leaving the scrubber will contain more ammonia and carbon dioxide than the original design and consequently this inert vapor will need further purification before it is released into the atmosphere. This purification of the inert vapor usually required retrofitting the plant and will result in a slight increase of the steam consumption and thus operating expenses of the plant.

For some clients it is recommended to transform the plant into an Adiabatic Flash Melt Plant Design, as shown in Figure 2. The high pressure scrubber will become obsolete and the low pressure absorber can be designed to be deflagration proof.

Mitigating the probability of an internal deflagration by design

Most urea plants operating under a Stamicarbon license are of the total recycle concept, with CO_2 stripping. For those plants the following safe operating conditions are required to mitigate the probability of an internal deflagration:

- Oxygen should be added to the synthesis section in the form of air. Enriched oxygen systems are prohibited in urea plants. Required passivation air dosing in the range of 0.6 vol% oxygen in the carbon dioxide feed for austenitic synthesis construction materials or 0.3 vol% oxygen in case of Safurex® as construction material in the synthesis, limit the hydrogen concentration in the feedstocks to the urea plant to safe values determined by calculation. In most cases, in order to meet these concentration limits, it will be required to operate a catalytic hydrogen removal system in the CO_2 feed. In exceptional cases a hydrogen reducing system (e.g. pre-flash) in the ammonia feed will be required.
- Hydrogen concentrations should be measured on a regular basis. In cases where the acceptable hydrogen concentrations are exceeded corrective actions must be taken (the plant must be stopped).
- To minimize potential ignition sources, all equipment (high pressure as well as low pressure, including piping) in which there is a flammable component (e.g. hydrogen, methane, or ammonia), that might contain oxygen or air, should be grounded and protected against direct hit by lightning. During operation, no "hot work" (welding, cutting) activities at or near such equipment is allowed. Such hot work activities may only be done when the absence of flammable components (including ammonia) in the equipment or piping has been ascertained.
- Continuous monitoring of the performance of the hydrogen converter is crucially importance to reduce the probability of an internal deflagration. However, it does not mitigate the effect of a deflagration!

Risk perspective of existing plant configurations

Before the introduction of the sphere type high pressure scrubber Stamicarbon designed cigar type high pressure scrubbers. Different types of cigar type scrubbers can be distinguished as the design developed over time. Some designs do not meet the criteria used as the basis for the design of the sphere type high pressure scrubber and do not meet the current safety standard of Stamicarbon.

As several customers who operate cigar type have requested to supply new (identical) cigar type high pressure scrubbers in Safurex® all types of cigar type scrubbers were assessed for possibilities to meet the current design criteria.

The conclusions can be summarized as follows:

1. Strictly identical high pressure scrubber replacement in kind is not possible. With such replacement, either modification of the design (scrubber and/or process), additional safeguarding, or operational constraints are always required.
2. Total condensation style cigar type scrubbers may have met Stamicarbon past design criteria but do not meet current safety design criteria.
3. Partial condensation style cigar type scrubbers may need relatively few (mechanical) design changes to meet the current standard design criteria.

Risk based decision making

When a customer decides that it is time to replace a high pressure scrubber (end of lifetime) the customer might have a strong preference to plan for a "replacement in kind". An end of lifetime replacement is often funded from a maintenance budget and requires few (if any) modifications to the existing plant structure and piping arrangements. In the case of a partial condensation style cigar type scrubber with a sufficiently large pressure relief volume this approach is perfectly acceptable. In all other cases it is preferred to transform the plant into an Adiabatic Flash Melt Plant Design.

If the customer persists in a replacement in kind, the unmitigated risk of such a configuration must be carefully evaluated. As the actual risk exposure is determined by many parameters, both in design as well as customer specific operating conditions, a comprehensive risk assessment of the present configuration is a pre-requisite. This initial assessment should be prepared in close cooperation with the operator/customer as it is unlikely Stamicarbon will be aware of all operational conditions or the availability and effectiveness of existing safeguarding provisions.

In this respect a full understanding of operating conditions and safety measures is required to ensure that the replacement scrubber can be operated within nationally (or internationally) accepted risk acceptance criteria. Risk Acceptance Criteria (RAC) are frequently used worldwide in many industries.

These criteria allow the authorities, or whoever adopts them, to detect activities with high level of risk for the society or an individual. Their goal is to ensure that the level of risk is acceptable with respect to safety, cost, and the environment and to balance and compare risk against the benefits.

Within the European Union the Seveso III-directive 2012/18/EU was implemented to provide the basis for legislatures to define the criteria in relation to societal risks. As directives and standards do not contain RAC figures, authorities often use F-N curves (Figure 4) which represent the function between the number of fatalities and their probability, and can provide a good indicator for evaluating societal risks.

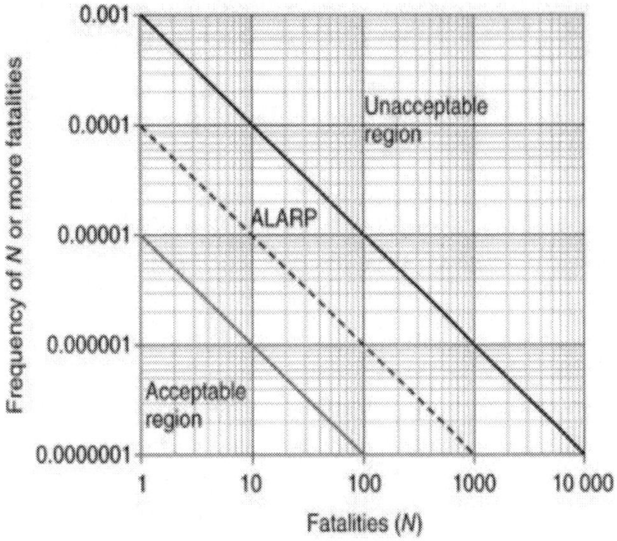

Figure 4. F-N curve example.

In the implementation of the Seveso Directive the Dutch legislature has considered the individual risk of fatal injury of somebody in the public outside of the plant perimeter to be acceptable at an individual risk probability of 10^{-6} per year. In general, the individual risk of plant population is considered acceptable at an order of magnitude higher than the public: 10^{-5} per year. Worldwide, RACs in this order of magnitude are commonly applied.

In the petrochemical industry it is impossible to retroactively bring existing facilities up to the latest design standards. Resources (manpower/budget) are scarce and will be allocated where they create the highest benefits regarding risk mitigation, quality improvement, or capacity gains. A risk based approach can help evaluate the tradeoffs between investment and potential risk mitigation.

RACs can be applied with regard to replacement projects in existing facilities. The initial risk assessment is the reference to determine what additional safety measures should be put in place to arrive at an acceptably low residual risk.

To determine tolerable risk for changes to existing facilities, in comparison to cost and other factors, industry often mitigates the risk to a level As Low As Reasonable Practicable (ALARP). ALARP demonstration is commonly practiced to document that the safest alternative requires disproportional investment or is not technically feasible. Figure 5 shows a graphic example of ALARP risk acceptance criteria.

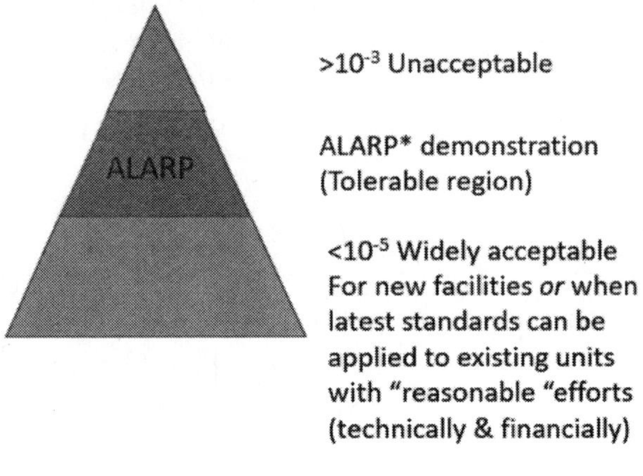

Figure 5. Example plant individual risk acceptance criteria

Although the licensor can advise in design solutions and facilitate the initial risk assessment, risk acceptance criteria are to be defined by the plant owner in conjunction with local and federal authorities.

Visualizing risk and mitigation opportunities

Stamicarbon can facilitate a risk assessment template by proposing a suitable methodology and drafting the parameters determining the actual risk level. However, populating this risk assessment template with actual process conditions and details concerning existing safeguarding remains the responsibility of the plant operator.

Initial risk

A widely used risk assessment methodology is Fault Tree Analysis (FTA). FTA is a graphical method that starts with a hazardous event and works backwards to identify the causes and enablers of the undesired event. Intermediate events related to the top event are combined by using logical operators such as AND and OR. This analysis method is mainly used in the fields of safety engineering and reliability engineering to understand how systems can fail, to identify the best ways to reduce risk, or to determine event rates of a safety accident. Failure modes of components being part of safety instrumented systems can be derived from industry databases or actual test records available at the operator of the facility.

Risk validation

The current risk level should be evaluated to determine whether it is in line with internationally accepted standards and federal/local legislative obligations. If this is not the case, additional risk mitigation measures should be put in place.

Fragmentation of an HP scrubber caused by an internal deflagration is estimated to result in a consequence causing 2-10 fatalities within the plant perimeter. Plotting this risk scenario on an F-N curve will result in a risk within the ALARP region. (Figure 6)

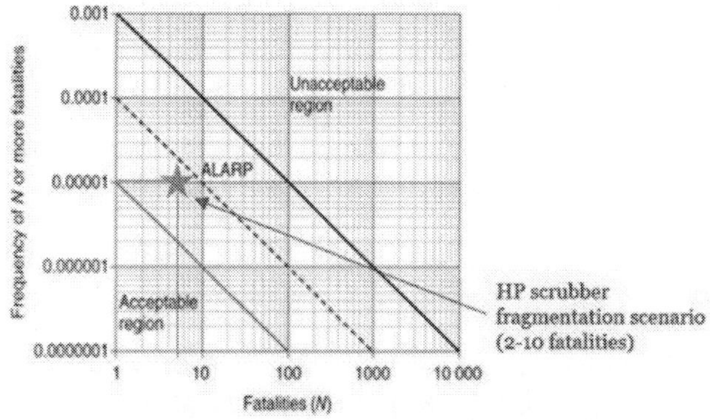

Figure 6. Probability of the risk scenario moving into the ALARP region

Figure 7 illustrates the potentially iterative process of assessing and evaluating risk and risk mitigations, in order to reach a solution with risk ALARP.

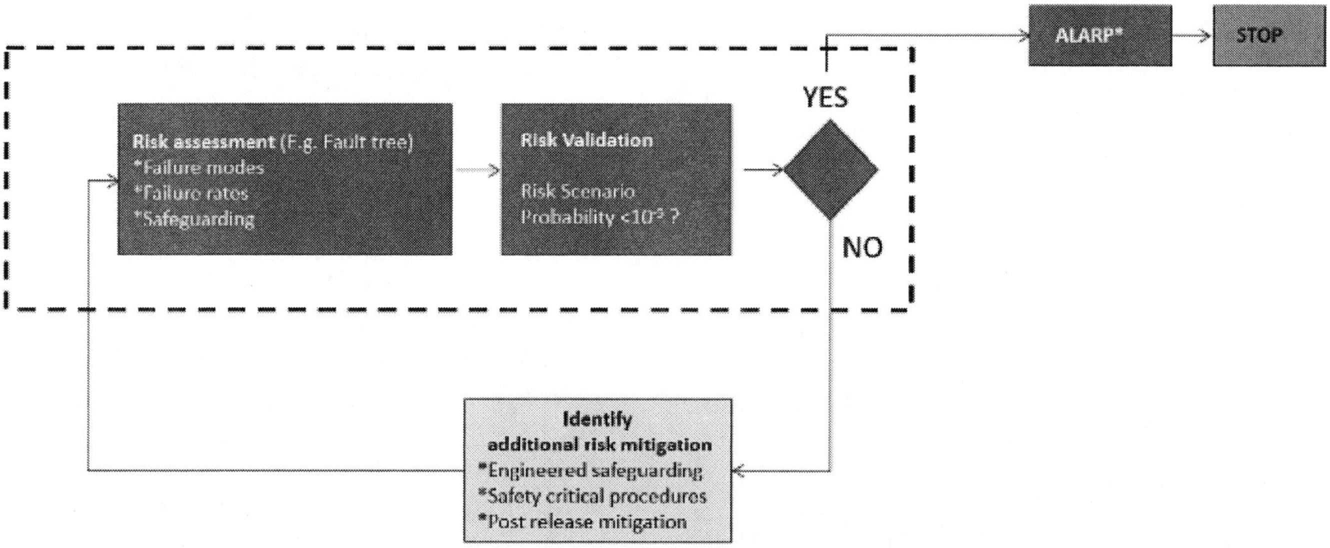

Figure 7. Sequence for risk validation

The effect of the implementation of additional mitigation measures should again be added to the risk assessment tool to determine if the target individual risk probability is achieved. If this can be confirmed the risk assessment should be documented by the customer in an ALARP document stating that the risk has been reduced to a level "As Low as Reasonably Practicable". This document may be required by regulators and should include the rationale why the safest solution was not chosen (In this example, a sphere type scrubber or Adiabatic flash revamp concept).

In the case of cigar type scrubber replacement projects, the most obvious arguments will be:

- Exceedance of the structure or foundation maximum load (sphere type scrubber adds weight).
- Disproportional costs to modify piping systems (to fit in larger sphere type scrubber)
- Additional equipment and piping in the Adiabatic flash concept revamp.

As the "disproportional cost" criterion is arbitrary, ALARP demonstration in relation to the replacement in kind of existing high pressure scrubbers may be difficult to defend to regulators, as affordable technical alternatives are readily available.

Risk mitigation opportunities

Under the header "Risk perspective of existing plant configurations" it is concluded that high pressure (strict identical) scrubber replacement in kind is not possible. This statement will be supported by the outcome of the initial risk assessment, which in most cases will call for additional risk mitigation measures to achieve the target risk probability.

Independent from the outcome of the initial risk assessment, the following design principles should at least be respected:

- For passivation of the synthesis materials, the application of air is mandatory. Application of enriched air or pure oxygen is prohibited for this application.
- Cigar type scrubbers based on partial condensation with a sufficiently large explosion dome volume are acceptable.
- Replacement of a cigar type high pressure scrubber on the principle of total condensation is only possible with the now standard sphere type high pressure scrubber. From a safety perspective the frequency of occurrence of the presence of a flammable oxygen containing process gas mixture is much higher compared to processes where partial condensation is applied. That makes the total condensation scrubber designs more complicated and requires a much larger pressure relief volume, due to the presence of the volume of the packed bed that adds to the volume in which a flammable oxygen containing gas mixture can be present.

The high pressure cigar type scrubber with total condensation will not withstand an internal deflagration. The residual risk (Probability x Consequence) when operating these types of scrubbers can only be lowered by reducing the probability parameter by lowering the probability of the presence of an explosive atmosphere in the scrubber. However, this will not mitigate the effect of a deflagration!

As already concluded under the header "Mitigating the probability of an internal deflagration by design"; Continuous monitoring of the performance of the hydrogen converter is of crucial importance in the mitigation of the frequency of occurrence of the event of an internal deflagration. For this reason, all additional mitigation is focused on the reliability of the hydrogen converter.

In cases where a properly functioning hydrogen converter and a high quality ammonia feedstock (low hydrogen content) can be sustained, there are not enough combustible components to create a combustible mixture during normal operation and/or plant upset conditions.

Possible mitigating measures to lower the probability of an internal deflagration include:

- Provide the hydrogen converter with adequate temperature measurements to monitor the temperature increase (ΔT), and thus the performance of the combustion process. Eliminate the human error probability (operator response to alarm) for the ΔT measurement by replacing it by a 2 out of 3 SIL2 temperature measurement and interlock.
- Provide the discharge of the hydrogen converter with a continuous analyzer with a

safety critical alarm (dedicated annunciator), alarming at a high hydrogen concentration. The units of the hydrogen analyzer should be indicated in ppm-vol to indicate a dangerous breakthrough of hydrogen. Periodic laboratory analysis of the hydrogen content in the treated CO_2 should be performed to ensure the accuracy of the analyzer.

- Lower Probability of Failure on Demand (PFD) of hydrogen analyzer by more frequent calibration.
- Avoid instantaneous failure of the hydrogen converter (catalyst poisoning) by providing clean CO_2 (no lubrication oil from reciprocating compressors or ammonia plant catalyst contamination).
- Ammonia feedstock preferably to be supplied from a refrigerated storage tank or flash section, to reduce dissolved hydrogen content.
- Analyze ammonia feedstock periodically to verify that hydrogen content is sufficiently low.

As the scenario of high pressure scrubber rupture is categorized in the highest consequence, risk mitigation should be proportionate to the risk. To effectively mitigate a risk of this type, engineered protective layers should be in place. Only engineered layers of protection can be proof tested to ensure their availability during the full lifecycle of the plant. Relying on human intervention (operator response to an alarm), will not effectively reduce the risk to an acceptable value.

If adequate engineering measures, as suggested above, are effectively implemented the probability of an explosive mixture in the HP scrubber will become very unlikely. The effect of these safeguards on a risk matrix is shown in Figure 8.

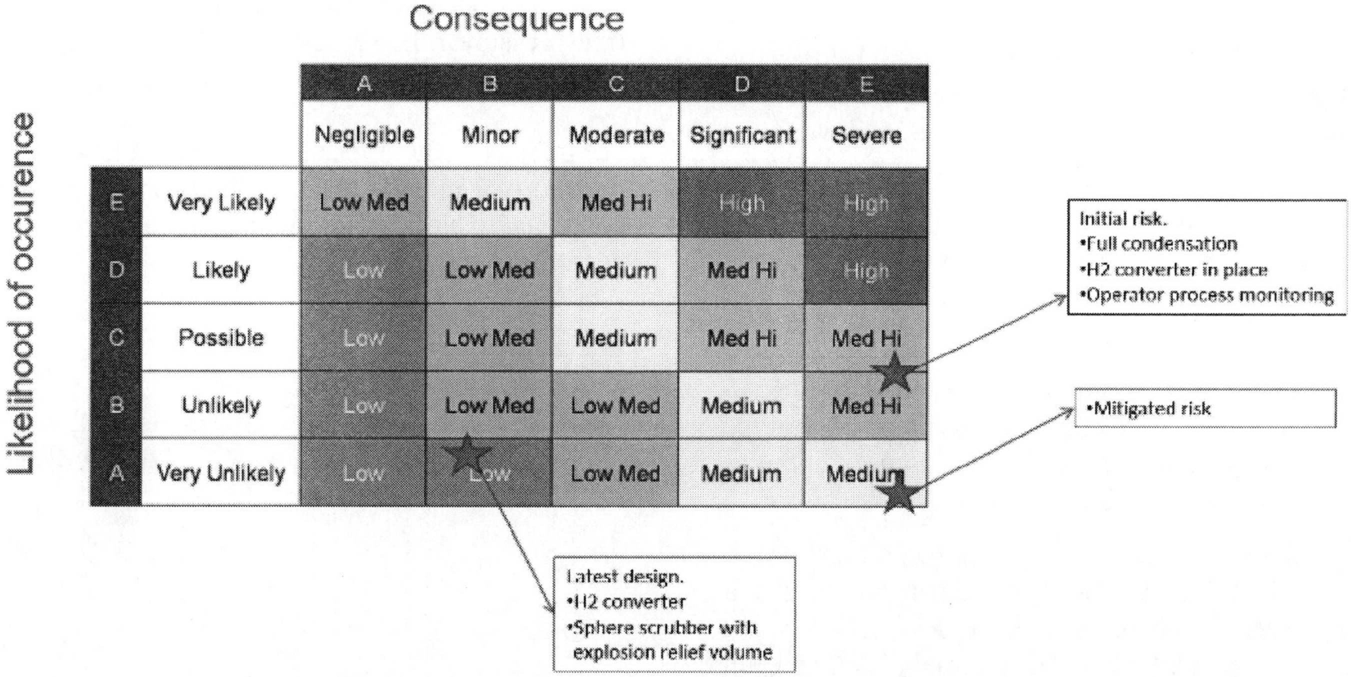

Figure 8. Plotted risk scenario after mitigation

Important note:

Engineering measures for which credit is taken as safeguards must be maintained during the lifetime of the facility. We must remember that the risk became acceptably low because of the presence and effectiveness of engineered risk mitigation. These scenarios must not be parked and forgotten, but will require continuous attention; periodic loop testing of interlocks, testing records, control of defeat, training of personnel, etc.

CONCLUSION

With respect to explosion risks, hydrogen is the dominant component contributing to an explosion in a urea plant. When urea synthesis vapors are fully condensed in a high pressure scrubber, the passivation air in combination with hydrogen and other flammables can form a gas mixture that may lead to a deflagration.

Recently, clients have requested Stamicarbon to replace their existing cigar type scrubber with an identical design. However, as a knowledgeable licensor Stamicarbon is expected to not 'blindly' copy past scrubber designs (which an equipment manufacturer may do) but to engage with the client to offer the "best possible design" from a safety perspective, as well as within the boundaries set by the client such as, but not limited to: budget, lead time, dimensions, weight, improved operating conditions, safety provisions etc.

Within the context of offering high pressure scrubbers in replacement projects the following prioritization sequence will be followed by Stamicarbon:

- Cigar type scrubbers of the partial condensation type are to be modified to accommodate an explosion dome volume as per latest standard.
- Cigar type scrubbers of the total condensation type are to be replaced by a sphere type scrubber with an explosion dome volume as per latest standard.
- Modify plant to the Adiabatic Flash Melt Plant Design. The high pressure scrubber will become obsolete and the low pressure absorber will be designed to be deflagration proof.
- In cases the client persists to identically replace a high pressure cigar type scrubber of the total condensation type, Stamicarbon can facilitate the clients' risk evaluation to achieve an acceptable low residual risk. To effectively mitigate a risk of this type, engineered protective layers should be implemented.

In this context, when ordering a replacement of a scrubber the client should be guided to make an "informed decision", fully understanding its operating conditions and safety measures in order to be able to operate the replacement scrubber within international (and local) acceptable conditions.

It should be considered that the owner of the Plant is responsible for the safe operation of its plant and is responsible for meeting requirements on process safety and mechanical integrity. The owner of the plant deals with local authorities and regulators.

In terms of international codes, it is the responsibility of the "user" of the equipment to identify scenarios that may lead to high pressure, to define pressure relief systems, and eventually define the design pressure of the equipment.

For plants of other types (e.g. conventional plants or non-total recycle plants), Stamicarbon can be contacted to clarify the applicability of these suggestions to the specific process.

REFERENCES

Messen, Donker, Vlake, and van den Aarssen, "Safety Aspects in Urea Plants: Y2K Update", AIChE Ammonia Technical Manual, 2001.

The #1 Safeguard for Any Urea Plant

As of January 2021, UreaKnowHow.com has collected and analyzed more than 135 serious urea incidents. Based on this analysis, what is the most important, or the number one, safeguard, for any urea plant? It is the leak detection system for loose liners of the high-pressure equipment. In just two recent years (2018 and 2019), at least six serious incidents have occurred whereby the available leak detection system failed! It is really a miracle no casualties were involved.

This paper explains why a proper leak detection system for loose liners of high-pressure equipment is crucial. This paper also explains why most existing systems are not acceptable anymore by illustrating their shortcomings in several incidents. The paper will conclude with how a modern state-of-the-art leak detection system, The #1 Safeguard for Any Urea Plant, should be designed. The goal of this paper is to convince urea plant owners to upgrade existing leak detection systems to state of the art standards and improve the safety and reliability level in our industry.

Prem Baboo
Dangote Fertilizers Ltd

Mark Brouwer, Jo Eijkenboom
UreaKnowHow.com B.V.

Majid Mohammadian
OCI N.V.

Introduction

High pressure equipment in a urea plant experience most safety incidents. High pressures and temperatures, the corrosive circumstances and the health risks of ammonia leaks place high demands on the integrity of these equipment items. The UreaKnowHow.com Incident Database ([1], [2]) already contains twenty serious incidents with high-pressure urea equipment items, where a failing leak detection system was the cause or one of the main causes. Of these twenty serious incidents, six occurred during in only two recent years (2018 and 2019). Serious is defined as the integrity of the carbon steel pressure-bearing wall being threatened, which could lead to a rupture of the vessel or a near miss of such a rupture, and could result in casualties and injuries and an extended, unplanned shut down with extensive repairs.

In seven of the twenty incidents the vessel ruptured, which is what one expects when corrosive ammonium carbamate is in contact with a large area of the carbon steel pressure-bearing wall. The rupture means a physical explosion of a heavy wall vessel with an internal pressure of some 140 barg-200 barg (2030 psig – 2900 psig) releasing a large quantity of toxic ammonia. In several incidents it was a miracle that such a rupture did not happen, which was a result of pure luck and not wisdom. For more details of these twenty incidents, refer to Appendix 1.

From the UreaKnowHow.com risk register of a high-pressure urea reactor with a 316L urea-grade liner [3], typical for most urea reactors still in operation today, one can conclude that 70% of the fifty identified safety risks could be prevented by a proper leak detection system.

These statistics explain why a proper leak detection system is crucial. This paper will further explain why most existing systems as well as currently offered systems are not acceptable (that is, not safe enough) anymore, by illustrating their shortcomings in several incidents. The paper will conclude that the minimum requirements of "The #1 Safeguard of Any Urea Plant" is a modern state-of-the-art leak detection system.

Why is a Proper Leak Detection System Crucial?

In order to avoid any risk that the mechanical integrity of the carbon steel pressure-bearing wall is endangered, one needs to assure that any leak through the protective layer, especially a loose liner, is detected immediately.

Without detection of a liner leak, the failure mode of high-pressure urea equipment has proven to be a catastrophic failure of the vessel. In case of a leak in the liner, a large area of the carbon steel pressure-bearing wall underneath will be exposed to the extremely corrosive ammonium carbamate. Experience has shown that ammonium carbamate can corrode carbon steels with very high rates (up to 1,000 mm (40 inch) per year).

The loose liner is attached to the carbon steel pressure-bearing wall and these enclosed areas form so called "liner compartments" as shown in Figure 1.

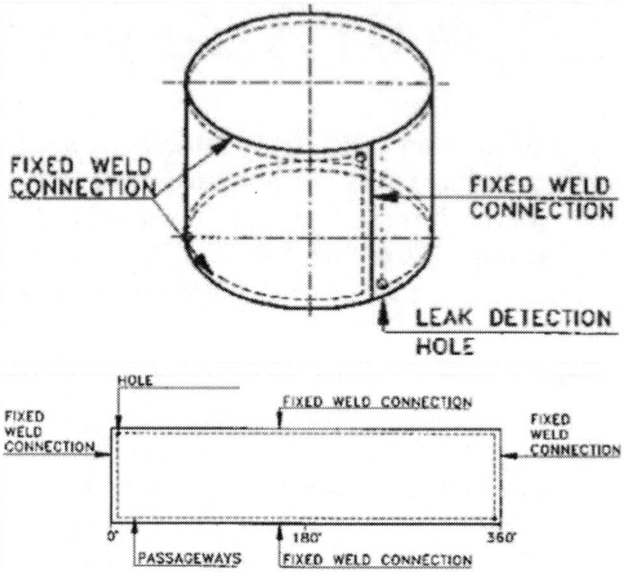

Figure 1. A typical liner compartment

The quality of the welds that attach the liner to the carbon steel (fixed-weld connections) cannot be checked by non-destructive testing methods, which makes these welds a higher risk for undetected leaks. Each liner compartment should have minimum two leak detection holes through the carbon steel.

During operation, loose liners continuously become thinner due to the unavoidable passive corrosion. Typically, after some 20-25 years a replacement of the liner is required. Besides thinning of the liner, several other failure mechanisms can cause a leak in a loose liner. Examples include material quality issues, design errors, fabrication/welding mistakes, condensation corrosion, strain induced intergranular cracking, fatigue cracks, chloride stress corrosion cracking, crevice corrosion, and active corrosion. One can be sure that during its lifetime every urea plant will experience leaks in loose liners.

A leak contains ammonium carbamate, water, and urea (in some process equipment). At the lower pressure present in the leak detection system, ammonium carbamate will dissociate into ammonia and carbon dioxide but only if the temperature is higher than 60°C (140°F). Urea, however, can easily crystallize or polymerize and clog the leak detection system. It is the major challenge for any leak detection system to avoid the clogging risk.

Existing Leak Detection Systems

Urea plants today operate various designs of passive and active leak detection systems.

In a passive leak detection system one waits until the leak shows up from the leak detection holes. Other systems connect the leak detection holes via tubing to a leak detector, for example a bottle with a phenolphthalein solution, which changes in color when the pH changes due to the presence of ammonia. An example of such a system is shown in Figure 2.

Figure 2. An example of a passive leak detection system.

The biggest risk with these passive leak detection systems is that one waits until the leak shows up at the detector and then relies on an operator who inspects these detectors during their plant tours. Experience has shown that in such a case clogging easily occurs before a leak is noted and serious threats to the integrity of the carbon steel pressure-bearing wall occur. Figure 3 shows a clogged leak detection hole.

Figure 3. Clogged leak detection hole.

The risk of clogging leak detection holes is also illustrated in a paper presented during the 2019 AICHE Ammonia Safety Symposium [4], where a leak detection hole was clogged with a solution that only contained ammonium carbamate and not any urea.

To operate safely one must understand that clogging can occur, and if so the pressure and temperature increases preventing the dissociation of ammonium carbamate into less harmful components. The trapped ammonium carbamate has a rapid corrosion rate for carbon steel, and can actively corrode stainless steel because oxygen is not available for passivation.

These are the reasons why, in our view, passive systems are no longer acceptable. But why do many urea plants still operate with these outdated, passive leak detection systems? Why, for instance, is a DCS operator warned when a pump may cavitate due to a low level in its suction vessel, but is not warned when the integrity of the carbon steel pressure-bearing wall of a high-pressure equipment item is threatened? A far higher threat!

We believe the answer is that most Hazard and Operability studies do not discuss in sufficient detail the case "failure of loose liners in high pressure equipment". The consequences of a carbamate leak through the liner can lead to a complete failure of the high-pressure equipment with major damage to the production facilities and several fatalities, possibly even outside the fence line. Using the F-N curve (Frequency – Number of fatalities) as societal risk acceptance criteria, the probability of such event must be lower than once every 100,000 years to be in the ALARP (As Low As Reasonably Possible) region, and meet The Netherlands risk criteria requirement. An example of LOPA (Layers of Protection Analysis) is showing that the PFD (Probability of Failure on Demand) requirement of the leak detection system must be 10^{-2} to reduce the probability of catastrophic failure, with fatalities outside the fence line, below 10^{-5} per year in that given example. [5]

Pressurized Versus Vacuum Systems

When talking about active leak detection systems, one can distinguish a **pressurized system**, in which an inert carrier gas stream flows through the leak detection circuits and a **vacuum system**, where one pulls vacuum pressure behind the liner.

A pressurized system circulates an inert carrier gas through the various leak detection circuits. This gas will flow only through the grooves behind the liner. Attention should be paid that the pressure is not too high when the equipment is not in operation, to prevent damage to the liner.

A vacuum system pulls vacuum behind the liner by means of a vacuum pump. On the end of each leak detection circuit an atmospheric ball valve is installed which is normally in closed position.

Figure 4 shows schematics of these systems.

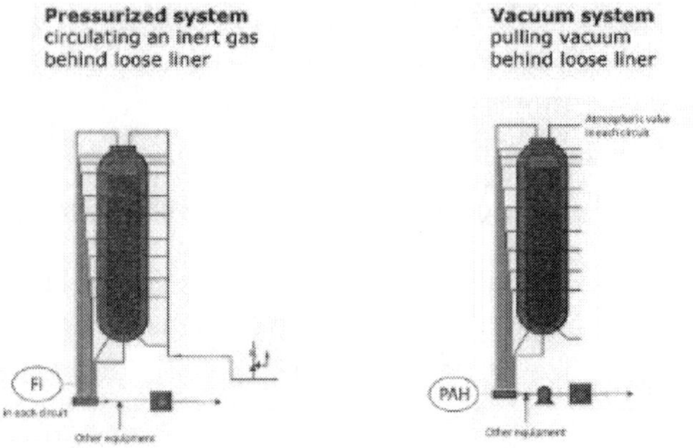

Figure 4. Pressurized (left) and vacuum (right) leak detection systems.

We recommend a vacuum leak detection system because of the following reasons:

i) A vacuum system always monitors the complete liner area, so including, for example, critical welds of clips of trays, even when parts behind the liner are clogged for some reason. In a pressurized leak detection system, one needs grooves through which a carrier gas flows. Typically, these grooves are along the liner welds to the carbon steel as one can see in Figure 5. Note there are two parallel routes from entrance to exit and if one groove is clogged for any reason, more than half of the liner area will not be monitored directly.

Figure 5. Grooves and leak detection holes in the carbon steel pressure-bearing wall.

ii) Once one faces the situation of a clogged leak detection system, a vacuum leak detection system is the only and thus best solution. For example, one can pull vacuum from both leak detection holes creating a reliable leak detection system again. Flushing with steam or water to dissolve the crystallized urea is not recommended. Introducing water behind the liner can lead to various kinds of stress corrosion cracking phenomena of the carbon steel, which one cannot see nor inspect, and which can lead to serious incidents. An example can be found in Incident 04-001 of the UreaKnowHow.com Urea Incident Database [1].

Figure 6 shows cracks in the pressure bearing carbon steel wall behind the liner. Also one can recognize boat samples for detailed analysis of the cracks.

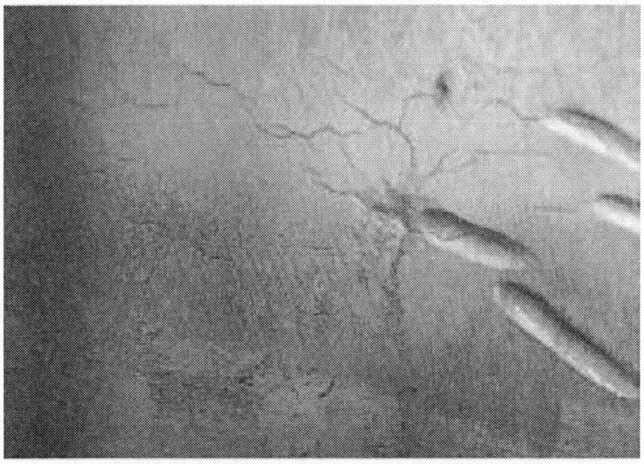

Figure 6. Cracks in carbon steel.

iii) A vacuum leak detection system can be applied for any kind of high pressure urea equipment, including when only one leak detection hole is available in a liner compartment or when there is a history of leaks and clogged areas behind the liner. This benefit is especially important for older equipment which was built without leak detection grooves or without compartmentalized liner sections.

iv) With a vacuum leak detection system there is no risk for liner bulging and damage due to high pressure in the leak detection system behind the liner, as indicated in Figure 7.

Figure 7. Example of a bulged liner

Some vendors avoid this risk of bulging by switching off the leak detection system when the pressure in the synthesis section is below a certain value. In our view this is not a good solution because during heating up and cooling down of the synthesis section one puts the highest stresses on the loose liner and during these operating phases a properly working leak detection system is very important.

v) With a vacuum leak detection system, one is able to significantly dilute the leaking fluid by lining up the leaking leak detection circuit with the vacuum pump and opening the atmospheric ball valve located at the end of the circuit. In a pressurized system one can only dilute the leak to a limited extend, which is determined by the maximum range of the carrier gas flowmeter.

vi) There are cases of failing active pressurized leak detection systems reported in the public domain. For example, Incident 19-005 of the UreaKnowHow.com Urea Incident Database [1] describes a case where a large leak in the loose liner is restricted by the flow meter present in the leak detection circuit of the pressurised leak detection system. The pressure increases and back flow occurs causing clogging of other leak detection circuits and even clogging the leak detection circuits of other high-pressure equipment items.

Some argue that a vacuum leak detection system has the disadvantage of giving a false alarm in case one experiences an ammonia leak in the ambient air, as this is drawn into the system. However, our AMMO Laser leak detection system [6] provides an ammonia level alarm at 100 ppm(vol); this amount of ammonia can only be reached in case of a real liner leak. It is impossible that 100 ppm(vol) ammonia will exist in the complete atmosphere around the high pressure equipment. A local highly concentrated ammonia release will be diluted to a level significantly below 100 ppm(vol) by the air sucked in via the other tube connections. The chosen alarm level of 100 ppm(vol) corresponds to a leak size big enough to find it with normal leak tracing techniques. Note that any liner leak contains at minimum 30%(vol) ammonia due to the free ammonia present and the dissociated ammonium carbamate.

How to Avoid Clogged Situations?

It is very important that the leak detector is very accurate and reliable. In high-pressure equipment any leak will contain ammonia. We prefer to use a tuneable diode laser analyser, which is specific for ammonia, self-calibrating, very accurate, and very reliable. An accurate and reliable leak detector is very important to detect a leak in an early stage, before any risk of clogging can occur. When a leak occurs it is very beneficial to dilute the leak to minimize the risk of clogging.

Proper tracing and insulation and a sufficiently large internal diameter of the applied tubing reduce the risks of clogged situations. Making use of a proper leak detection system beginning with the first day of operation reduces the risks of clogging situations.

How to Detect and Solve Clogged Situations?

To detect a clogged situation with a pressurized leak detection system, one can install a low flow alarm in each leak detection circuit. Similarly, in a vacuum leak detection system one can install a high-pressure alarm and detect a clogged situation, as in this case of a clogged situation the vacuum will get lost due to the unavoidable leaks along the tubing connections.

In case of a clogged situation the best practice is to divide the leak detection circuit first into sections to determine the location of the clogging.

- In case the clogging is present in the tubing, one can disconnect the tubing and flush / replace the tubing.
- In case the clogging is in the leak detection hole itself one can remove it mechanically by means of small rod (use one with a blunt end in order not to damage / puncture the liner) or with help of a flat bottom drill.
- In case the clogging is behind the liner, there are almost no options to solve this problem. Do not flush with steam or condensate for the abovementioned risks of cracks in the carbon steel wall. In a vacuum system, one detects the maximum liner area directly, even when clogging exists behind the liner. In case there is no open connection between two leak detection holes, we advise to connect both leak detection holes with the vacuum pump. In that way one re-installs a proper leak detection system again.

Requirements of a State of the Art Leak Detection System

It is very important that a leak is detected as soon as possible to reduce the risk of clogging of the leak detection system.

An active vacuum leak detection system actively transports the leaking ammonia to the detector.

An accurate and reliable detector, which can detect the smallest leak at any time.

DCS alarms to immediately warn the operator a liner leak is present or the leak detection system is not working properly. In the AMMO Laser leak detection system, the following DCS alarms are included:

i) A high pressure alarm,

ii) an ammonia detector failure and

iii) a high ammonia level indicating a liner leak.

In order to minimize downtime it is recommended that the leak detection system is able to:

i) Identify the leaking leak detection circuit,

ii) introduce a leak detection gas, and

iii) distinguish false leaks from real leaks.

The overall reliability of the leak detection system should meet national or international standards. Reference 5 states that the Probability of Failure on Demand of a leak detection system must be 0.01 to reduce the probability of catastrophic failure with casualties outside the fence below once every 100,000 years, which is a safety standard valid in the Netherlands. The AMMO Laser leak detection system fulfils this reliability requirement.

We recommend to check the complete leak detection system at regular time intervals and thus to include the leak detection system in the field operator checklist. In this way, operators continuously pay attention to the leak detection system and realize its importance to ensure the safety and reliability of the high-pressure equipment. Best practice is to appoint one person as the "owner" of the leak detection system in order to assure it does get proper attention.

Conclusions

Urea incident databases and risk registers clearly prove the need to a proper leak detection system for loose liners of high pressure urea equipment as The #1 Safeguard for Any Urea Plant.

A proper leak detection system is in our view an active vacuum leak detection system with an accurate and reliable ammonia detection system.

We believe the reason that still many urea plants operate with less reliable leak detection system is because Hazard and Operability (HAZOP) studies do not discuss and recognize in sufficient detail the case "failure of loose liners in high pressure equipment" including its proper safeguards.

An active, vacuum and accurate leak detection system, which has been in operation since the first start-up of the plant, is of utmost importance to secure the mechanical integrity of the carbon steel pressure bearing wall of high pressure urea equipment.

Appoint one person as the "owner" of the leak detection system in order to assure it does get proper attention.

We urge all urea plant owners to upgrade its existing leak detection system to a state of the art standard and together improve the safety and reliability level in our industry.

References

[1] UreaKnowHow.com Incident Database: www.ureaknowhow.com/ukh2/round-table-1/33.html

[2] 2018, AMMONIA TECHNICAL MANUAL, Prem Baboo, NFL, "The Lessons Learned from the First 100 Safety Hazards In Urea Plants".

[3] Risk Assessment of a Urea Reactor: www.ureaknowhow.com/ukh2/images/rr/rr_urea_reactor_v7.pdf

[4] 2019, AMMONIA TECHNICAL MANUAL, Alex Scheerder, Stamicarbon, "Successful Repair of a Severely Damaged HP Scrubber Sphere".

[5] 2014, AMMONIA TECHNICAL MANUAL, Shane Roysum, Yara, "Failure and Repair of the Liner of a High Pressure Carbamate Condenser".

[6] https://www.ureaknowhow.com/ukh2/services/1252-ureaknowhow-com-leak-detection-system.html

Appendix 1: List of Incidents With Failing Leak Detection Systems

1. Incident 19-006 Severe liner cracks in bottom HP scrubber not indicated by pressurized LDS

   Cause: The pressurized leak detection system fails due to backflow and clogging.

   Consequences: No active leak detection is available for all high pressure equipment items, extensive work needed to clean / open all circuits, relining required for bottom HP scrubber

2. Incident 19-004 Damage of reactor CS wall

   Cause: Failure in tray clip weld

   Consequences: Unplanned shut down and extensive reactor repair

3. Incident 19-003 Damage of urea reactor CS wall

   Cause: Failure in tray clip weld

   Consequences: Unplanned shut down and extensive reactor repair

4. Incident 19-002 Damage of urea reactor CS wall (outer CS layer coil ruptured)

   Cause: Failure in tray clip weld

   Consequences: Unplanned 30 days shut down and extensive reactor repair

5. Incident 18-005 Serous corrosion of CS wall HP scrubber

   Cause: Passive leak detection system did not alarm in time

   Consequences: Leak through wall, extensive repair, future replacement

6. Incident 18-004 Damage of reactor CS wall (outer CS layer coil ruptured)

   Cause: Failure in tray clip weld

   Consequences: Unplanned shut down and extensive reactor repair

7. Incident 15-005 Severe corrosion CS wall urea reactor

   Cause: Active pressurized leak detection system did not work properly

   Consequences: Leak through insulation was noted by operator and plant was shut down, serious repair work was required

8. Incident 13-002 Urea reactor damage after 35 years in operation

   Cause: Active corrosion during blocking in

   Consequences: Plant shut down, serious repair, no casualties, no damage to other plant parts

9. Incident 12-004 Damage CS wall urea reactor

   Cause: No proper leak detection system

   Consequences: Repair work 15 days 24 hours

10. Incident 12-003 Damage CS wall urea reactor

    Cause: No proper leak detection system

Consequences: Repair work 20 days 24 hours

11. Incident 12-002 Urea reactor liner severely damaged

    Cause: Unknown but no active leak detection system present

    Consequences: Relining required, long unplanned shut down

12. Incident 05-001 Urea reactor rupture after 5 years in operation

    Cause: Stress corrosion cracking of carbon steel behind the liner likely due to steaming out leak detection holes

    Consequences: 4 casualties, 32 injured, 4.3 mln US$ financial loss

13. Incident 04-001 Urea reactor rupture after 14 years in operation

    Cause: Corrosion of carbon steel from behind the liner likely due to steaming out leak detection holes

    Consequences: no casualties, complete damage of the 200 mtpd total recycle urea plant

14. Incident 03-002 Urea reactor damage in sleeve nozzle

    Cause: Unknown but no active leak detection system present

    Consequence: Shut down, no casualties

15. Incident 98-003 HP scrubber rupture after 16 years in operation

    Cause: Leak detection hole was still plugged with transportation plug

    Consequences: Plant shut down for 2 months, no casualties, significant damage to other plant parts

16. Incident 95-001 Urea reactor rupture

    Cause and consequences not exactly known but no active leak detection system present.

17. Incident 92-001 Urea reactor rupture

    Cause: clip weld failure in liner, no active leak detection system, no proper actions after leak was noted

    Consequences: 10 people injured + 90 citizens injured

18. Incident 90-001 Urea reactor rupture

    Cause: liner leak, no active leak detection system

    Consequences: Plant shut down

19. Incident 79-001 Urea reactor leak after 8 years in operation

    Cause: wrong design nozzle liner

    Consequences: repair, no casualties, no damage to other plant parts

20. Incident 77-001 Urea reactor rupture after 14 years in operation

    Cause: lead liner damage, corrosion from outside

    Consequences: no casualties, likely damage to other plant parts

Integrity Assessment and Restoration of a Urea Reactor Following Repeated Failures

This paper describes the rehabilitation of a Urea Reactor installed in 1977 and operating at a urea plant located in Russia. Frequent leakage from the liner weld seams was reported in the last period of service and Casale SA was requested to inspect and provide solutions. The following internal inspection confirmed a generalized deterioration of the liner base metal and seam welds and this situation could have led to potential catastrophic failure in the worst-case scenario. The repair plan was developed to assure the safety of personnel employed during the repair work-scope and compliance with Code and local rules. The in-situ repair was successfully completed, allowing plant operation to continue prior to equipment replacement.

Roberto Gorza
Casale SA

Introduction

In early 2019 Casale was informed of repeated process leaks from a high pressure reactor operating at a urea plant located in Russia. A meeting was immediately set-up to discuss the details and develop a path forward to determine the cause of the leaks and develop an effective repair plan.

The urea reactor, shown in Figure 1, was fabricated by an Italian company and commissioned in 1977. This is a mono-wall, vertical, high-pressure vessel. It is 36 m (118 ft) tall tangent-to-tangent, with an inner diameter of 2.3 m (7.5 ft), and a 95 mm (3.7 in) thick carbon steel shell for pressure retention. There is also an 8 mm (0.3 in) AISI 316L UG corrosion resistant liner on the interior surface of the pressure shell.

Figure 1. Urea reactor view from ground level.

Background

Casale was informed that during the last period of service, repeated leaks were experienced at different cylindrical shell elevations, due to selective corrosion of the liner seam welds.

The reactor had undergone various modifications and repairs in the past, namely the following:

- Two liner belts were re-lined in 2004 and 2006 respectively due to accelerated wall loss.
- Several weld repairs on longitudinal and circumferential liner seam welds due to selective corrosion attack.
- The last tray from top (i.e. tray#14), was removed and replaced with special devices at the very bottom section of the vessel.

During February 2019, the urea plant was taken off-line to repair a leak on the bottom cylindrical shell section and Casale was invited to perform an internal visual inspection and provide recommended repairs.

Inspection Findings

The upper section of the vessel was difficult to inspect because the inspection could only be performed from the entry ladder.

The overall condition of the top head's AISI 316L UG liner was satisfactory. No scaling or accelerated corrosion marks were observed. The relined section was found in general good condition. The minimum recorded liner thickness was 3.6 mm (0.14 in) against nominal 4.0 mm (0.16 in) (Figure 2).

Above tray #1, the longitudinal and circumferential weld seams were found to be badly corroded and severely attacked in the heat affected zone (HAZ). The minimum liner thickness recorded was 5.6 mm (0.22 in), compared to the original 8 mm (0.3 in) thickness.

As a whole, the liner courses throughout the vessel were corroded, but still fit for reliable service. The measured thicknesses were between 5 and 6.7 mm (0.20 in and 0.26 in).

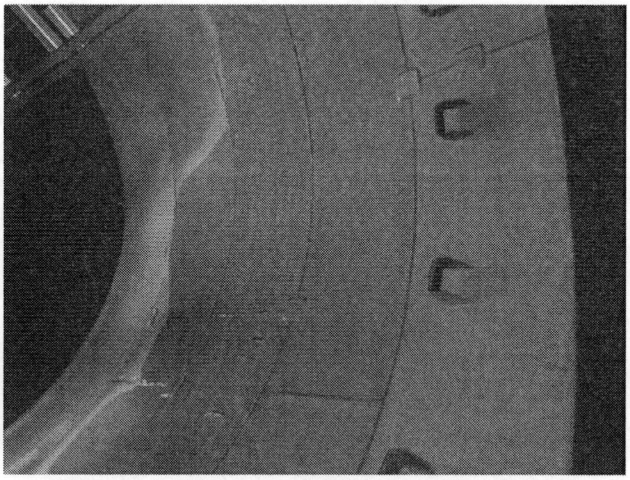

Figure 2. Upper cylindrical relined shell section.

The liner weld seams were badly deteriorated, and significant toe-detachment was witnessed on both side of the welds (Figure 3).

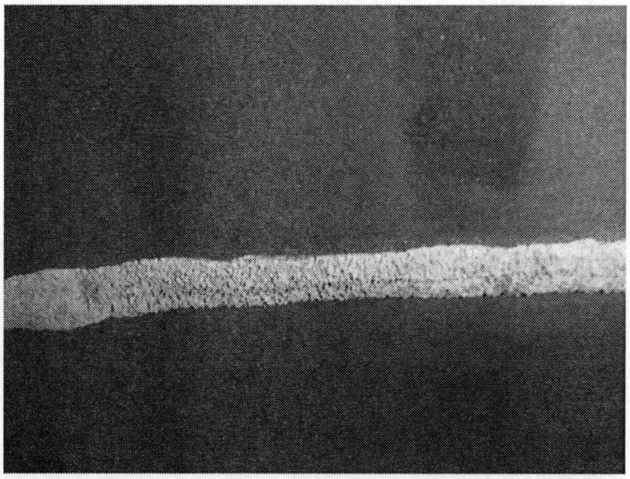

Figure 3. Corrosion on longitudinal liner seam weld.

The liner belt between tray #2 and tray #3 was found severely corroded compared with others, and the minimum thickness recorded was 2.5 mm (0.10 in). This liner belt was already partially re-lined during 2006 maintenance plant shutdown.

The overall condition of the reactor trays was satisfactory, whereas the existing support were generally corroded, and metal deterioration was visible on the base metal and seam welds (Figure 4).

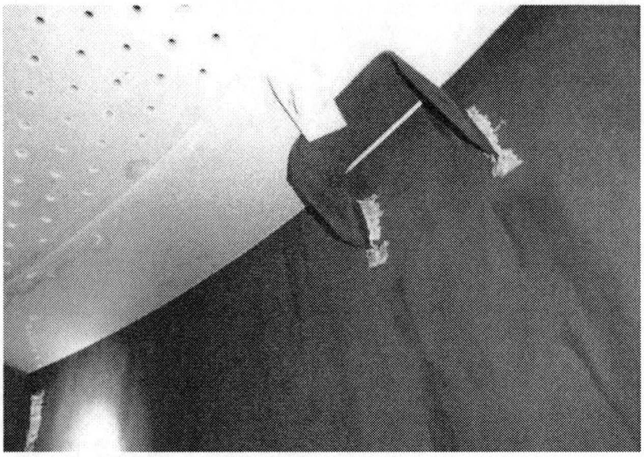

Figure 4. Trays and support cleats.

Access to the bottom section, underneath the special devices, was somewhat limited by the ongoing repair work. This situation hindered the ability to complete a thorough visual inspection.

As reported, a new liner belt had been installed to replace the existing liner just above the round weld connecting the hemispherical head to the cylindrical shell course (Figure 5).

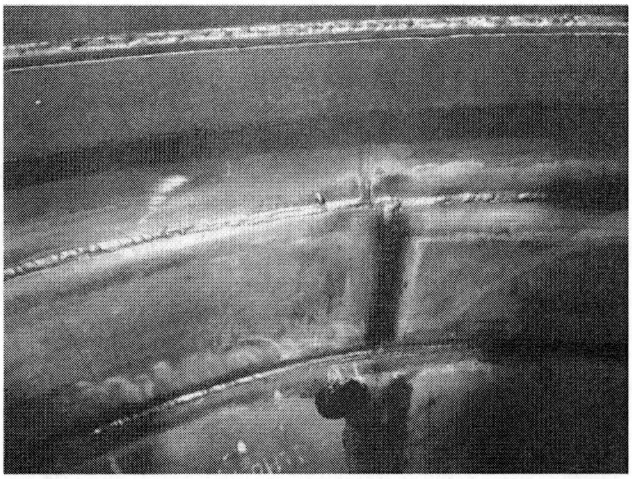

Figure 5. New liner belt at bottom shell section.

Repair Process

The repair process basically consisted of installing liner bands in 25Cr22Ni2Mo material on top of the existing circumferential liner seam welds, installation of new tray support cleats and refurbishing the remaining corroded liner seam welds by weld build-up. The new liner bands were connected to the existing leak detection system behind the liner to be able to convey unforeseen leak toward the weep tubes.

The in-situ repair was planned for September 2019, so work preparation began immediately. Early work preparation improves job quality and reduce time of execution.

A specialized Italian contractor was engaged to carry out the repair work, since skill and expertise are required to successfully implement repairs of this type.

The new liner bands were pre-formed to suit entry through the manway. Within the scope of supply, we decided to include all tools and equipment necessary to carry out all of the activities in the work scope. Three high frequency welding machines with torches and umbilicals were required to weld inside the reactor, while keeping the generators outside, on the top platform of the reactor. A low-voltage illumination system made of a 50 m (164 ft) cable with 15 lamps was used to illuminate the entire reactor interior. Angle grinders, pencil grinders, general consumables, personal protection equipment, and other items were dispatched to the site in a 20 ft (6.1 m) ISO shipping container.

Any inspection and maintenance work inside the urea reactor requires specific attention to the safety of personnel. Safety can affect the work execution.

The Urea reactor is long and narrow with a single point of access, a 500 mm (19.7 in) diameter manway, located at the top head. There are also two nozzles at the bottom which were used to supply fresh air to the vessel. Also, the plant was shut down for maintenance but six similar plants were operating in the surrounding area, which was considered while preparing the job safety analysis.

An electric hoist was supplied and installed on top of the reactor. This hoist was specifically adapted for working inside the reactor, and was provided with cylindrical basket for use as a man-lift.

Aluminum round scaffolding, suitable for fast installation over reactor trays, were assembled at each working level to ensure the safety of personnel inside the vessel. (Figure 6).

Figure 6. Urea Reactor view from inside.

Before mobilizing the team, a site visit was scheduled to meet with the Owner's project team, to discuss and confirm the scope of work and relevant responsibilities. Based on this discussion, a detailed work procedure for the shutdown repairs was issued, including requirements for work permits, qualification of welders and welding processes, and the preparation of as-built documentation in accordance with Russian rules and regulations.

The estimated job duration was 28 calendar days, with two shifts of 12 hours each. The work scope was performed during the agreed turnaround schedule by a team of total 6 welders, 2 fitters, and a supervisor.

Repair Execution

After the hoist was installed, all of the trays and internals were removed and stored at ground level. An internal visual inspection was performed throughout the length of the vessel.

The repair work-scope was revised based on the inspection findings. One change was the removal of most tray support cleats, which were replaced and installed over a cover plate.

All longitudinal and circumferential liner weld seams were rehabilitated using weld build-up or partial refurbishment, as necessary (Figure 7). The work proceeded from the bottom-up by using internal temporary aluminum scaffold installed at the working level.

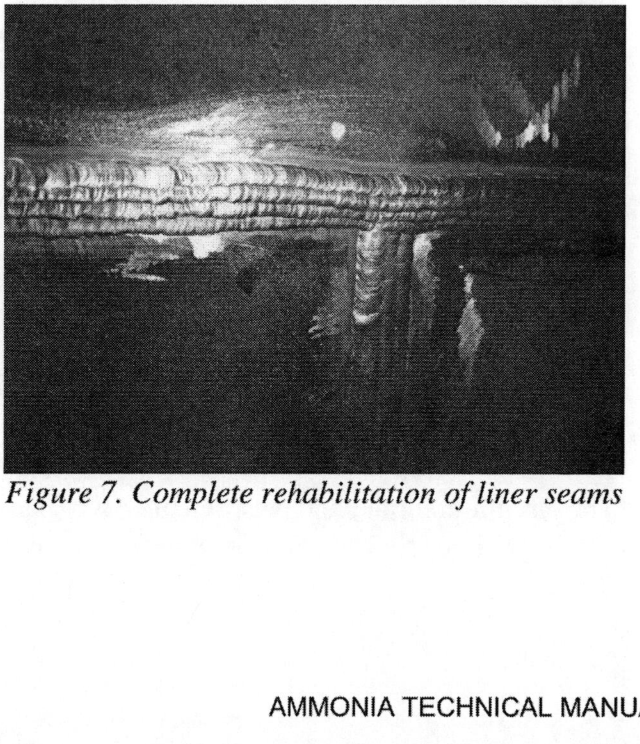

Figure 7. Complete rehabilitation of liner seams

Once the cleats were removed severe corrosion marks and cavities were found on the liner base metal, hence new liner bands were installed at multiple elevations in the reactor over these affected area. The trays were re-installed at the same elevation (Figure 8).

Figure 8. Corrosion marks on liner base metal

Two holes were drilled through the existing liner for communication with the existing leak detection network. Afterwards, a new liner band was installed by tacking and then welding it into place. New cleats were welded to the new liner band and the existing ring support was installed to match the original orientation (Figure 9).

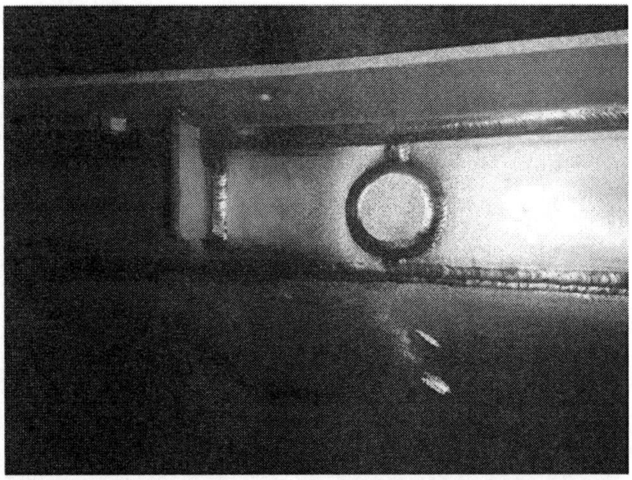

Figure 9. New liner band with tray support rib

The circumferential weld between the cover pads and liner of nozzle 1N (solution outlet) was found to be badly corroded. It was repaired over its entire length using weld built-up (Figure 10).

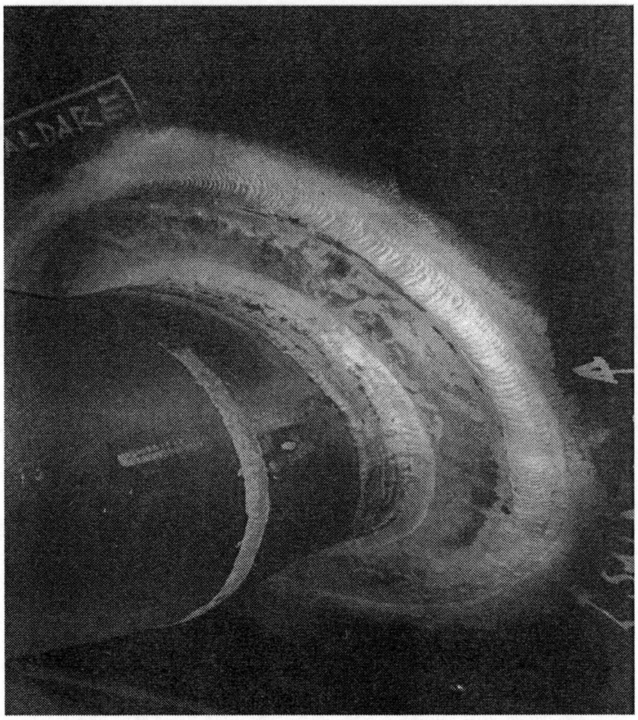

Figure 10. Extent of repair on nozzle 1N

All weld repairs in the work scope was completed as per schedule. The special devices at the bottom of reactor were placed in position. All trays were installed except one sector of each tray to facilitate the access for NDT inspection.

Following was the expected path forward:

- 100% Dye Penetrant Test (DPT) + Ferrite content check (Fe%) + Positive Material investigation (PMI)
- Hydrotest (HYT)
- DPT (spot check)

The hydrotest failed a couple of times. Water leakage was observed from weep tubes in the bottom section, which was not included in Casale work scope.

The special devices installed at the very bottom section of the reactor were again removed to ensure free access to the leaking area (Figure 10).

Figure 10. Limited access to the leaking area

An internal visual inspection and air/soapy water leak test was performed on the plate section the Owner installed during the previous maintenance shutdown, which was monitored by the weep holes that leaked during the hydrotest. The testing revealed an indication alongside the top longitudinal weld of the plate section. This test also flagged additional suspect areas, because the overall quality of these weld was poor (Figure 11).

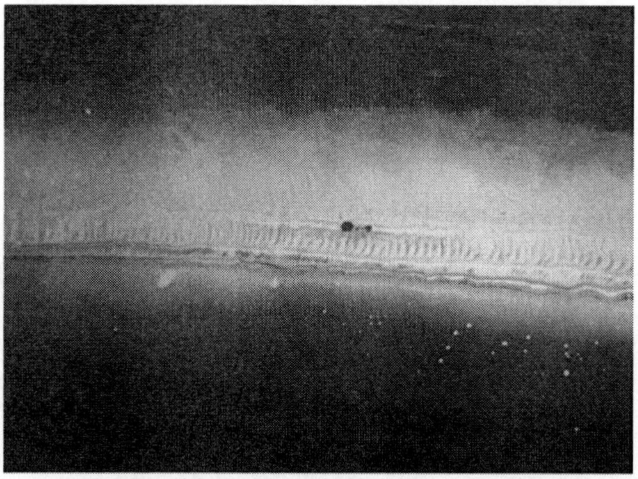

Figure 11. Defect on liner seam weld

The Owner and Casale agreed that the top and bottom circumferential seam welds of this plate section should be completely refurbished. This repair was completed by weld build-up and DPT tested.

Following this repair the reactor was hydrotested again and failed. Water was observed leaking from the same two weep tubes in the bottom section.

An internal visual inspection confirmed a large fracture beside the top circumferential weld of the same plate section that failed during the first hydrotest. The fracture was approximately 550 mm (21.7 in) long and 3 mm (0.1 in) wide (Figure 12).

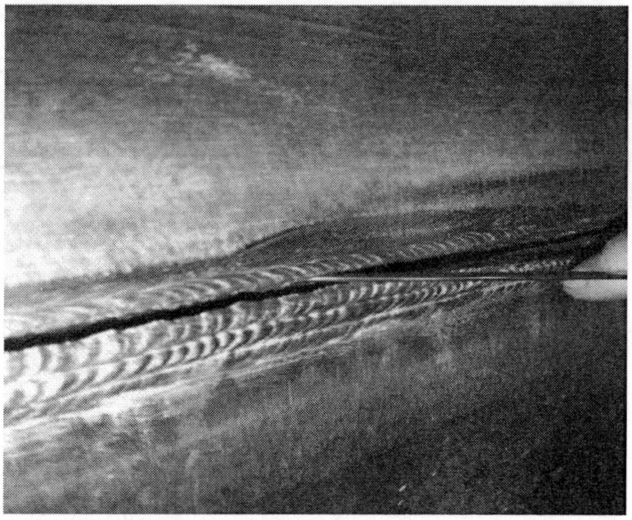

Figure 12. Mechanical fracture on seam weld

The Owner and Casale immediately met to discuss the potential root cause of this repeated failure and determine the path forward.

The Owner reported that repeated leakages had occurred from the same area as these hydrotest failures. The plant was shut down for inspection and repair several times during the last year of service. All the repair attempts failed and finally this defective portion of liner was replaced. The details of this repair were not recorded nor reported on dedicated inspection report hence the root cause of these failure were not determined because of lack of information.

Based on the inspection results and discussion with the Owner, a repair plan was issued. If seal weld type is confirmed, the following procedure is used:

- Remove insulation in the bottom cylindrical section and provide an ultrasonic thickness survey for mechanical integrity confirmation;
- Remove all reactor internals, including bottom special devices, along with relevant grid support to assure free access to the leaking area;
- Grind off the top circumferential seam weld of the plate section, which was confirmed to lack full weld penetration;
- Prepare welding bevel to assure full weld penetration;
- Provide internal visual inspection;
- Weld new top circumferential weld seam, pushing the plate section against the inside surface of the reactor with a hydraulic jack;
- Grind off a portion of the bottom circumferential seam weld to verify the penetration of the existing weld seam.

If full weld penetration was confirmed, no action would be required, only the sample repair would need to be repaired. If the weld was found to be a seal weld the procedure applied for the top circumferential seam weld would be used.

An additional team of two welders and one fitter from the Italian contractor were engaged to speed-up the repair work in a quality, safe, and effective manner. A total of four welders, one fitter and one supervisor were available at site.

The detail of the original circumferential weld seam between the liner cylindrical barrel and bottom hemispherical head is shown in Figure 13.

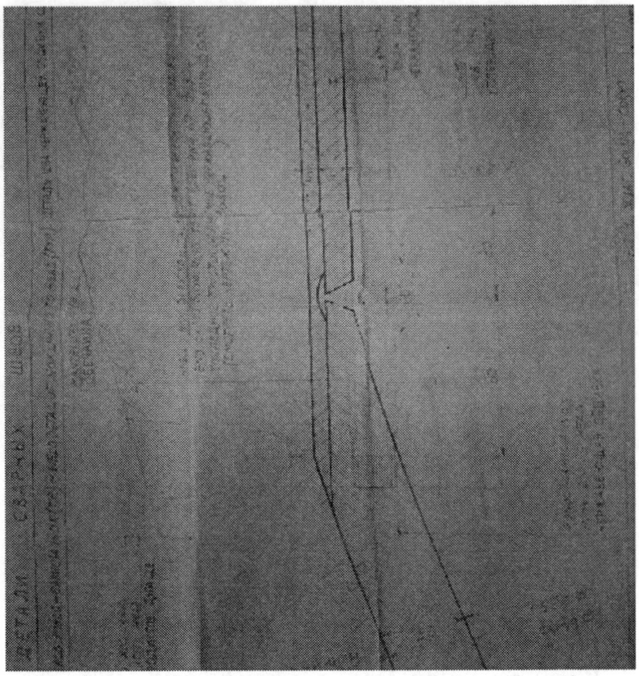

Figure 13. Original construction detail

The original shell barrel liner was 8 mm (0.3 in) thick and is installed over a 5 mm (0.2 in) carbon steel buffer layer, with stainless steel inserts at the end of each liner section. The hemispherical head liner is made from a single formed 25 mm (1 in) thick plate.

The sketch in Figure 14 indicates the design solution applied by the Owner for the previous repair.

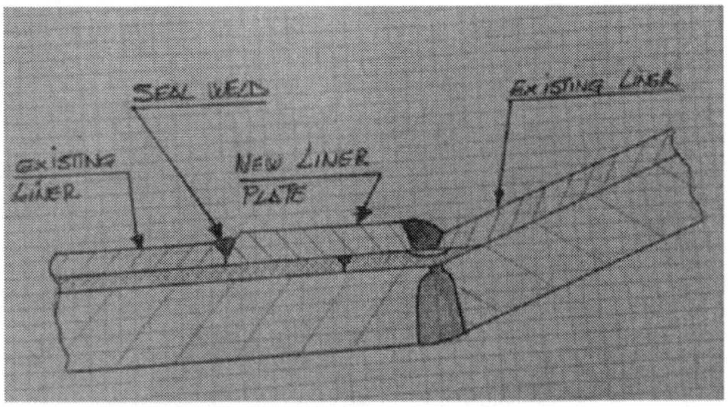

Figure 14. Repair scheme

As can be seen, a defective segment of liner, which was approximately 250 mm (9.8 in) high, had been cut and replaced with new liner plate. The new plate was 10 mm (0.4 in) thick, which was thicker than the existing 5 mm (0.2 in) liner.

Both the upper and lower circumferential weld seams were not anchored to the backing layer. The hemispherical head liner consists of single formed plate, and is anchored to the pressure vessel through the two nozzles welds. Considering this, Casale determined the hemispherical head liner could have shifted upward during cutting and welding activities, creating a gap between the liner and the inside surface of the pressure shell.

The plate section was partially removed for inspection (Figure 15).

Figure 15. Plate section under removal

The hemispherical head liner shifted down approximately 8 mm (0.3 in) when the upper circumferential weld seam was partially removed by grinding. The plate section was then prepared, tacked in position and welded (Figure 16).

Figure 16. Welding of the plate section

Finally, an additional cover strip was installed over the upper circumferential weld seam of the plate section (Figure 17).

Figure 17. Final repair configuration.

The hydrotest was completed successfully after these repairs, with a test pressure of 160 barg (2320 psig).

Conclusion

The reliability, efficiency, and working life of this critical equipment can be significantly affected or even compromised by misunderstanding the relevant corrosion propagation mechanism and associated inspection and maintenance strategies.

It is very important to develop and manage detailed inspection strategies, with attention given to all components being evaluated. These include:

- Inspection and maintenance activities should be performed by qualified and experienced contractors, since skill and expertise are required to properly execute repairs of this type.
- Every inspection and maintenance activity should be documented with a comprehensive inspection report, supported with drawings, sketches, and photographs.
- The design of vessel internals should account for the safety of personnel working inside the vessel for inspection and maintenance work.
- Good work preparation generally improves job quality and reduce time of execution.

Safety can also affect the execution of work in confined spaces, such as the reactor. Some of the critical factors that affect confined space safety include:

- Purging
- Blinding
- Vessel entry preparation
- Hole watches with rescue team
- Proper ventilation and gas monitoring
- Internal and external illumination of the work area

It is also very important that all personnel, even those not directly involved in the confined space work, are well informed about the job in progress to avoid dangerous situations.

In conclusion, a plant can remain reliable and safe only if it is properly maintained, a target that can be achieved by turning to professional experts and specialists with wide experience and knowledge. Casale has put in place a specific program – PAMS (Pro Active Maintenance System) – as an opportunity to share Casale's experience and technical knowledge in the urea field to benefit plant owners directly in their operations. It is a way to manage inspection and repair strategies in a cost-effective manner.

Regular communications between Owners and urea specialists is a strategy to prevent costly equipment failures.

Explosion Inside HP Urea Reactor

A client was operating a urea plant already for decades when during a restart an abnormality was observed in the form of a rumbling sound not heard before. No actions were taken at that time and the reactor was operated as normal with barely any difference in operational parameters. Upon opening of the reactor during a planned turnaround it was observed that all the trays were displaced from their supports and deposited in the bottom of the reactor. Most of the trays were heavily deformed.

A Root Cause Analysis found that an explosion must have occurred between the 3rd and 4th tray during startup with sufficient force to displace all the trays from their supports, depositing them in the bottom of the reactor. Analysis found that the most likely cause is the formation of a combustible mixture which exploded due to self-ignition or due to ignition caused by a static discharge. Before this occurrence no scenarios were recognized which could lead to such an incident.

In this paper we try to derive the conditions and timeline which would explain this occurrence. As not all the information was readily available, the conclusion is based upon a most likely scenario. Measures taken to eliminate the occurrence of such formation in the future are presented.

Freddy Buitink, Hans van den Tillaart
Stamicarbon

Introduction

The urea plant was designed in the 1970s and based upon TOYO technology later on revamped by Mavrowich to improve the energy consumption figure and capacity of the plant. The synthesis section includes two reactors in series in which one is used for heat exchange and the other is provided with trays to improve the conversion of ammonia and carbon dioxide to urea. As the conversion is limited due to the equilibrium, urea and water together with the unconverted ammonia and carbon dioxide is being let down in pressure where the ammonia and carbon dioxide is separated from the urea/water mixture and recycled back to the synthesis.

The remaining water in the urea solution is removed in the evaporation section and the urea melt is consequently prilled and stored in a warehouse.

The synthesis conditions are typically between 200 to 250 barg (2900 to 3625 psig) with a temperature ranging from 185 to 195°C (365 to 383°F).

During a startup in 2012, a rumble was heard, and vibrations were felt in the urea plant. The plant was started up without any further problems.

People checked the plant for abnormalities, but could not find anything indicating the source of the noise nor any substantial deviation in operating conditions. No further action was therefore taken. During inspection in 2014 of the high pressure (HP) reactor, the inspection team found all the trays displaced from their supports and piled up in the bottom of the reactor. (Figure 1)

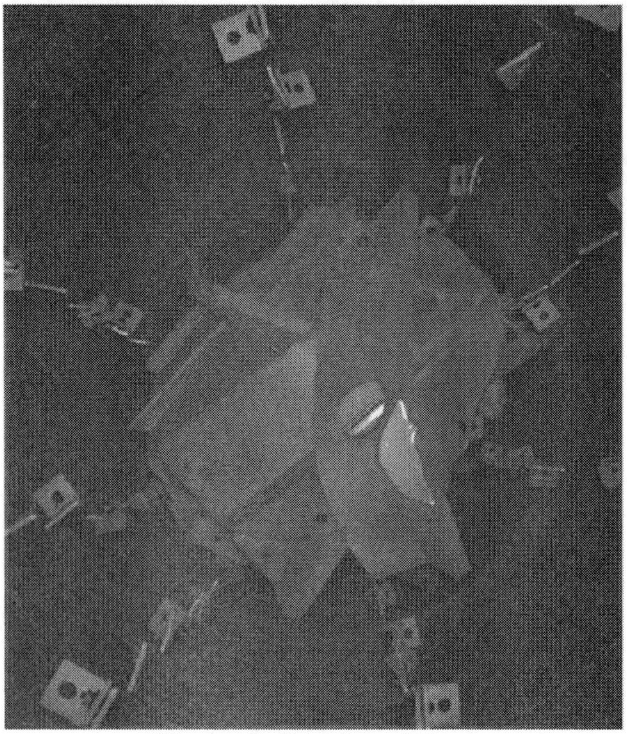

Figure 1. Displaced Trays

The preliminary conclusion was that this occurrence likely happened during the start up in 2012. A Root Cause Analysis (RCA) investigation was organized to identify the causes of and avoid reoccurrence of this incident. Stamicarbon was asked to provide inspection services of this reactor and bring in external expertise for the RCA.

RCA analysis

Looking at the extent of the damage to the trays and tray supports it was concluded that this was caused by a substantial incident which could have been either a sudden liquid displacement or an explosion. As the incident happened two years earlier, retracing what happened during the startup proved to be a challenge. It was theorized that if the incident was caused by a pressure surge and as a consequence fast liquid displacement, all the trays would be bent in one direction, either upwards or downwards. In the case of an explosion, this would not be the case. At the point of explosion, the upper trays would bend upwards and the lower trays would bend downwards if the explosion initiated in between the trays.

It was therefore decided to remove all the trays and place them in order of original installation on the ground to investigate the cause of deformation of the trays.

The top trays of the reactor were severely bent upwards (Figure 2) whereas all other trays are less damaged and are all bent downwards.

Figure 2. Upwards Bent Tray Segment

The urea reactor is equipped with fourteen trays. Each tray consists of four segments. All trays were lined up according to their actual position in the reactor. The amount and orientation of the deflection was measured for each tray. Figure 3 presents a table of the tray deflection for the damaged tray segments.

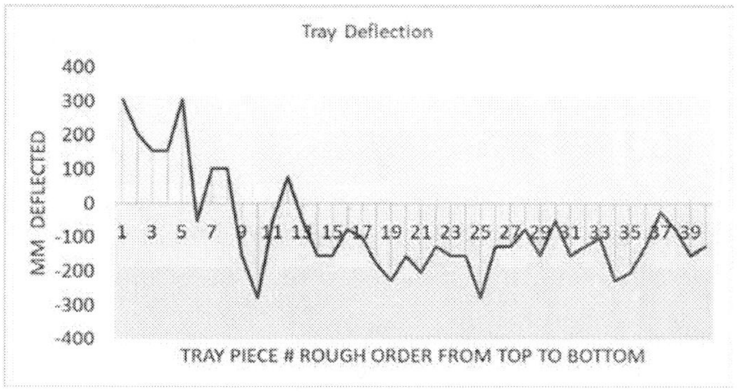

Figure 3. Tray Deflection

The evidence matched the explosion theory. The explosion occurred in between the third and fourth tray from the top of the reactor (see Figure 4).

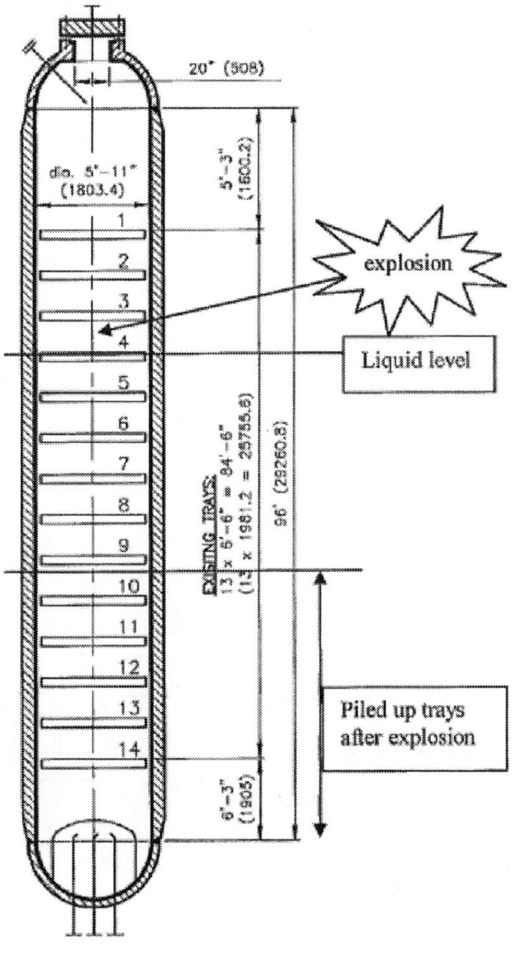

Figure 4. Reactor Diagram

At the moment of the explosion, the reactor was most likely filled with liquid up to the fourth tray.

Based upon the damage the explosion occurred in the vapor phase just above the liquid level. The pressure wave deformed the trays above the point of explosion in an upward direction, whereas the trays below the point of explosion are deformed downwards. The damage done to lower trays was found to be less severe than that of the top trays, due to the buffering effect of the liquid.

During the two years of operation of the urea plant hardly any difference in consumption figures and conversion rate were noticed while all tray parts were lying in the bottom of the reactor. The trays were piled up on top of each other to the level of the 5th tray from the bottom.

All trays were replaced during the turnaround in 2014. Identical trays were installed with the exception that 316L material was used rather than the originally specified material, 316L Urea Grade, which was not available on such short notice. It was decided to replace the trays identical as there was simply not enough time to re-design and approve a new tray design during the turnaround.

These trays are considered a temporary solution and will be replaced in the near future.

Damage assessment HP reactor

All trays and tray clips had to be replaced due to the severe deformation as a result of the explosion. The supports of the gas divider in the bottom of the reactor was replaced preemptively, but the gas divider was reused. The liner and weld overlay were not seriously damaged due to the explosion.

Mechanical damage due to the impact of tray segments and bulging of the liner at the position of some of the tray clips due to the mechanical forces exerted by the trays were minor.

From a corrosion point of view the reactor is still in good condition given its age. The main reason is the choice of material used (X2CrNiMo 25-22-

2) combined with the chosen thickness of the weld overlay and liner which was 12.5 mm (0.5 in).

Sequence of events

The typical start up procedure is to first fill about 1/3 of the reactor volume of the HP reactor with hot condensate. Both carbon dioxide feed and ammonia feed are then started slowly pressurizing the system at a N/C ratio of 4 to 1. During this period, the vent valve of the HP system is closed to reach normal operating temperature and pressure as quickly as possible, to avoid too low conversion at overflow.

Due to the heat of carbamate formation (exothermic) the temperature and pressure of the reactor and its content quickly rises. The increasing pressure causes more gases to condense releasing more heat of carbamate condensation increasing the temperature and pressure even further.

The carbon dioxide fed to the HP reactor also contains some nitrogen, methane, hydrogen and oxygen. Oxygen is added to the carbon dioxide to passivate the stainless-steel liner and internals.

The carbon dioxide and ammonia react with each other and are increasingly removed from the gas phase hereby increasing the relative concentration of the inert components in the gas mixture.

Depending on the composition of inert gases in both the carbon dioxide and the inert gases dissolved in the liquid ammonia an explosive mixture could be formed at certain conditions.

Figure 5 includes a graph showing the increase of hydrogen in the gas phase as function of the level of condensation of reactants. For passivation 0.3% oxygen as air is added. A typical hydrogen concentration is 0.6 vol%. In some cases, this can go up to 1.5%. In the graph both lines have been plotted. The two red lines indicate the upper and low explosion limits.

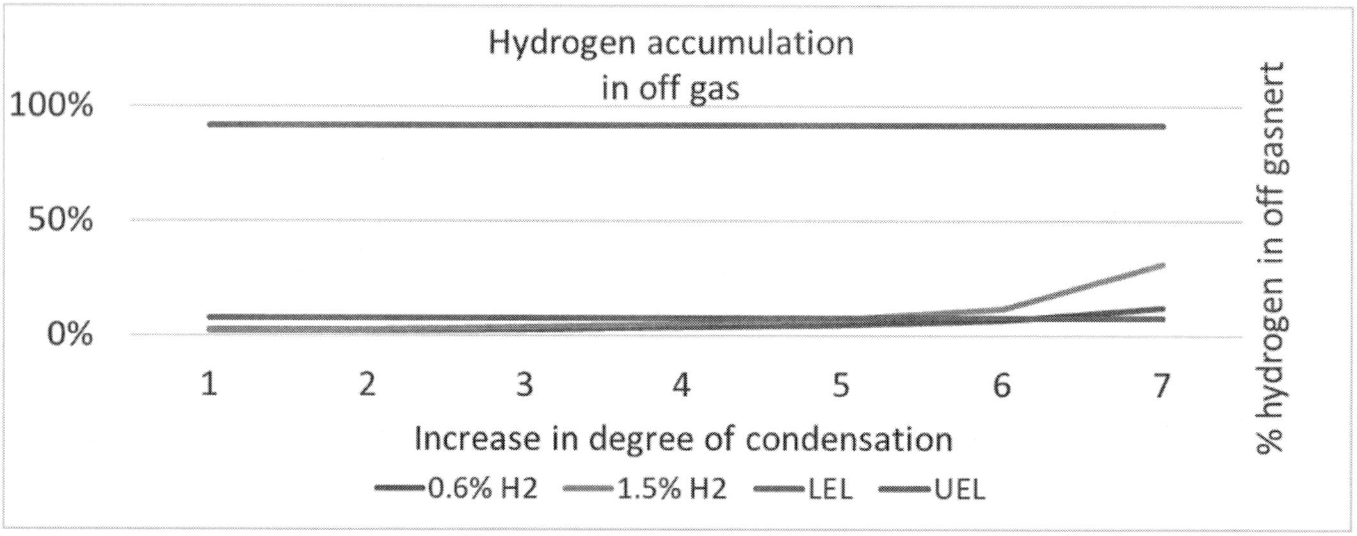

Figure 5. Hydrogen in Gas Phase

The ignition source can either be a static discharge or reaching the auto ignition temperature. Reaching the auto ignition temperature in a well-mixed environment is unlikely since as the concentration of inert increases, the partial pressure of ammonia and carbon dioxide in the off gas drops and with it the condensation temperature. During start up, steam is supplied to heat up the equipment to avoid crystallization of carbamate before feeding the ammonia and carbon dioxide to the HP Reactor. When dosing the reactants (ammonia and carbon dioxide) a thermo-compression effect can be observed which means that steam starts condensing at the colder reactor wall and the trays at vapor phase conditions that follow the condensation curve of steam at the reactor pressure. Temperatures quickly increases to well above 200°C (392°F) even at low pressures. Figure 6 shows the condensation curve simulated as a function of pressure for both steam and for carbamate gas. As the reactor wall can dissipate the heat easily due to the mass of metal, this is not the case for the trays which can retain this temperature for some time and hence be a potential source for (auto) ignition of the gas mixture.

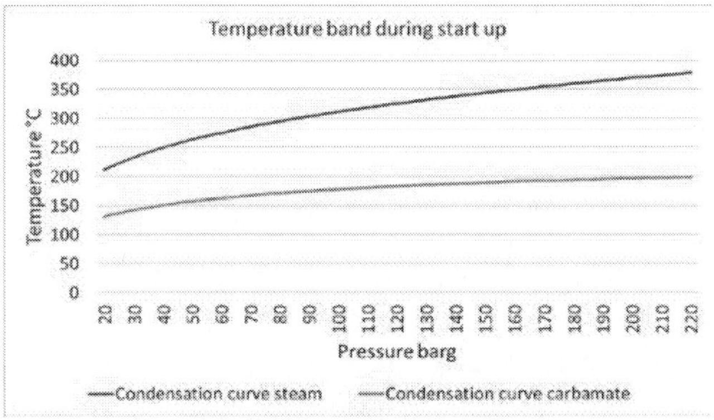

Figure 6. Condensation Curves

It is known that the self-ignition temperature of a gas is dependent on the composition and pressure of the gas mixture. Increasing the pressure decreases the auto ignition temperature. An increase of hydrogen content has a similar effect. During start up the pressure and temperature increase while at the same time, due to condensation of the reactants, the gas mixture enters the explosive region.

At the time of the startup in 2012, the ammonia plant had some known problems causing the carbon dioxide to contain more hydrogen than usual. A higher concentration of hydrogen means a wider explosion triangle (explosive region increases) but also a lower auto-ignition temperature of the mixture. Figure 7 shows the influence of increased hydrogen on the explosion region.

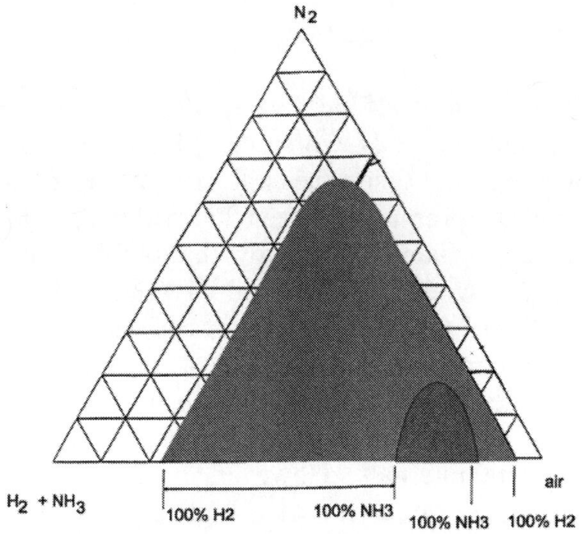

Figure 7. Flammability Diagram

In this figure the blue indicates the explosion area for an ammonia air mixture where ammonia is the only combustible. The red indicates the explosion area if hydrogen is the only combustible. A mixture of ammonia/hydrogen will have an area in between these two limits. This figure shows clearly the large influence of the hydrogen concentration on the explosion area of a mixture.

It is assumed that during start up, due to an abnormal high hydrogen concentration, the temperature of the gas/liquid crosses the self-ignition temperature of the gas mixture which in itself enters the explosive region due to condensation of carbon dioxide and ammonia. As the inert concentration and temperature would be the highest just above the liquid makes it plausible that the explosion occurred just above the liquid level. The pressure wave would then push the liquid down where the impact is limited due to the inertia of the liquid which is not the case in the gas phase. The found evidence of the damaged trays supports this theory which makes this scenario the most likely root cause of the incident.

Summary root cause of the incident;

- High hydrogen content in carbon dioxide
- Accumulation of inert

Of the above only the first was deviating from the normal start up situation.

The incident itself luckily did not lead to any loss of containment or worse, but this could easily have been the case. The explosion occurred most likely at a pressure between 20 to 50 barg (290 to 725 psig). Calculations show that in this case the pressure would have surged to 71 to 179 barg (1,030 to 2,600 psig). As the vessel was designed for 250 barg (3630 psig), the vessel and its appendages where able to contain the sudden increase of pressure which damaged in this case only the internals. However, if the explosion would have occurred at a higher pressure, the outcome could have been different. Catastrophic failure of the reactor was a possible scenario resulting in high material and immaterial damages.

Mitigating measures

The scenario described only occurs when inerts get the chance to accumulate in combination with a high hydrogen content in the carbon dioxide. To prevent this scenario in the future the following measures have been taken:

- A hydrogen analyzer has been installed in the carbon dioxide feed to the reactor.
- The startup and shut down procedures have been adapted to prevent accumulation. The vent valve will not be fully closed any more during these periods to allow inert to be vented to the downstream sections.
- All personnel will be instructed with respect to the risks of inert accumulation in general and specially during startup and shut down activities.
- All work instructions have been adapted to reflect the findings and to avoid this scenario to occur again.

Conclusion

Operating a plant for more than 40 years in a similar way without any major incidents does not mean that it is always safe doing so. Transient process operating modes such as startup and shutdown could especially pose unrecognized risks. Our advice is to repeat a safety analysis of these situations in regular intervals to not only investigate potential new safety scenarios, but to also teach personnel on the potential scenarios associated with operating the plant under all conditions.

Furthermore, do not presume that the absence of an obvious ignition source prevents explosions from occurring if a mixture enters the explosive region.

Installation of an Early Warning Solution for Gas Leaks for the OCI Urea and Melamine Complex at Chemelot Industrial Park

Ammonia is one of the most widely used chemicals and requires safety measures for handling and storage in order to prevent hazardous releases. However, no reliable measures have been developed for early warning and situational assessment of such incidents. Warning for ammonia releases becomes an ever more urgent problem considering increasing worldwide demands for ammonia., and the prospect of ammonia playing a major role in the transportation and storage of hydrogen as green fuel will eventually catapult the demand for ammonia production and storage to new heights.

Automated early detection of gas leaks and tracking of moving gas clouds in case of an incident would be critical for leak response and mitigation. This combination of risk prevention and incident mitigation was the design task for the early warning solution that was installed at CHEMELOT chemical park in Geleen, The Netherlands. The site is near a densely populated campus, requiring warning of a release incident within minutes. Once ammonia is in the air, the propagation of the moving cloud must be tracked and mapped live until it reaches the fence. The remote monitoring solution covers large areas and can locate a gas anywhere within sight. The early warning solution is based on spectroscopic analysis (Fourier-transform infrared (FTIR) spectroscopy) that allows safe identification and mapping of hundreds of different chemicals – both flammable and toxic – over distances up to 4 km.

René Braun
Grandperspective GmbH

Peter Schmitz
OCI Nitrogen B.V.

Introduction

The Chemelot industrial park in the Netherlands is located near Maastricht on the Dutch and Düsseldorf on the German side. It is one of the largest chemical parks in Europe with numerous companies such as OCI, Arlanxeo, DSM, and SABIC. The major product lines onsite are naphtha/gasoil through hydrocarbons to plastics and natural gas through ammonia to organic and specialty chemicals. Chemelot is connected via its own port, the main road network to nearby motorways, and a rail terminal. One of the specialties of the Chemelot industrial park is the Brightlands Campus, a hub for startups, R&D facilities, and education with more than 3.900 researchers, entrepreneurs, and students.[1] The campus is situated within close range to the production facilities, of which the OCI melamine- urea complex is less than 400 m

(1300 ft) from the campus. OCI Nitrogen B.V. has three plants on site: OMM (OCI Nitrogen B.V. Manufacturing Melamine), OMA (OCI Nitrogen B.V. Manufacturing Ammonia), and OMF (OCI Nitrogen B.V. Manufacturing Fertilizers).

The scope of the first phase of the project focuses on the OMM production unit, in particular the reduction of risks due to unexpected airborne releases. A technical feasibility study for a future scale-up to monitor the entire Chemelot north side has also been performed. There are four stacks within the OMM plant that must be monitored for a real-time detection of an accidental ammonia release. Once a release of ammonia is detected, the operator must be warned immediately. Situation assessment requires determination of the location of the release, the compound that has been found in the air, the total amount of gas that has been released, and the dimensions and the concentration of the gas cloud. In case of a spontaneous, large-scale emission, the gas cloud will not be diluted within close range of the incident and a gas cloud is formed that moves over the compound. This cloud bears a hazard potential, especially for low to medium wind speeds.

Our question is: To what extent will the risks from an unexpected airborne release be reduced at OCI's Melamine and Urea plants at Chemelot using the scanfeld™ remote monitoring solution, utilizing Fourier-transform infrared (FTIR) spectroscopy?

OCI's main requirements were:

- A swift detection at levels of 10 ppm and above,
- visualization of the gas cloud in the control room in real time,
- no false alarms,
- continuous operation for at least 99.9% of the time,
- at least four hours operating time in case of power outage,
- on site data storage,
- internet independence,
- 24/7 support in case of outages.

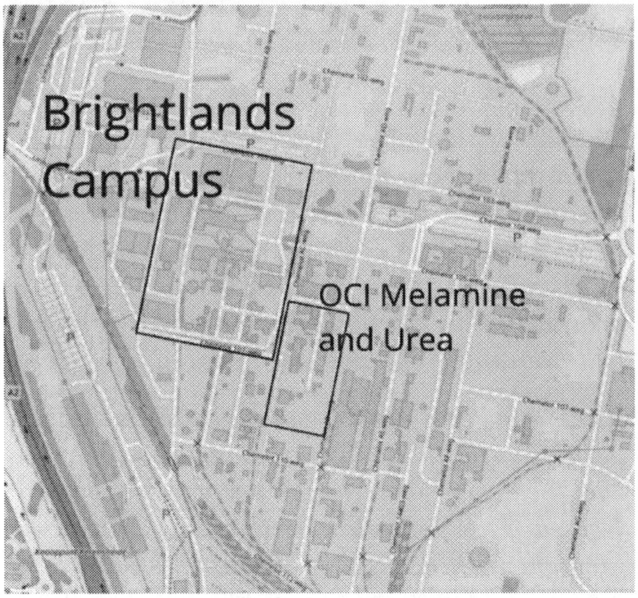

Figure 1. Map of Chemelot north showing the location of the Brightlands campus and the OCI production complex.

Methods

Monitoring large industrial sites with FTIR based remote sensing

This report introduces a permanent monitoring solution for the mitigation of hazardous incidents due to an unnoticed release of a hazardous gas. The principle of this solution is to monitor critical infrastructures in real time with a small number of measurement units. This is made possible by an optical remote sensing technique that is based on FTIR spectroscopy.

FTIR spectroscopy is a well-established optical measurement technology for analyzing remote objects. In principle, the distance between the object and the measurement device is irrelevant, which allows chemical analysis of a gas without the need to place the sensor inside the gas cloud. Astronomy uses this technology for chemical analysis of remote planets or galaxies. Since the 1960s remote sensing within the atmosphere became a scientific subject. Various remote sensing systems have been developed and since 2006 remote monitoring technology has become standard equipment for European early response task forces.[2, 3]

The measurement principle is based on the spectroscopic analysis of infrared radiation. Due to their temperature all objects emit infrared radiation, which allows chemical analysis from the distance around the clock. Within the infrared spectrum, one wavenumber range, called the fingerprint region (from 500 to 1500 cm^{-1}), is very interesting for remotely analyzing the chemical composition of a gas. In this spectral region, many chemicals interact with radiation in a specific manner. The effect of this interaction is a spectral absorption feature in the spectrum of infrared radiation that is unique for the chemical compound – a fingerprint.

The FTIR remote sensor can detect a gas cloud instantaneously from several kilometers distance without any form of background calibration. The sensor can analyze the chemical composition of the gas and measure the gas concentration.

Covering a whole industrial site from a distance has significant benefits. Most importantly, it will allow the detection of a gas release without prior knowledge of the exact release location. It can also assess the entire gas cloud, locate its position, and track its propagation.

Figure 2. The remote monitoring principle: detection and localization of gas clouds from a long distance. The optical measurement technique uses FTIR spectroscopy to identify the chemical composition.

The scanfeld™ remote monitoring solution

The scanfeld™ monitoring solution uses scanning imaging remote sensing units. Covering the entire fingerprint region of the infrared, they detect and identify over 400 different chemical compounds, including ammonia and methane. The detection limit is a specific value for each chemical. For example: A 10 m (32 ft) wide ammonia cloud under normal ambient circumstances and at 1 km (0.6 mi) distance, this detection limit is in the order of 4 ppm. The identification algorithms and quantification techniques that can automatically identify and quantify a chemical compound in the infrared spectrum are certified according to VDI4211. Due to the spectral region in the infrared over long distances the measurement technique is not affected by weather conditions such as rainfall or fog. Without the need for ambient light, the measurement technique also works day and night.

Figure 3. The scanfeld™ sensor unit is a scanning imaging FTIR spectrometer. The picture shows the sensor (white) with the large aperture infrared telescope head. The sensor is moved continuously by a pan-tilt unit (black) to scan the area of interest.

The scanfeld™ sensor units are designed to automatically scan predefined scan areas from distances of up to 4 km (2.5 mi). Mounted in elevated positions they overview large areas. The large aperture infrared telescope of the sensor unit focuses the field of view to 3 mrad (0.2°). The maximum spatial resolution that a sensor unit can cover is thus approximately 0.3 m (1 ft) at 100 m (330 ft) distance (or 3 m (10 ft) at 1 km (0.6 mi) distance, respectively).

The sensor unit moves along a predefined scanning pattern and constantly analyzes the chemical composition within the field of view of the sensor. The units can be positioned in a 360° horizontal and ± 90° vertical angle to point in any direction. They automatically scan predefined areas at an arbitrary scanning speed within wide speed range. Slower scanning increases the spatial resolution while higher scan speeds cover larger areas in less time. With a constant spectral acquisition rate of six measurements per second, a predefined scan area can be scanned within seconds. Once the sensor unit detects a known target gas, it is used to measure the location and dimension of the gas accumulation. The scan area and scan speed are adjusted automatically to map the entire gas cloud.

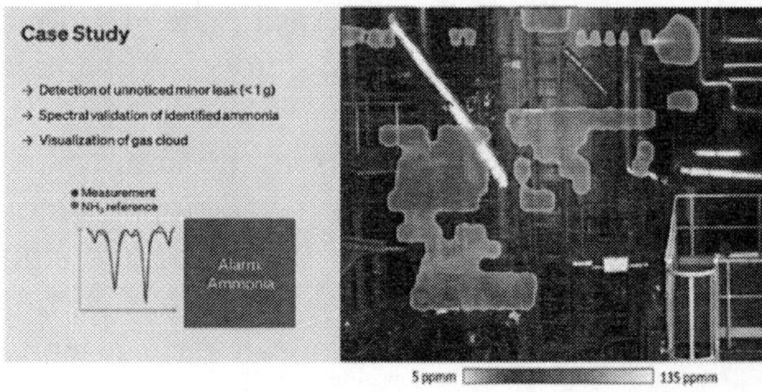

Figure 4. Measurement example: Detection of minor ammonia leak from 100 meters distance. On the left: The area where the gas is located is automatically displayed as a red overlay to a live video image of the scene. Ammonia is identified by spectral analysis and automatic comparison to a proprietary library of hundreds of gaseous compounds. On the right: The concentration distribution is displayed in false colors. The color-code maps the column-density of the cloud. From the point of view perspective of one sensor unit the column-density is the product of concentration and cloud depth

The scanning sensor units measure the dimension and the concentration distribution of the gas cloud. In terms of risk management, this point-of-view visualization of a gas cloud, in combination with the automatically generated alert, provides two barriers for the escalation of the release event. First, the operator is warned of an unexpected gas release with the specific information of how much of which gas is in the air. Secondly, the point-of-view visualization of the gas distribution provides concise information about the leak location and the near-field propagation of the gas accumulation. Immediate action can be taken based on this specific information.

Real-time tracking of gas clouds with scanfeld™

To localize a gas cloud, two or more scanfeld™ sensor units are needed for cross bearing and tomographic reconstruction of the gas cloud distribution. The scanfeld™ monitoring solution automatically coordinates the positioning and defines new scanning areas for all available sensor units to automatically determine the location and shape of the gas cloud.

Figure 5 shows how two sensor units map the concentration distribution of a small ammonia cloud in a release experiment. The concentration is depicted directly on a map that allows of immediate assessment of the situation in the field. Emergency response can be coordinated effectively.

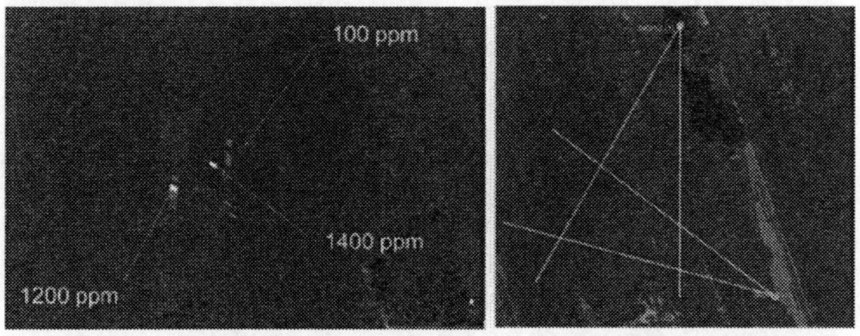

Figure 5. Measurement of an ammonia cloud from two measurement positions. The concentration distribution is displayed on a map.

The alert generation and alarm log help to assess how the event evolves. For a better understanding of the situation and 24/7 observation, both the visualization of the gas and the alerts are mapped on a timeline with time-shift and fast-forward features to move back and forth in time to quickly understand how the situation evolves over time (Figure 6).

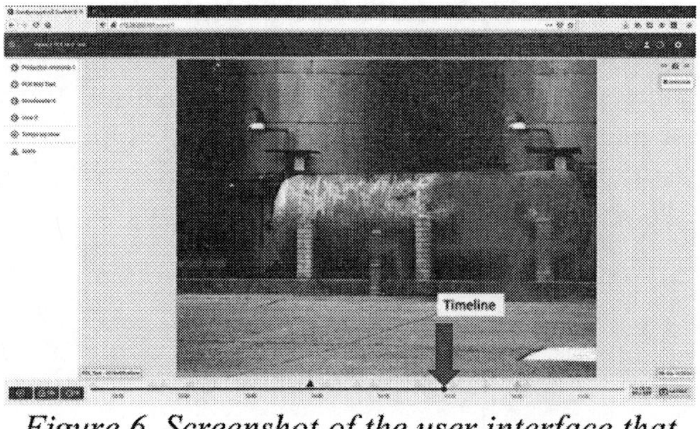

Figure 6. Screenshot of the user interface that shows how the point-of-view visualization of a gas cloud is organized together with the incident alert in a time line.

Topology of a typical scanfeld™ installation

The general setup of the monitoring solution is shown in Figure 7. The sensor units are connected via ethernet to an on-premises server. The server controls the sensor units and evaluates the measurements automatically. The user interface displays the events in a timeline display with alert logging. A dedicated software module analyses the alerts in time and space and classifies the incidents with alert levels. The classification is based on the evaluation of the total amount of gas in the air, the cloud size, and the persistence of the event. The alert levels with reporting can be connected to the DCS.

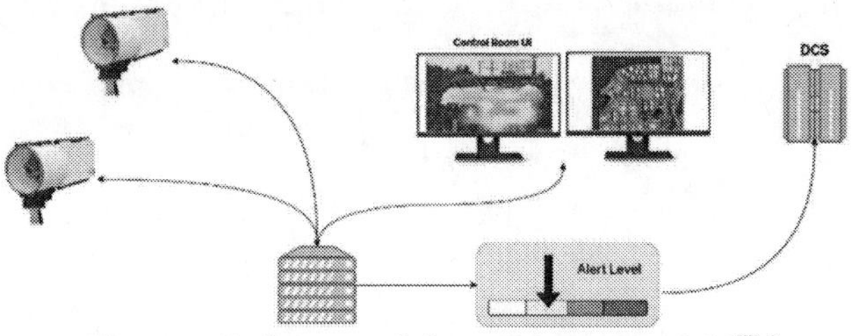

Figure 7. Setup of the entire scanfeld™ monitoring solution. The measurement units are controlled by the server. Alerts are generated that classify the incident. The user sees the POV distribution of the gas cloud and the localization on the map.

Result: Installation at Chemelot industrial park

The number and the positioning of the sensor units is driven by the scope of the monitoring plan. Each individual sensor unit can cover a radius of 4 km (2.5 mile). Large area monitoring of an entire facility can thus be accomplished by two or three sensor units. The installation at Chemelot focuses on fast early warning within 1 min and a quick monitoring of emissions onto the Brightlands campus. For the 3D localization of a moving gas cloud, two sensor units are installed, autonomously monitoring the production area at OMM. The two units are coordinated in real-time to track the event.

The scanfeld™ solution sequentially monitors the four stacks of the OCI melamine plant OMM, including the urea plant for real-time detection of an accidental gas release. The two sensor units also oversee the entire Brightlands campus area to monitor any emissions onto the campus. The total monitoring area is approximately 350,000 m² (0.14 square mile).

The first unit is installed at the highest point within the production complex that oversees the entire facility. Figure 8 shows the installation points for the two sensor units and the viewing areas that map the four potential emission points of OMM. The Brightlands campus is partly shown in the satellite image and marked yellow.

Figure 8. Installation point for sensor unit #1 at an elevated point within the production complex (the viewing areas are marked magenta) and sensor unit #2 at close distance outside the production site (blue viewing areas). The relatively close distance of the sensor unit to the target allows the fastest release detection and a high spatial resolution.

Each unit must be positioned according to the early detection principle and the requirements for real-time situation assessment. For the first criterion for the selection of an installation point maximum overview of areas of interest is essential. The distance to the area of interest is a balance between oversight and spatial resolution.

To meet the second criterion the sensor units must have a clear view on the moving gas cloud along the propagation path over the chemical park. Figure 9 shows the area over the Brightlands campus that is visible to the two sensor units.

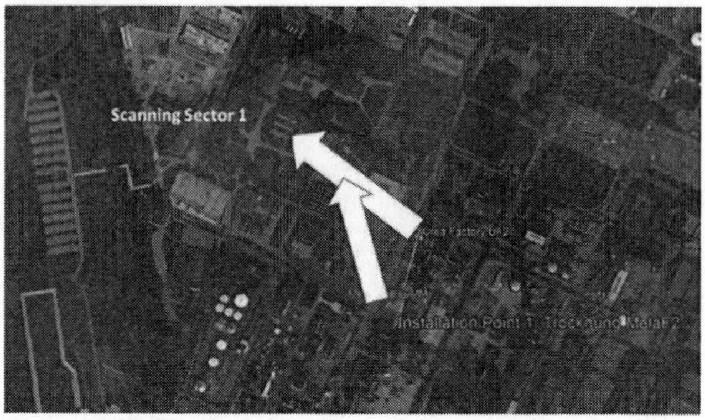

Figure 9. The image shows the installation point and the viewing area that effectively oversees the entire campus. The white arrows indicate the potential propagation direction of a release from one of the four stacks. The location of the sensor unit enables the continuous monitoring of a gas cloud from the point of release on over the entire monitoring area.

A precise localization of the gas cloud requires an appropriately wide angle between the two perspectives on the gas cloud. **Error! Reference source not found.** displays three possible scenarios for a gas cloud in the moment of entering the perimeter of the Brightlands campus, and the resulting angles under which the two sensor units measure the gas cloud. In all three scenarios, the angle between the two perspectives is wide enough for localization precision within the spatial resolution of the instruments.

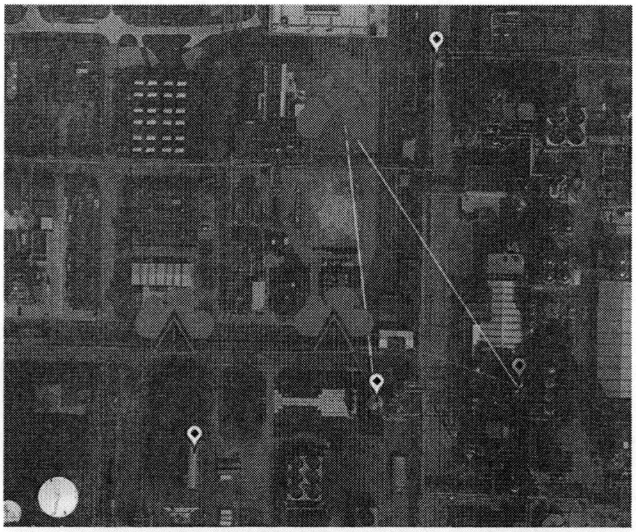

Figure 10. The image shows three scenarios for a gas cloud entering the perimeter of the Brightlands campus. In both cases the gas cloud can be monitored from both installation points under a wide enough angle to allow for precise localization.

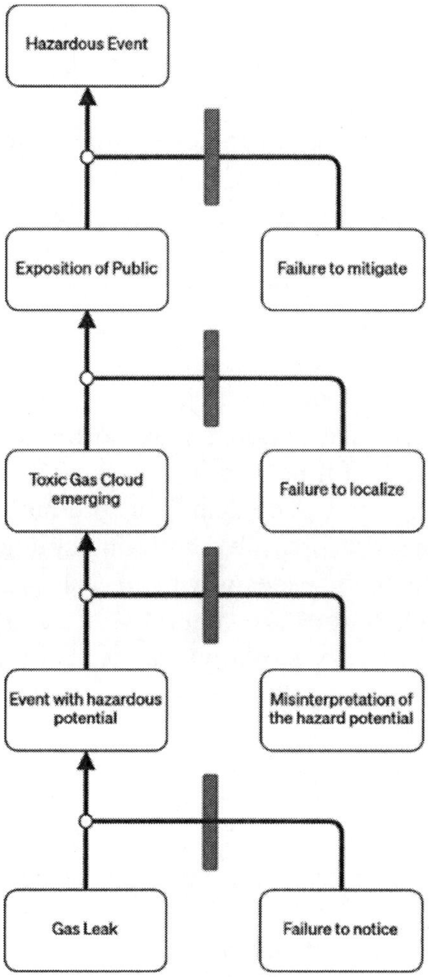

Figure 11. Fault tree model of a hazardous event with a gas release as root cause. Large area gas monitoring can provide barriers in all stages of the incident. They are drawn in red into the model.

Discussion

Creating barriers

A hazardous event that has its root cause in an accidental gas release is often the result of subsequent failures of preventive and mitigation measures. Risk analysis must take all stages of the incident into account and define safety measures to minimize risk. Figure 11 displays a fault tree model of the various stages of an incident. To prevent further escalation, each stage of the incident needs its own barrier. scanfeld™ establishes a barrier for each of the stages by providing warning levels and detailed information when needed. Conversely, scanfeld™ helps to rapidly scale down alert levels once the immediate threat is gone and safe conditions are confirmed.

Early Warning

One thinks of the release of a hazardous gas as a rare event. In normal operating mode the process is under control and therefore safe. Turnaround and startup phases are considered more dangerous and are under closer observation. In contrast, statistics show that more than half of all large emissions do not occur in the startup-phase. It is during normal operation that most large-scale releases occur, and with operator error leads to most of catastrophic events.[4] Therefore, the first principle of a reliable monitoring solution is permanent monitoring of the production and storage areas in normal operation mode and early warning in case of an unexpected gas release.

In the early stage of the event, quick notice and correct situation assessment are crucial. Early warning of the gas release can prevent further escalation of the situation if the operators in the control room accurately recognize the situation at its inception. Measures taken at an early stage of an incident are most effective, but from an operator's point of view this is a very consequential decision to make if measures taken lead to down-time and loss of production. The need for rapid action and the pressure to take moderate measures are conflicting options for the operating team, especially in an early stage of the event. Operators are naturally hesitant to shut down until the situation is clear. In order to prevent potentially hazardous situations, a significant gas release needs to be detected quickly and properly assessed.[5]

To fulfill the need for a solution that warns reliably in case of the unpredicted event, early warning monitoring must cover all relevant areas within the production site, the loading areas, and all other critical infrastructure (e.g. pipelines). In principle critical infrastructure can be monitored with local sensors. However, monitoring all flanges, valves, compressors units, and storage tanks with local gas sensors would require considerable investment and, more importantly, incur significant operating costs due to maintenance and regular calibration of the sensors. Remotely monitoring critical infrastructure is a more effective way of covering large areas with a small number of remote sensing units.

Based on the evaluation of the total amount of gas and an analysis of an event over time, an incident is automatically classified in three alert levels, the first of which merely indicating the presence of a gas and the highest level indicating the persistent presence of a large amount of gas. For each measurement in a scan that results in a pixel in the mapping of the gas cloud, the system's algorithms identify the presence of each target gas within the target compound library. The algorithm is based on a radiation transfer simulation that takes into account the spectral signatures of interfering atmospheric and known potentially present technical gases. The incident level is calculated based on the amount of gas that is detected in each scan and a quality criterion for the identification including the signal-to-noise ratio and the confidence value for the target gas identification. The indication levels are also subject to a persistence criterion that filters single identifications which are subject to cross-sensitivity or fugitive emissions. These calculations are specific for both the target gas and the known behavior within the viewing areas. This feature enables the monitoring solution to learn the normal behavior of the plant over time and self-adapt.

The high-level information on the incident is visualized in the timeline as well as the alarm logs, giving the operator a clearly readable aid to monitor the event. Prior events can also be brought back onto the screen, so in case an event is not recognized immediately it can be compared to historical incidents. The high-level incident information can be directly routed to the DCS via an OPC UA interface. This integration into the safety architecture of the control room is scheduled for the installation at Chemelot.

Mitigation measures

There are generally two potential hazards from an ammonia release: Ignition of the gas cloud and exposure to the toxic gas.[6] Ignition of all kinds of flammable gases, including ammonia, becomes possible, when the local concentration of the flammable gas exceeds the lower flammability limit (LFL). Mitigation measures must therefore detect gas accumulation in both toxic and flammable concentration ranges. The flammability limit of ammonia is between 14.8 % and 33.5%. The gas is also immediately dangerous to life or health (IDLH) in much lower concentrations. The IDLH level for ammonia is

500 ppm. Toxic hazards are usually present prior to flammable hazards, as toxic concentrations, especially in the case of ammonia, are much lower.

Sensors must therefore work over a wide concentration range, from the low ppm range up to double digit percentage levels, for reliable mitigation tools. More importantly, the gas concentration information must reflect the highest possible concentration. Therefore, any concentration information should always refer to the center of the gas cloud in order to assist situation assessment. FTIR spectroscopy, as used in the remote monitoring solution presented in this paper, has the benefit of identifying hundreds of different chemicals in a wide concentration range from single digit ppm to double digits percentage values.

Near-field mapping

Gas releases in the open are fundamentally different from gas emissions in confined spaces. The gas coming from the release point is diluted more quickly in open fields than in confined spaces, which reduces the risk. On the other hand, it also makes it harder to detect and map the distribution of a gas cloud. As propagation is subject to wind dynamics, there is the risk of local gas accumulation and the areas of high likelihood for accumulation are difficult to predict. This is especially true among dense infrastructure like production facilities, as wind dynamics are extremely complex. [4, 7] As a result, even the assumption that a gas heavier than air accumulates near to the ground is not necessarily true in the area around the release point. Also, a gas lighter than air does not necessarily rise until complete dilution has occurred. The 2005 incident at the Yara facility in Rostock-Peetz, Germany shows how a cloud of ammonia, which is lighter than air, was carried far before coming down again when the gas emissions could be smelled in a town 12 km (7.5 mi) from the incident. [8]

Near-field monitoring (Figure 12) of the accumulation of a flammable or a toxic gas must cover the entire area around the point of release to ensure accurate monitoring of the entire gas cloud. With respect to the dynamic gas distribution, it must also cover the near field in real-time.

Figure 12. Measurement of a release of a small amount of methane (20 g/min). The example shows how the gas cloud follows the changing wind direction. Measurement taken from [11].

Compared to field measurements with hand-held gas sensors or monitoring vehicles, remote sensing offers the ability to map the gas concentration in the near field, without prior knowledge of the location and the propagation of the gas cloud. Monitoring near field distribution can be performed in seconds. It is thus quick enough to map the dynamic propagation behavior of gas accumulation. [3]

Wide area mapping of toxic gases

A study has been published comparing the potential hazards of an ammonia release due to fire and the toxicity of the gas. The study finds that danger by the fire jet is imminent in a radius of approximately 300 m (1000 ft) around the release point after ignition. Hazard potential due to the toxicity of the gas is less local around the release point. The study focuses on a release scenario with approx. 81 kg (180 lb) of ammonia and concludes that the concentration of the gas cloud at 1 km (0.6 miles) from the incident can still reach lethal levels of 1000 ppm. [6]

Measures to mitigate the effects of an accidental gas release are most effective if they accurately present the gas concentration across a wide radius of the incident. Without information on the location of a gas accumulation, wide areas of the compound need to be monitored continuously. Assessment of the gas concentration becomes more difficult as the affected area grows larger, making reliable situation assessment impossible with common gas sensors.

With the dilution of the gas cloud, it spreads over the facility. But in a dynamic situation, especially shortly after the release, the concentration at a certain location may change dramatically over time. The concentration of the gas can be low at a certain location but rise dramatically when the wind sweeps the gas cloud in another direction. Therefore, the reading of a local gas sensor can create a false interpretation of safety when the location of the gas cloud is yet unknown. It can miss the cloud entirely or pick up only the wide-spread diluted parts of the gas cloud which can lead to a false interpretation of the situation. This is especially concerning if one considers that gas concentration measurements in the field can only be taken a certain height, usually at ground level, whereas the propagation of the gas cloud can occur at greater heights. It must be noted that the general understanding of the propagation of a gas often leads to a false interpretation of the propagation of a gas cloud in the field. It is generally assumed that a gas lighter than air will travel upwards, while a gas heavier than air stays at ground level, but this assumption does not take the effect of air flow into account. Air flow can carry a heavy gas up, while a gas lighter than air can be blown down, even at locations far off the release point.

The propagation of the gas cloud is determined by the local wind situation which on the ground is far more complex to evaluate than just assuming that the general wind direction is also valid on ground level. The calculation of the distribution of a gas cloud is discussed at great length in the literature. [4] Mitigation measures rely on real time information of the location and distribution of a moving gas cloud. As most hazardous gases are invisible, the assessment must rely on measurement technology. A wide variety of distribution simulation software is available to fill the information gap. The general approach of distribution simulation takes the initial amount of gas release into account, but more sophisticated software can also simulate the distribution of a known amount of gas among plant structures by simulation the air stream that carries the gas with it among the structures located in the plant. Local situation assessment takes gas concentration measurements into account.

Both measures are limited in their ability to provide correct situation assessment. The precision of the simulation of the air stream depends on the number of wind sensors that feeds the simulation tool. It narrows down the area of higher likelihood for a high concentration of gas. For a correct simulation of a local concentration distribution at a distance from the release point, knowing the total amount of gas that has been released into the air is essential. In case of an incident the knowledge of this Figure is vague, especially within the first minutes of the incident. The larger the distance to the release point, the uncertainty becomes even greater. This is especially true when the gas concentration is likely to be hazardous when the gas cloud reaches the perimeter of the compound and the risk assessment must include the hazard potential for the public. Gas concentration measurements are the only option for effective warning and response.

Remote sensing early warning systems in comparison to common, local sensing technology provide both information on the location and concentration distribution of the gas cloud. In contrast to the deploying of gas sensors and filling the information gap by distribution simulation, remote sensing can be applied over wide areas within the field of view of the optical sensors. Remote sensing can continuously monitor and track the propagation of a gas cloud. [9] Due to their ability to continuously monitor large areas, a gas accumulation can be located and tracked from the moment of the release forward. The gas cloud propagation can be tracked over vast areas, up to several square kilometers, and the concentration distribution measured in real-time. Remote monitoring of the moving gas cloud also works at high elevations that conventional gas sensors cannot reach.

Information on the propagation behavior of the gas cloud and the location of the gas while the event is unfolding are essential for a precise and timely understanding of the hazardousness of the situation and the coordination of mitigating measures. Fire brigades must know where to set up a water curtain to rain down a hazardous gas. Workers must know which gathering points or buildings are safe and within reach. [10]

Conclusion & Outlook

The scanfeld™ monitoring solution is a highly automated monitoring tool that scans predefined scan areas automatically in a programmed order. In normal operation mode the scan pattern is optimized for quick scan rate in order to map the status of the plant in minutes. The scan-pattern is specific for the scope of the installation with a free definition of scan-speed and spatial resolution. Thus, monitoring areas with the potential release of highly toxic or flammable gases can be optimized for speed, whereas large area monitoring can be optimized for better spatial resolution. Once a gas release is detected, the scan pattern shifts to automatic gas cloud distribution mapping. The operator is notified of all events with an indication on the user interface. As permitted emissions naturally occur in production facilities and fugitive emissions from other areas periodically come into view, an automatic situation assessment aids to better distinguish hazardous situations from harmless emissions.

At OCI Chemelot a very short response time to unexpected releases at the OMM is achieved by the careful choice of two installation sites in the near vicinity of both the OMM and the Brightlands campus. With a minimal time-to-alert the scanfeld™ installation will provide the exact knowledge about What, Where, and Level of unexpected gas emission incidents. It thereby creates four levels of barriers for the prevention of hazardous events and protects the Brightlands campus. Furthermore, it allows the speedy post-incident release of affected areas as soon as the dispersion of the gas cloud is confirmed by the scanfeld™ sensors. Compared to conventional gas sensor networks, scanfeld™ is cost effective and provides more detailed information, both from the near and wide field point of view. In the future, the coverage of the scanfeld™ solution can be extended to the entire Chemelot North site (Figure 13). Since the scanfeld™ system is highly scalable, just two additional sensor installations will allow for the coverage of the complete area and provide additional safety to the site and its neighboring areas.

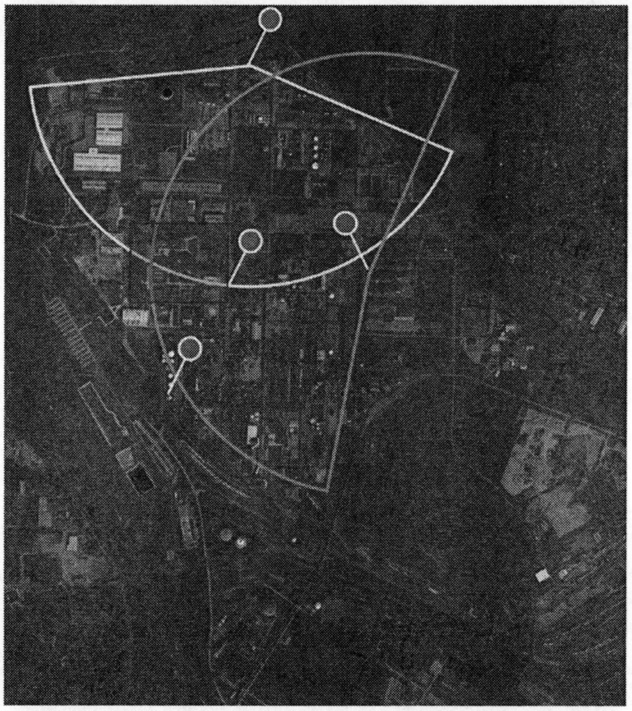

Figure 13. A sketch for an installation of two additional sensor units to monitor the entire Chemelot north chemical site. The monitoring area covers 1.6 km² (0.6 mi2). The monitoring solution covers 8 toxic or flammable gases.

The scanfeld™ monitoring solution as presented in this paper was installed in March 2021 with a full on-site acceptance test scheduled for May 2021.

References:

[1] University, M. (17th. Sept 2020). AMIBM: University collaboration at Brightlands. Source: https://www.maastrichtuniversity.nl/news/amibm-university-collaboration-brightlands

[2] Matz, G., & Harig, R. (kein Datum). Infrarot-Fernerkundungssystem für die chemische Gefahrenabwehr. (B. Bund, Hrsg.) Zivilschutzforschung, 58.

[3] Sabbah, S., Harig, R., Rusch, P., Eichmann, J., Keens, A., & Gerhard, J.-H. (2012). Remote sensing of gases by hyperspectral imaging: system performance and measurements. (SPIE, Hrsg.) Optical Engineering, 51(11).

[4] Lees, F. (2012). Lees' loss prevention in the process industries: Hazard identification, assessment and control. Oxford: Butterworth-Heinemann.

[5] McGillivray, A., & Hare, J. (2008). Offshore hydrocarbon releases 2001-2008. Executive, Health and Safety Laboratory for the Health and Safety. Derbyshire: Buxton.

[6] Karimi, H. A. (August 2019). Modeling of Ammonia Emission in the Petrochemical Industry. Jundishapur J Health Sci.

[7] ProcessNet-Fachgemeinschaft „Anlagen- und Prozesssicherheit". (2017). Auswirkungsbetrachtungen bei störungsbedingten Stoff- und Energiefreisetzungen in der Prozessindustrie.

[8] Bakli, K., Versteele, W., & Swensen, B. (2006). Safe Ammonia Storage. AMMONIA TECHNICAL MANUAL.

[9] Harig, R., & Rusch, P. (March 2010). 3-D Reconstruction of Gas Clouds by Scanning Imaging IR Spectroscopy and Tomography. IEEE SENSORS JOURNAL, VOL. 10, NO. 3.

[10] Hofstra, R. J. (2002). Minimizing Facility Siting Risks. AMMONIA TECHNICAL MANUAL .

[11] Feitz, A., Schroder, I., Philips, F., Coates, T., Negandhi, K., Day, S., Griffith, D. (2018). The Ginninderra CH_4 and CO_2 release experiment: An evaluation of gas detection and quantification techniques. International Journal of Greenhouse Gas Control, Vol. 70, S. 202-224.

25 Years of Experience With a Retrofitted aMDEA CO_2 Removal System

In 1997, an OCI Nitrogen plant converted its CO_2 removal system from Benfield solution to aMDEA. In the years since, various operational and integrity issues with the aMDEA CO_2 removal system developed. In 2001, material of several sections was changed from carbon steel to stainless steel. In the meanwhile, plant capacity increased and more operating and integrity issues developed, which had to be solved in recent years. These problems include excessive vibration, cracked welds, and process upsets due to foaming. This paper describes these issues and the steps taken to resolve them.

Mart Grooten
Sitech Services

Werner Ploumen
OCI Nitrogen

Introduction

OCI Nitrogen operates two ammonia plants in Geleen, The Netherlands. Both plants have a design capacity of 1360 MTPD (1500 STPD). One plant was designed and constructed by Bechtel and was commissioned in March 1971. This plant was designed for a specific energy consumption of 34 GJ/MT (LHV) (29 MMBtu/ST), a typical value for plants of that vintage. The plant used a hot potassium carbonate system (Benfield) for CO_2 removal. A major revamp was executed in April 1997 which improved the energy efficiency by 8.6%. This revamp is described in detail in an earlier presented symposium paper in 1998 [1]. An important part of the revamp was a switch from Benfield to aMDEA solution for CO_2 removal. The change to aMDEA and the first operating experiences are described in detail in another symposium paper from 2002 [2]. In the years following the 1997 revamp various operational and integrity issues developed with the aMDEA system. In 2001, the material of construction in several sections was changed from carbon steel to stainless steel. In the meanwhile, plant capacity increased further to 1650 MTPD (1820 STPD). Additional operating and mechanical integrity issues had to be resolved in recent years, including excessive vibration, cracked welds, and process instability due to foaming.

This paper describes the issues faced by this unit after the revamp. OCI Nitrogen shares these operational experiences to help improve safety in our industry. Experience with retrofitted installations and operation above nameplate capacity may lead to better design guidelines for new process units or replacement projects.

CO_2 removal section

The CO_2 removal section is briefly described below. An overview is shown in Figure 1.

The process gas leaving the Low Temperature Shift (LTS) is cooled subsequently in H252, passes through the V202 process condensate separator, before additional cooling in the aMDEA reboilers H204A/B/C/D and the vacuum condensate heaters H206A/B. The gas temperature is reduced to approximately 80°C (176°F) before entering the CO_2 absorption column C203S.

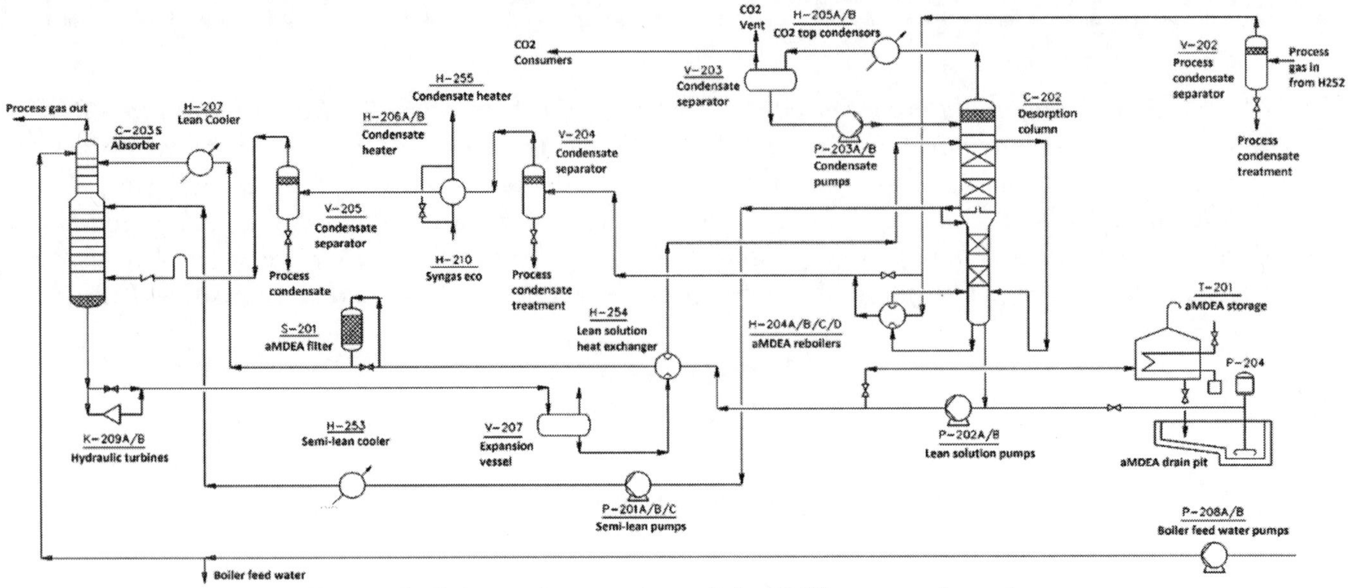

Figure 1. Schematic overview of the CO2 removal section

The CO_2 is removed from the process gas at 30 barg (435 psig) and the cleaned process gas leaves C203S at the top at about 60°C (140°F) and a CO_2 content of ~50 ppm. The absorbent solution contains 35-40%-MDEA (methyldiethanolamine), 3-5 % activator (piperazine), with the balance water. Small amounts of Amerel, a silicone based defoamer, are dosed to suppress foaming. Amerel can be added to the process at the lean solution pumps P202A/B or near the control valves in the rich solution piping downstream of the bottom outlet of C203S. The pressure of the rich MDEA solution from C203S is reduced over two hydraulic turbines, K201A/B, before entering the expansion vessel V207. The V207 pressure is controlled at ~7.5 barg (109 psig). Dissolved gases, mainly hydrogen, leave the high pressure flash vessel V207 and can be recovered as fuel gas or vented. The rich solution continues to the lean solution heat exchanger H254, before entering the top of the desorption column C202 at a pressure of 0.5 barg (7 psig). CO_2 leaves the top of stripper C202 at a temperature of ~75°C (167°F) to two CO_2 overhead condensers H205A/B. Condensate is removed in the V203 separator and returns to C202 via condensate return pump P203. The cooled CO_2 is sent to downstream customers, such as urea and liquefaction plants, or vented. The desorber has two packed sections, a low pressure flash at the top section and a bottom desorption section. The rich MDEA flow to the desorption column is up to 2200 m^3/h (9700 gpm). The bulk aMDEA flow (semi-lean) leaves the upper section of the desorber with a CO_2 content of ~9%. Two 60 barg (870 psig) steam driven semi-lean pumps P201A/B convey the semi-lean solution through semi lean cooler H253 and to the bottom section of the absorption column. The solution temperature at this point is ~70°C (158°F). An electrically driven P201C is available as a spare pump or to use for startup. H253 has a bypass to control the semi lean flow over the heat exchanger and the temperature to the absorber C203S. The bottom section of desorber C202 further removes CO_2 from the MDEA solution with the aid of heat recovered from process gas using four thermosiphon type aMDEA reboilers H204A/B/C/D.

The regenerated aMDEA solution (lean solution) is pumped by P202A/B from the bottom of the desorber and cooled in H254, then further cooled in the lean solution cooler H207, before entering the top of the absorber at about 45°C (113°F). A side-stream of the lean solution passes through a filter to remove impurities.

Issues from 2001 to 2021

The gradual increase in plant capacity to 1650 MTPD (1820 STPD) has resulted in a CO_2 removal system becoming one of the bottlenecks for further capacity increase. The semi lean cooler H253 was found to be undersized, resulting in vibration damage and erosion in the bypass line. Other major problems were mainly corrosion related. As a result, the top section of desorber C202 and the rich and semi lean MDEA solution lines were replaced with stainless steel material in 2001 [2].

Unfortunately, several problems occurred after these revisions:

1. Erosion corrosion in condensate return vessel V203 leading to its replacement in 2018.
2. Corrosion under insulation (CUI) at V204 and galvanic corrosion at welds of the stainless steel inner liner.
3. Corrosion fatigue and Chloride Stress Corrosion Cracking at tubes on the shell of overhead condensers H205A/B, starting from the inside. Internal cooling water leakages due to low cooling water flow and under deposit corrosion.
4. Corrosion at desorber C202. Galvanic corrosion at a stainless steel-lined tube in the carbon steel conical shell. CUI and heavy atmospheric corrosion at the skirts, requiring reinforcement repairs.
5. Vibration, leakages due to fatigue, and broken supports and dampers in rich solution feed line to desorber C202.
6. Baffle hammering in reboilers H204A/B/C/D, likely due to high inlet flow velocities. Also CUI of the reboilers.
7. Erosion corrosion in absorber C203, leading to its replacement in 2014.
8. Leakage in 2019 at welds made in 2014 in semi-lean solution feed line to absorber C203S.
9. Recurring leakages in semi lean cooler H253 due to leaks along old tube plug repairs and severe fouling of cooling water side.
10. Vibration at the rich solution lines at the absorber bottom outlet.
11. CUI of the high pressure flash vessel V207.
12. Heavy wear and leakages at lean solution and semi lean solution pumps.
13. Damage at condensate return pump P203.
14. CO_2 breakthrough after absorber C203S due to foaming and limited absorption in the lower section.

Some of the more relevant and significant failures are detailed below.

Erosion corrosion in absorber C203

Between 2005 and 2013, several planned and unplanned repairs, such as adding stainless steel cladding, had to be executed to maintain integrity of the absorber C203. In 2014, the column was replaced. A 304 stainless steel liner was installed for the entire column and the sieve trays replaced with structured packing in the lower section and random packing in the upper section. A pressure drop decrease from 650 mbar (9.4 psi) to 100 mbar (1.5 psi) allowed plant capacity to increase by ~30 MTPD (33 STPD). Corrosion in the column is now under control. However, problems with foaming and peaks in CO_2 slip remain.

Issues in rich solution feed line to desorber C202

In 2003, the C202 rich solution inlet was modified to a tangential entry design. Following this change the top section of C202 exhibited low frequency shaking which was visible from ground level. This vibration was likely caused by process conditions, as the movement was not noted during windy conditions. The movement was calculated to be within acceptable static and dynamic loads limits, and confirmed with strain gauge vibration measurements. Leakage due to fatigue and broken supports and dampers remain an issue.

In 2006, a leak developed in the rich solution feed line to C202, near support PS28. The location of the leak is shown in Figure 2.

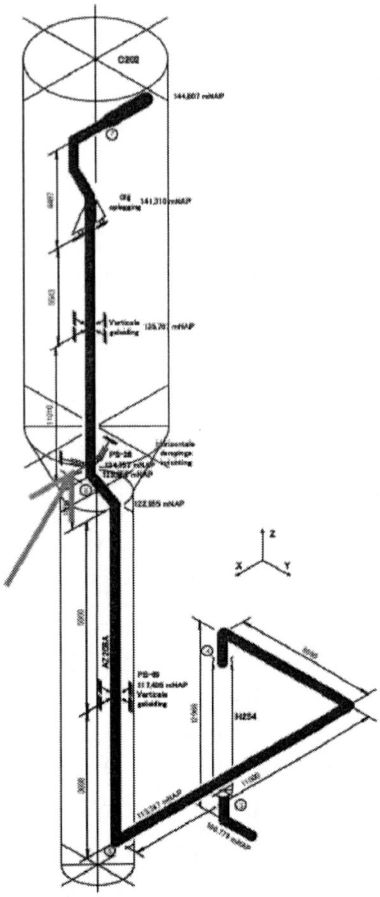

Figure 2. Isometric representation of rich solution line AZ208a to desorber C202. The affected support is indicated with the arrow.

The most probable cause of the leak was fatigue due to vibration. The crack initiated next to the weld where the support is connected to the line. The visual crack length was 45 mm (1.8 in) into the rich solution line and 120 mm (4.7 in) into the support. Additional clamps were added to avoid further crack growth.

Later inspections revealed no further crack growth. The damaged piping and support were replaced at the next opportunity to complete a full repair.

In 2014, construction of the support was reinforced and two dampers were added. This resulted in reduced vibration.

In 2018, one of the two dampers broke off, due to a combination of faulty installation and increased vibration. Vibration was found to increase with increasing plant load, and vibration levels were within acceptable design limits after replacement of the dampers.

The permanent solution would be a redesign of the piping and the supports. Simulation showed that rich solution in the horizontal and vertical lines downstream of heat exchanger H254 have a plug / slug flow profile developing into a slug / churn / annular flow profile while moving vertically up to C202.

The random vibration in the vertical line section along the column at support PS-28 was likely caused by deflections of the flow in the 45° curves. Inspection of the supports was added as a specific checkpoint during turnarounds.

Vibration in the rich solution lines at absorber C203S outlet

In 2014, a new absorber C203S was installed. As a consequence, modification to piping and supports was required. In the 2018 turnaround, the control valves in the C203S outlet were replaced by a cage type due to excessive wear.

After the 2018 turnaround, the piping downstream of the control valves was found to vibrate heavily under specific flow rates and valve positions. The pressure drop over the valves results in degassing of the aMDEA solution, causing a resonance at certain conditions. This was investigated by vibration measurements and the use of a motion amplification camera. For the short term, valve positions have been limited to diminish risks. Long term solutions under investigation include redesign of control valves, redesign of piping with supports, and installing a bypass valve to ensure that the control valves operate in a range where vibrations are acceptable.

Leakage at welds in semi-lean solution feed line to absorber C203S

In 2019, an aMDEA leakage occurred at the DN500 semi-lean solution line AY2501 near a T-piece, as shown in Figures 3 and 4.

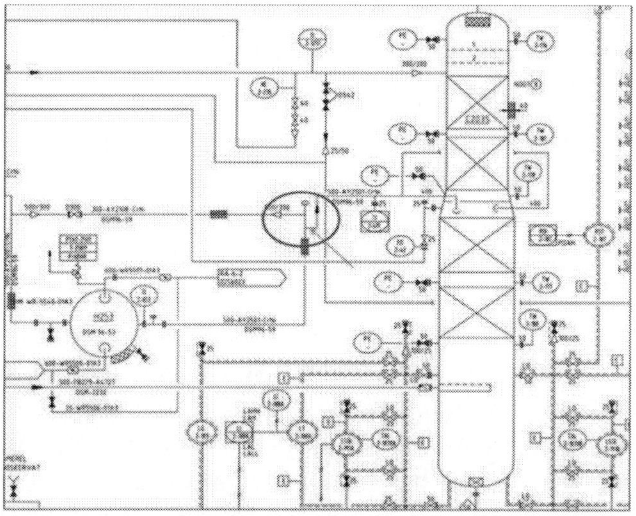

Figure 3. Schematic of the position of aMDEA leakage

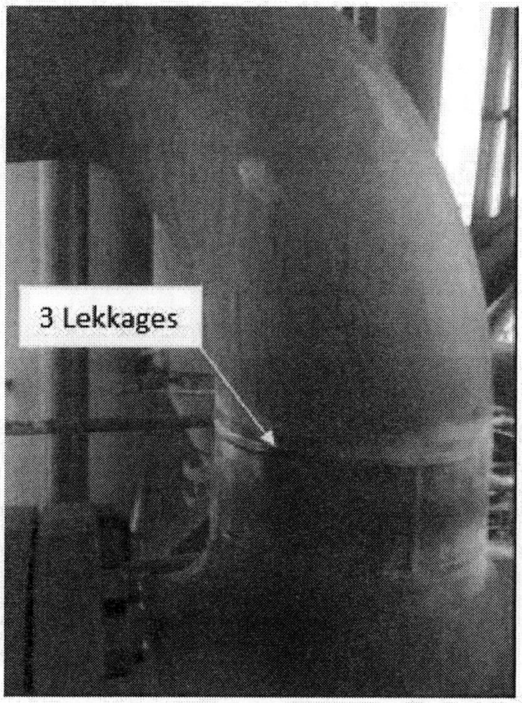

Figure 4. Photograph of the position of aMDEA leakage

The 304L stainless steel line was installed in 2014 along with the installation of a new absorber C203S. The process pressure in this line is ~30 barg (435 psig) and the temperature is ~70°C (158°F). A circumferential crack developed in a weld, starting from the inside. Operators found droplets during the night shift leaking from the crack. In a couple of hours the leak developed into a serious trickle. The plant was immediately shut down and investigation and repair was initiated. The plant was shut-down for 18 days in order to understand the cause of the leak and decide on the repair method. This was the longest single cause of downtime in 2019 for this plant.

304L stainless steel should not be susceptible to CO_2 corrosion. The welds were prepared with great care and within specifications during the installation of the line in 2014. Material certificates and welding consumables showed no deviations. All comparable welds in the system were X-rayed and the same damage (circumferential cracks in the weld propagating from the inside) was found in multiple welds, each in different stages of development. Seven welds were replaced: four were repaired by replacing the piping and three were repaired in the field.

Additional investigation was started in order to identify the exact damage mechanism and to determine if system conditions such as pipe stress, thermal expansion, vibrations, mechanical loads, etc. were within acceptable limits.

Calculations on static as well as dynamic loads showed that all parameters were within acceptable ranges. Also the alloy was acceptable for the process parameters. There were nine optional damage mechanisms left to be considered. Only True Corrosion Fatigue (TCF) could not be ruled out. All others were ruled out based on factual evidence. Therefore TCF was identified as the root cause.

Damage caused by mechanical fatigue was excluded because:

- The maximum permissible stresses were not exceeded;
- The simulated stress requirements as well as occurring forces are amply within the permissible design limits;
- No mechanical fatigue has been seen in the microscopic pictures.

Damage caused by 100% corrosion due to degradation products or CO_2 from semi-lean aqueous solution is also excluded because:

- The protective chromium oxide skin of the 304L and 316L materials was not damaged by corrosion (inspection revealed a nice, blank surface);
- In addition, no visible corrosion components have been detected by energy dispersive x-ray spectroscopy (EDX) microscopy;
- Concentration of decomposition products as well as the temperature (<100°C, <212°F) is too low to achieve such corrosion rates in 304L and 316L stainless steels.

On the inside surface, various types of notches were visible such as high-lows (due to welding as well as wall thickness differences), metal smearing, grinding grooves, etc. These are initiation points, where it is relatively easy to breach the protective chromium oxide skin.

TCF is a damage mechanism where two other contributing mechanisms, Vibration Induced Fatigue (VIF) and Corrosion, amplify each other in a way that leads to failure even though the individual mechanisms themselves are not strong enough to cause the failure. Some features of TCF in this case include:

- The fatigue share in TCF is caused by VIF with a milder amplitude than usually associated with mechanical VIF.
- VIF crack growth can grow or stop growing temporarily, depending on vibration amplitude;
- Chromium oxide skin has been breached at the various types of notches on the inner surface by VIF;
- The corrosion share in TCF is caused by a combination of very low amounts of amine degradation products, residual CO_2 + organic acids + heat stable salt anions in aqueous semi-lean solution, which have only actively caused damage after the chromium oxide skin has been compromised. The temperature increase due to cooling water side fouling in H253 (from 62°C (144°F) to an average of 80°C (176°F)) contributed to this;
- TCF was further initiated as well as propagated with some indications stopped and others resulting in a positive leak;
- TCF makes no distinction between 304L or 316L grade stainless steels. As soon as the chromium oxide skin is breached at the location of the notches or where VIF can occur on a sufficiently strong level, TCF initiates and propagates further.

Over time, various projects gradually increased the capacity of the plant by approximately 7%. This resulted in an increase in the operating range of the semi-lean temperature from 62°C (144°F) to 75°C (167°F), with peaks up to 80°C (176°F).

In 2014 a new absorption tower C203S was put in operation. This tower was prefabricated close to the old tower, and cut-over during a turnaround. This resulted in an illogical and unfortunate piping layout. All piping was designed and installed within acceptable limits, but the location where the crack developed into a leak had a relatively high stress load.

The licensor was not able to provide adequate advice to mitigate this problem and maintain mechanical integrity. All current applicable piping design rules could not prevent the damage mechanism from developing. New or additional design rules have to be developed in order to be able to build a safe, reliable and durable system.

Equipment and piping in aMDEA service should be treated as an exception because static as well as dynamic damage mechanisms amplify each other beyond conventional design limits. Further study and investigations must lead to additional provisions in order to operate the process without unwanted and unexpected surprises. Sharing of experiences with peers may help. At this time, a selection of welds is regularly inspected using phased array ultrasonic testing to find defects in a timely manner, before new cracks develop or existing cracks propagate.

Damage at a semi lean pump

In 2012, heavy damage in the semi lean pumps occurred due to mechanical overload. Positive Material Identification (PMI), visual assessment, and Raster Electron Microscopy (REM) were performed. Figure 5 shows the damage at the 347 stainless steel guide vanes and Figure 6 shows the damage at the 304 stainless steel impeller. All cracks have similar characteristics, such as: irregular crack flanks, without crack initiation and that the cracks spread in various directions. In addition to stepwise crack growth, the fractured surfaces have the appearance of split wood, which is characteristic of ductile violent fracture. The impeller material has been changed to Duplex 1.4470 with a Peek impeller wear ring, and the diffuser material and casing wear ring were changed to Duplex 1.4462.

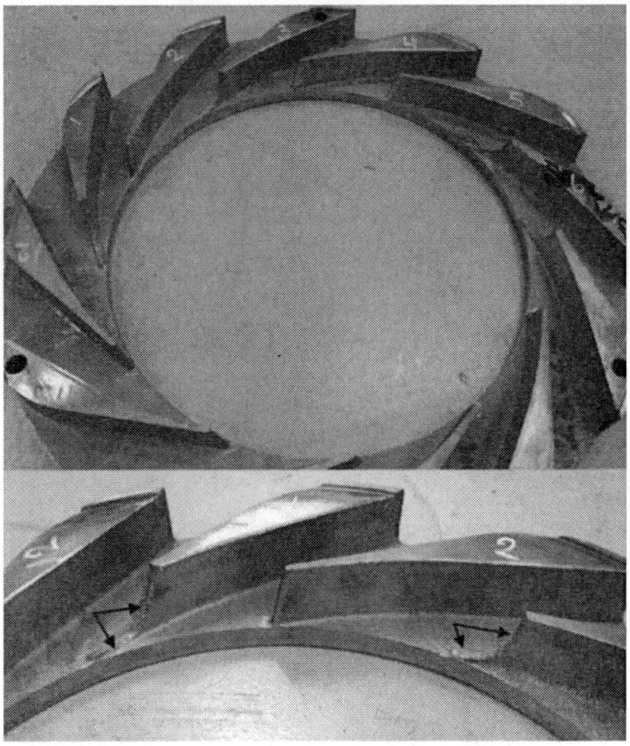

Figure 5. Overview and detail of the guide vanes. Four tips are ripped off.

Figure 6. Overview of the damaged impeller 1st stage. Material has broken off.

CO_2 breakthrough after absorber C203S

With increasing plant capacity, performance stability of the CO_2 removal system became problematic. Especially, CO_2 breakthrough (spikes) observed as a result of foaming and limited absorption in the bottom section of absorber C203S. (See Figure 7.)

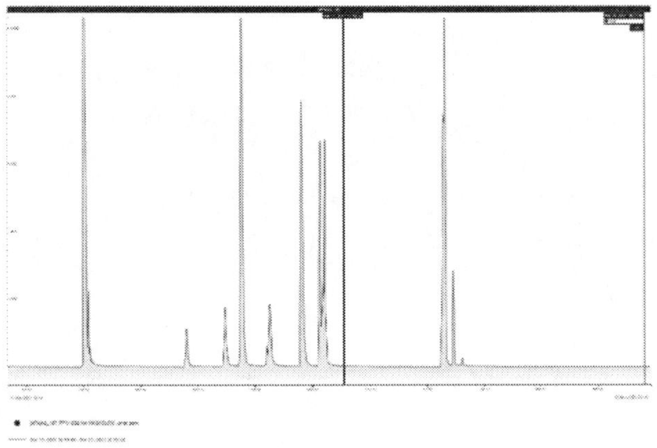

Figure 7. Typical pattern of CO_2 breakthrough at C203S over 11 days, with peaks to approximately 1000 ppm.

These occurrences are indicated immediately by gas analysis, absorber column liquid level, and pressure drop fluctuations. The new absorber, with new internals, installed in 2014, as described above, did not solve this issue. Although the hydraulic load of the absorber and desorber are within the acceptable limits, the lean absorption has reached its full CO_2 capacity. Regular quality checks on the MDEA solution show no deviations from allowable limits. The process gas inlet temperature is too high, in summer, due to cooling capacity limitations, at about 90°C (194°F), instead of 80°C (176°F). This negatively affects the absorption of CO_2 into the amine solution.

In the future, lean solution pumps P202 will be upgraded to increase lean solution flow for better absorption. Unfortunately, erosion corrosion risks in carbon steel parts is expected to increase as a consequence. Higher capacity packing in the bulk absorber (semi lean) section, heat exchanger modifications, and installing packing within flash vessel V207 are other measures being considered.

Conclusions

This paper showed a number of examples of operational and integrity issues in the CO_2 removal system that have occurred over the period of 2001 to 2021 at one of OCI Nitrogen's Geleen plants. Revamping from a Benfield CO_2 removal system to an aMDEA CO_2 removal system in 1997 resulted in corrosion issues. Unfortunately, these issues were not fully solved by the metallurgy upgrades of various process parts in 2001. New or additional design rules have to be developed in order to be able to build a safe and reliable CO_2 removal system. Over the years, the continuous production push has contributed to many of these issues. The CO_2 removal system has again become the bottleneck in this plant.

Acknowledgements

The authors want to thank S. Schepers, P. Drissen, W. Hupperetz, R. Dubois, J. Veenendaal, G. Smeets and G. Decramer for their input.

References

[1] J. de Wit and A. Riezebos, "Upgrading a 25-year Old Ammonia Plant", AIChE Ammonia Technical Manual, Volume 39, 1998

[2] T.L. Huurdeman and S. Goole, "Operating Experience with the Change-Over from Hot Potassium Carbonate Solution to aMDEA in a CO2 Removal System", AIChE Ammonia Technical Manual, Volume 43, 2002

Review of Layer of Protection Analysis Results After Process Safety Incidents

Layer of Protection Analysis (LOPA) is a well-established and valuable tool to assess process risk in a plant. Due to the complexity of the method, the composition of the team evaluating a safety scenario bears the inherent risk of subjective judgment and may influence the outcome significantly. A factor influencing the results of the analysis is the experience and knowledge of team members, but also national regulation and standards. Thus, LOPA gives a good basis for process safety risks but needs to be accompanied by additional considerations.

The aim of this paper is to share experiences of practical application of LOPA in existing plants, and review scenarios of Process Safety Incidents. The comparison of results between five Borealis ammonia plants located in France and Austria has revealed significant differences in approach and result.

Hans Juergen Essl
Borealis Linz

Daniel Sandholzer, Leonhard Werner, Georg Zimmer
Borealis Nitro

Introduction

Assessing the hazards of chemical processes and determining the necessary safety protection layers to fulfill safety requirements has always been a challenging task for chemical as well as engineering companies. The methods used for this purpose have been refined and standardized over the years. Currently HAZOP (Hazard and Operability Study) and What-If are widely spread in chemical industry.

However, for the next step in assessment – to define the amount and quality of safety protection layers needed for safeguarding against a certain scenario –a large variety of approaches exist. These differ on the one hand in terms of application effort (i.e. time and data needed) and on the other hand on the degree of objectivity of the method.

The two extremes of these methods can be defined as such:

- Applying safety protection layers based on scenario consequence only. This leads to a fast and easy definition of safety protection layers as no discussion on probability of occurrence is necessary. However, this may also lead to a large amount of protection layers safeguarding very unlikely scenarios, and increasing the resource effort in inspection and maintenance. In addition, the definition of the consequence is highly subjective and depends on the experience of the assessment team.

- Assessing the need of safety protection layers with a Quantitative Risk Analysis (QRA). As the term quantitative implies, this method is based on calculations and statistical data of incident consequence and probabilities. It is very accurate and gives a good objective vision of hazards, but it is extremely time consuming.

To balance effort and objectivity Borealis, like many other companies, has decided, in general, to apply the Layer of Protection Analysis (LOPA) method to define required layers of protection for a given scenario. This method is nowadays not only a widely used assessment tool, it is also formally acknowledged, being part of the standard IEC / EN 61511 "Functional safety – Safety instrumented systems for the process industry sector".

LOPA is a semi-quantitative method that on the one hand is based on the assessment of the team to define out of a HAZOP or What-If analysis the consequence of a scenario – the subjective, qualitative part. On the other hand, it uses standardized values for initiation probability as well as conditional modifiers, like the probability of ignition or human presence – the objective, quantitative part [1]. These values are based on probabilities calculated from large databases of failure modes and incident occurrences [2].

The consequence will define what remaining risk of occurrence is acceptable, the higher the potential consequence the lower the acceptable remaining risk. For example, if the consequence would be one fatality the acceptable remaining risk often is defined as one occurrence in 10.000 years.

With the initiation probability and the conditional modifiers, a so-called naked risk – the risk of the scenario to occur without the consideration of any safeguards – can be calculated. In case the naked risk is higher than the acceptable remaining risk, the existing gap needs to be closed by introducing independent protection layers.

The gap an Independent Protection Layer (IPL) can close depends on its type. For example, a safety valve in clean service can close a larger gap than an alarm with requiring operator intervention.

Borealis has been using the combination of HAZOP for identifying critical scenarios and LOPA to define the required safety protection layers and their quality now for more than 20 years. During this time, the application has been used for all relevant production plants as well as a majority of pilot and lab scale installations. Experience in correctly applying these methods has grown, and the scope has been expanded as well. Where once scenarios were cut off after identifying a loss of primary containment, now the ultimate consequences are elaborated in more detail, considering harm to people, but also damage to the environment.

During the years of applying the combination of HAZOP and LOPA, experience has shown that there are some points that need to be considered when applying this tool for protection layer analysis to ensure a conclusive risk assessment.

We will describe some of the key points that need to be taken into account before, during, and after a risk assessment using LOPA. The review of two incidents will show what issues can arise when all considerations are not consequentially put into implementation.

Experience

The secret of every tool is that you need to know how to apply it. Experience and thoroughness are necessary in the team conducting a LOPA. This is also a requirement for the step before the LOPA, the process hazard analysis, which needs to identify relevant scenarios in the first step. Whenever a potential loss of containment or other event that may cause damage to people and environment has been identified a LOPA will be performed. In scenarios with low probability, the team can also skip a scenario.

In current times, the workload of experienced plant personnel has increased to a level that people are trying to spend time only on "really necessary" tasks. Unfortunately doing a HAZOP and the following LOPA is often seen as a tedious work, whose benefits are not fully understood and accepted (at least until the first major accident happens). The main risks of the plant are seen as obvious and well known. The willingness to send the experienced (and therefore often overloaded) employees to attend such studies is low. Often young and inexperienced engineers are chosen as team members; as such, a study is seen as good opportunity for them to learn and understand the process better (which it definitely is). It is also known that inexperienced teams tend to overestimate the consequences of small scenarios and underestimate the consequences of large scenarios.

An experienced hazard study leader will help in such cases, but will only be able to facilitate the process by asking proper questions; as he is not – and even should not be- an expert in the chemical process, he cannot provide expert judgement.

Correct Risk Assessment

Generally, the team benefits from having experienced people in the hazard study. Sometimes, this can also have negative effects. Personnel that has been in the plant for many years can tend to negate or underestimate certain risks due to their own limited experience. This phenomenon of underestimating risks because they have not been experienced personally or in the near surrounding is known as risk homeostasis.

In such cases during the discussion, it is often said: "but this has never happened in our plant" or "I have never heard of this in our industry". The statements themselves may be true but statistics do not care if you have heard of the result or not. It will happen and it is only a question of time. To accept this general paradigm of process hazard analysis and LOPA sometimes is hard for experienced and practical operations staff. Such a situation – especially when the hazard study leader is inexperienced or not able to assert his expertise – can lead to risk assessments that exclude probable scenarios because the occurrence of the initiating event is not believed by the study team.

Effect

Mathematically, LOPA simply multiplies the frequency of an initiating event by the probabilities that each independent protection layer will fail to perform its intended function. The application of LOPA is targeting to close the gap between the naked risk and the acceptable remaining risk by adding independent protection layers. When this gap – typically expressed as required Risk Reduction Factor (RRF) - is lowered to one or below, the process is considered safe and the assessment team often considers that their job has been done.

However, in a broader context - and very much depending on the location of the plant– it might not be enough to limit the consequences to that extend. For example, noise or smell, even without potential to harm people or environment, should be prevented in order to avoid complaints, authority focus, or media attention.

For this purpose, it can be necessary to add additional protection layers to the process, like interlocks that act before a PSV opens to atmosphere or that at least limit the release time. Such protection layers do not necessarily have to be defined as safety critical, but from a society point of view they may be very important indeed. These facts need to be understood by the LOPA team so that the assessment does not stop when the method itself has been satisfied.

Some countries, such as France, have even made this aspect part of their legislation. The "Mesures de Maîtrise des Risques instrumentées" (Instrumented Risk Control Measures) – in short MMRI – requires safeguarding against scenarios that lead to substances leaving the area of a chemical process facility, no matter the concentration.

Implementation

After doing the process hazard analysis the job is only half done. Now the measures to close the gap need to be engineered as well as implemented. In case a safety instrumented function (SIF) is foreseen to close a certain gap, this may mean installing new, certified valves or instruments. For existing systems, components need to be checked for their suitability (effectiveness in all plant states; reliability). If these do not have formal certificates, they need to be checked if they can be further used via the "prior in use" approach (as defined in IEC / EN 61511). The correct location of instruments needs to be verified for measuring of unwanted deviations.

Thoughts have to be given about the application of new components. They not only have to fulfil the requirements regarding their setup and certificate but primarily they need to fit to the purpose. A valve with a given Safety Integrity Level (SIL) certificate does not serve all process conditions by default.

Doing these checks and implementing new systems requires effort and costs. This sometimes leads to a lagging implementation, especially when the criticality is not understood by all players. In addition, the resulting independent protection layers need to be submitted to a thorough Life Cycle Management (LCM) to guarantee reliability. This does not only mean suitable inspection and testing intervals, but also marking in the field and operating procedures regarding bypassing and general handling of protection layer components. Without this last step the safety improvement of new systems is not realized.

However, a LOPA gap does not necessarily need to be closed by a SIF, according to IEC / EN 61511. This is not new and is well described in literature. However, experience demonstrates this to be one of the most difficult steps in the whole process, especially when a LOPA gap requires a Risk Reduction Factor (RRF) of 10.

According to IEC / EN 61511-3:2016 Basic Process Control System (BPCS) can be used to close a gap up to and including RRF 10 (nevertheless, the IPL needs to be independent, which is in general difficult to realize [3]). The same is true for an operator response to an alarm, where a RRF of 10 can be taken into account as well. Both, the BPCS and the alarm with human intervention, might be used in order to "substitute" a SIF. IEC / EN 61511-1 excludes in its Scope Definition the applicability for SIFs with an Integrity Level lower than SIL 1. No doubt, such safety functions do exist. TÜV Austria describes in their publication pre-conditions to use SIF functions without SIL certificates in existing plants [4]. It is required to operate and maintain those with the same principles (bypassing, modification, testing, etc.) as SIFs with defined SIL.

Learning from incidents

As part of the Borealis Operational Excellence (OPEX) program, a formal learning from incidents process was established. Whenever an incident with a production loss, a (potential) loss of containment or the activation of a critical safety device (e.g. last protection layer activated) has occurred the risk of the incident is evaluated. Those incidents, which are charged "high severity" are then investigated by corporate departments of Borealis. The learnings are used to drive the implementation of mitigation actions, knowledge sharing and training, but also to review the effectiveness of procedures.

The following pages contain two remarkable process safety incidents that will be reviewed. These occurred in the last two years and triggered a discussion on the overall effectiveness of the process hazard analysis approach of Borealis Group.

Process Safety Incident 1

The first incident has occurred in one of the ammonia plants in Linz, Austria. On March 26, 2019, the control room operator of the synthesis section was alarmed by a sudden pressure increase in the ammonia flash tank and a level decrease in the ammonia separator. The setup can be seen in Figure 1.

As the pressure increase was very sharp, the safety valve was activated in less than a minute. Consequentially the level control valves from the separator LC-704 A and from the condenser LC-702A to the flash tank were closed. Field operators were ordered to close the bypass of the control valves as this was thought to be the cause of the pressure increase. When they arrived at the staircase to the top of the tank, the men found a significant ammonia smell.

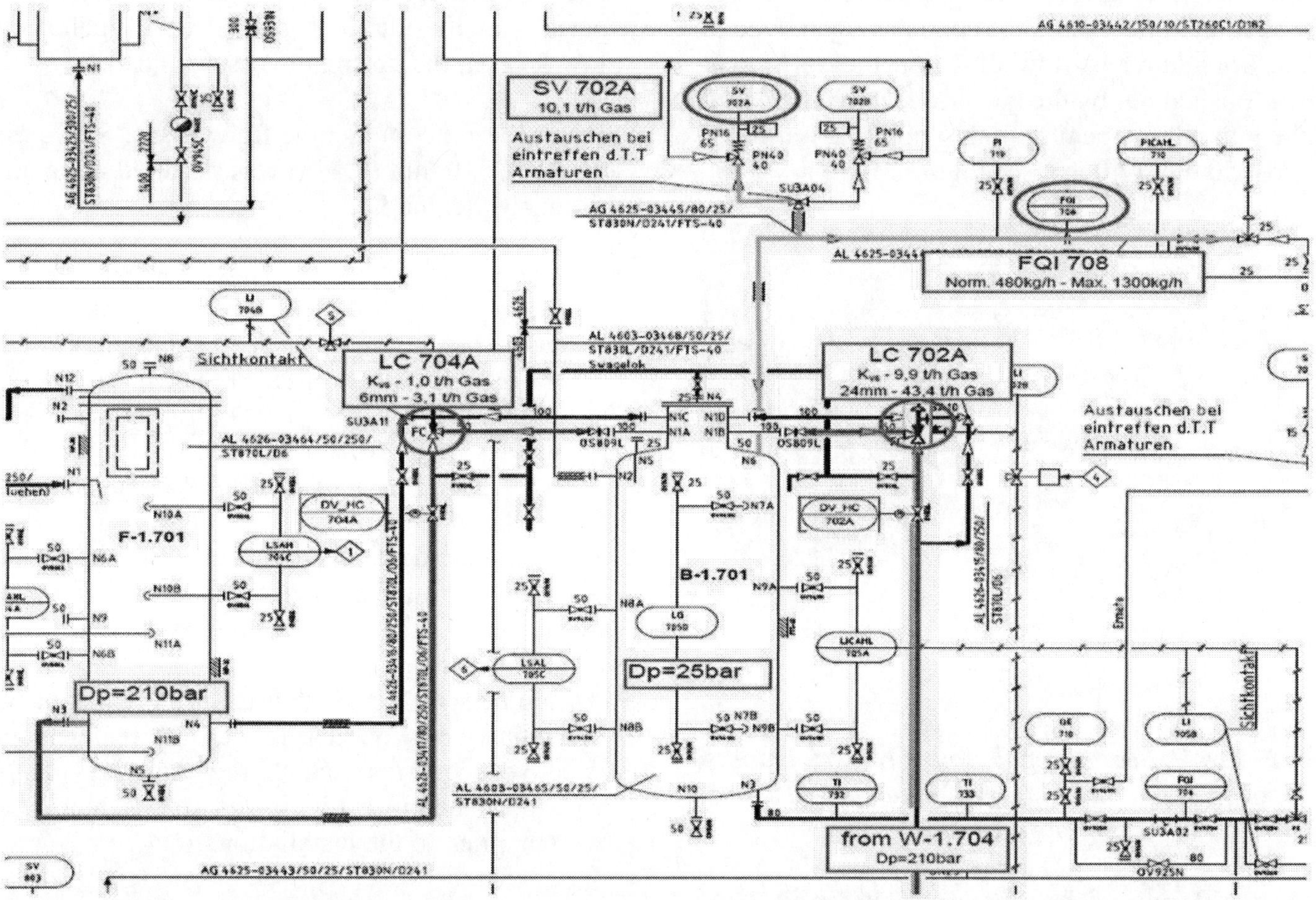

Figure 1. Configuration of separator F-1.701 and flash tank B-1.701 in synthesis section.

After contacting the control room by radio, the motor driven valve HC 704 between separator and flash tank was closed by the control room operator four minutes after the pressure increase, and the tank pressure began to decrease. The synthesis section was stopped immediately after the incident. As the pressure had dropped back to operating level there were no signs of leakages in the plant. Flash tank B-1.701 was decommissioned and internally inspected. It was found deformed and several cracks were detected in welding seams.

As the cracks were not deep, those could be removed by grinding. Pressure test, ultrasonic and visual inspections were performed together with the notified body. After recalculation of the areas, which were exposed to the highest-pressure stress, the tank could be requalified for twelve months of operation with additional external inspections (phased array testing) of affected longitudinal welding seams every four months. A new vessel was ordered immediately and installed during a short outage in November the same year.

The level control valve LC 704A was dismantled and investigated. The threaded sleeve of the seat was found broken and the cone damaged. Due to the orientation in flow direction the valve seat was pushed out by the gas flow, which increased the maximum opening from 6 mm (0.24 in)to almost 50 mm (2.0 in) (see Figure 2).

Figure 2. Seat of regulation valve was found broken at the thread (right). Undamaged spare of the seat (left).

As the design of the pressure safety valves was based on the maximum opening of the control valve, the pressure relief valves could not discharge the increased gas flow, so that the pressure was exceeding design. Additionally, the interlock on "pressure high" in B-1-701 was acting on the level control valves. As the cone of LC 704A could not close the broken seat of the valve, the interlock was not effective.

The Retroactive Hazard Analysis of the plant had investigated four scenarios dealing with pressure increase in the flash tank. The design of the safety valves was seen as sufficient to deal with increased gas flows. Consequentially no LOPA was conducted. For the high pressure interlock in B-1701 (no doubt – the aim was a safety function!), the basic principles for an IPL were not respected: Neither independency from the initiating event was given, nor effectiveness was verified. The question needs to be raised, if a high pressure interlock is fast enough to avoid the scenario – or the interlock should be triggered by low level in the upstream vessel instead.

Because of the incident, a flow orifice with a diameter of 10 mm (0.4 in) was installed as an additional safeguard.

Plant	LOPA	IPL for this scenario
Austria 1	No	PSV + No Interlock
Austria 2	No	PSV + DCS Interlock*
France 1	Yes	PSV + SIL 1 Interlock
France 2	Yes	PSV + SIL 2 Interlock
France 3	Yes	PSV + SIL 1 Interlock

Table 1. Evaluation of scenario before incident

When the incident was reviewed at the corporate level, it was found out that there was a difference in the result of the safety analysis between the different plants.
As can be seen from Table 1, the Borealis plants in France were conducting a LOPA for their respective scenarios, which the Austrian plants were not. Therefore, high-pressure SIL interlocks were applied to the installations.

Process Safety Incident 2

The second case has occurred in one of the French ammonia plants in Borealis on September 12 2019. During reconnection of the 5.5 kV drive of a Benfield circulation pump, an electrical arc was produced in the switchgear compartment. Consequently, this shut off two 5.5 kV to 400 V transformers. This caused the carbon dioxide removal section of the plant to trip and the respective interlock logic was activated.

The emergency shutdown (ESD) sequence caused by high carbon dioxide concentration on the outlet of the absorber of the Benfield section activated a motor driven valve on the inlet of the Methanation section. Due to the failure of both transformers, the valve did not close according to interlock design. Although shift operators in charge realized the interlock fault and were closing the hand valve to the Methanator, they were not able to prevent the ingress of carbon dioxide into the Methanator.

According to the reading of the temperature probes, the inlet temperature reached 711°C (1311°F) (approximately 400°C (720°F) increase) and 600°C (1112°F) (approximately 300°C (540°F) increase) on the outlet of the first bed within four minutes after the incident. In contrast, the temperatures in the second bed were hardly influenced as the inlet temperature increased from 303°C (577°F) to 343°C (649°F) and the outlet from 307°C (585°F) to 334°C (633°F) (see Figure 3). Approximately seven hours after the incident the insulation on the reactor had been removed and members of the local inspection department began investigating shell temperatures of the Methanator with mobile temperature measurement devices. As shown in Table 2, the measurements were significantly lower compared to the temperature probes in the first bed of the reactor.

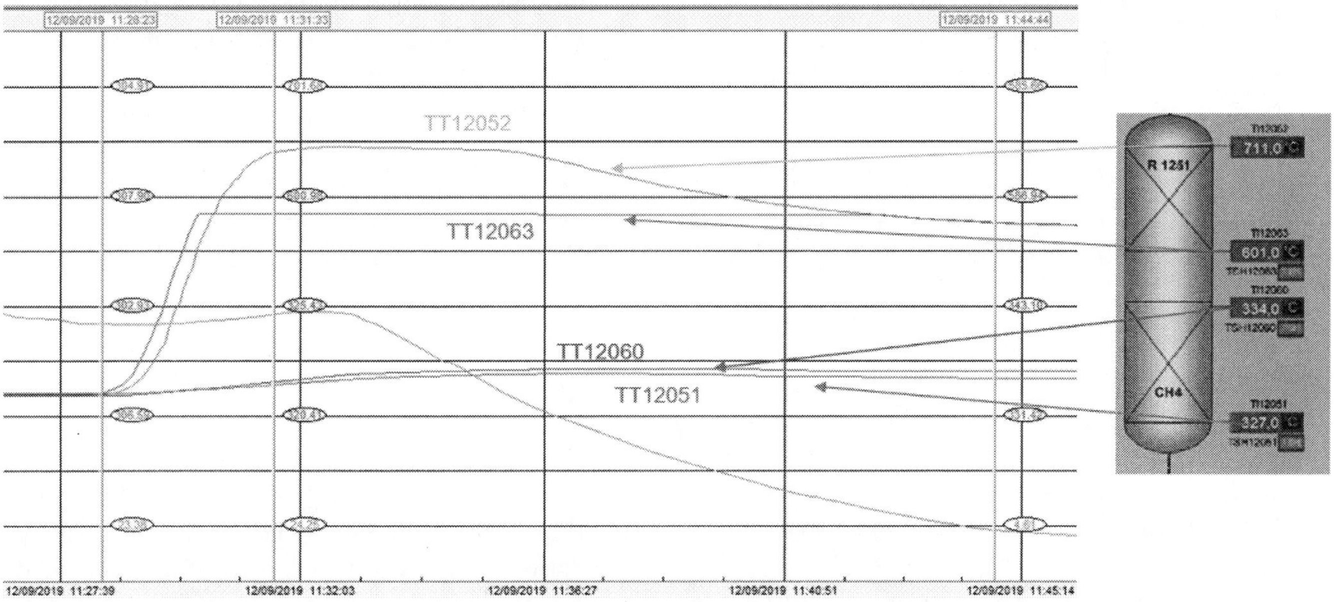

Figure 3. Temperature increase on Methanator temperature measurements

Point	Temperature [°C] / Time after incident [h]		
Time [h]	7	20.5	32.25
Location 1	250	177	288
Location 2	381	330	287
Location 3	330	270	285
Location 4	323	272	285
Location 5	315	225	280

Table 2. Local Temperature Measurements

The location of the measurements can be seen on Figure 4. As can be seen from Table 2 the temperatures measured on the surface of the pressure vessel were far below the readings on the temperature probes in the first bed of the reactor. These external measurements were used to estimate the thermal impact on the material of the vessel.

After investigation, it was found the design limits of the apparatus were likely not exceeded. A damage assessment (based on historical data) according to API 579 was performed together with Bureau Veritas. Mechanical testing (tensile test, hardness test, Charpy test) on representative plate of same material was performed (i.e. same thermal heat treatment and thermal cycle).

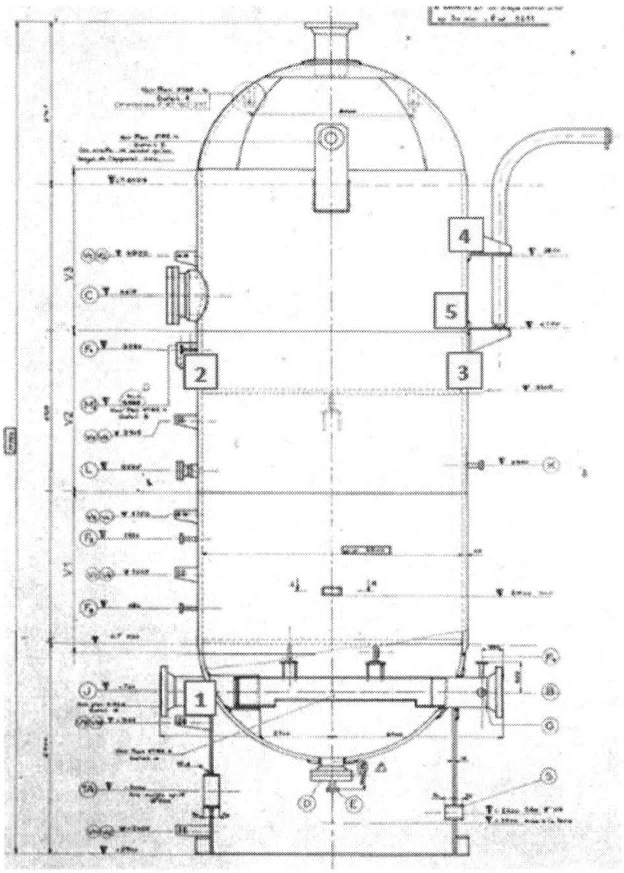

Figure 4. Position of local temperature measurements on the outside of the equipment

Finally, it was concluded that the reactor could be operated until the next planned outage.

Apart from an inspection program for the vessel it was also decided to replace the motor driven valve of the high temperature interlock with a pneumatic driven valve in failure safe close configuration.

In addition, the interlock scenario had to be reviewed. When the situation of all plants was compared at the corporate level, a large difference on approach was discovered (see Table 3). Two of the units had classified the interlock SIL 2, but both were experiencing problems with the tightness of the certified valves. The interlock of one plant was designed according SIL 1, but the formal certificate for the valve was missing, as it was manufactured in the 1990s. Finally, another plant had shown exactly the same setup as the unit where the incident has occurred.

Plant	LOPA	IPL for this scenario
Austria 1	Yes	SIL 2 interlock, flap valve.
Austria 2	No	Proven in use ball valve, fail safe close
France 1	Yes	SIL 2, ball valve
France 2	Yes	Motor driven valve
France 3	Yes	Motor driven valve

Table 3. Evaluation of scenario before incident

After consulting literature and discussion of the incident on group level Borealis is proposing a SIL 1 interlock as the minimum standard for a methanator high temperature scenario. In addition, in this case it was demonstrated that it is not enough to conduct a LOPA, focus on a SIL certificates of specific components, and do scientific PFD calculations. The safety function must also meet the specific need and the very basic principles of IPLs to be fulfilled. In this case, reliability (in terms of robustness) and efficacy of the safety function itself.

Discussion

The investigation of process safety incidents has clearly shown remarkable differences regarding the results of HAZOP and LOPA at different locations. The result of the same scenario was differing from no SIL to SIL 2. The result was depending on several factors like HAZOP study leader and experience of team members. With the definition of the required risk reduction factor out of the LOPA, the requirements for the different IPLs are set. Other companies have experienced the same deviation and have proposed a "typical" SIL level for a specific scenario in all their units [5].

Borealis, however, has concluded that not every LOPA gap needs to be closed with a SIF. Once the required SIF level is defined by the LOPA team, a dedicated automation and instrumentation team should start with the evaluation of the most attractive solution considering the technical requirements out of the applicable standards (e.g. IEC / EN 61511), the timeline, and overall effort. Typically, the result of this evaluation is formulated in a so called "gap closing" project.

In addition, another trap often appears on the way towards a state-of-the-art SIS, the "proven in use" or nowadays more precise "prior in use", and in line with the latest release of the IEC / EN 61511. Especially in the fertilizer industry the plants were erected in the 1970's and the embedded safety systems are sometimes still in operation, which is typically seen on relay based ESD systems. Once IEC / EN 61511 was published and the discussion around the implementation complexity started in the branch, "prior in use" was seen as an attractive approach to overcome bigger system changes and modifications. The results were, once again, a high allocation of internal as well as external resources to gather formal evidence that the systems fulfill the stringent requirements of the new codes and standards.

However, a couple of years later a lack of spares or knowledge, internally as well as from the system supplier, was experienced and an exchange became unavoidable. Considering the experience of different fragmented projects and failing attempts to get a formal certification the prior in use approach is the choice of selection on the actuator side and here especially for shutoff valves. As a kind of unwritten rule if the costs for a single valve exceed $10,000, "prior in use" should be seriously considered. Furthermore, in the case of shutoff valves with special construction requirements, many times there is no better alternative. Otherwise, unavoidable modification in terms of piping, not mentioning the impact on the fluid flow etc., needs to be considered and the effort accepted.

Another factor of the issues observed is national law and regulatory requirements. Here, the focus area of French units is to avoid a release of harmful substances beyond the limits of the site, whereas in Austria priority is given to the protection of the equipment integrity. Of course, this observation immediately raises the question, why priority is given to one and not to both topics. The answer is simply that the high implementation effort as well as the fact that the introduction of new equipment into the plant bears risks as well.

As the ammonia industry can be considered conservative in general, it is not surprising that responsible managers try to avoid abandoning proven, installed equipment and to exchange it towards a new setup, which might bring along unknown or new risks, like sticking or leaking valves. As such, a certificate does not really change the availability of a safety critical interlock, but a proper life cycle management as well as a proper engineering description will significantly improve its functionality.

Other companies have proposed a general common rating based on a risk matrix, but this might conflict with local authority requests, as well as immediately requiring a high number of changes to a "proven in use" setup. In general, vintage plants are still struggling to change the design of interlocks according to IEC / EN 61511 as well as related standards like IEC / EN 746-2 or VDMA 4315 (a standard of the German association for machinery and plant engineering), because of the high implementation effort and complexity. Because of this, gaps might remain for quite a while until those are fully mitigated.

Conclusion

After applying LOPA in existing plants for several years, Borealis can highlight the main learnings that this powerful tool provides in factual, veritable safety improvements for the operation of plants:

Company specific procedure and qualified Hazard Study-leaders

Although the methodology is very well described in literature for many years and various guidelines for practical applications exists already, some specific agreements and interpretations need to be safeguarded in a company procedure. Beyond that, it is key to develop a pool of educated LOPA leaders to ensure a high degree of objectivity and consistent outcomes. Ideally, those experts should regularly meet and align in a network. Also, participants need to receive training so they can understand and accept the approach used.

Define company standards for typical scenarios

In ammonia plants there are some basic, well-known process safety risk scenarios. As discussed above in detail, one of those is high temperature in the Methanator caused by high amounts of carbon dioxide entering the vessel. The evaluation of such scenarios should not be left to the evaluation by the plant team.

Typical requirements for safeguarding this scenario should be provided by corporate functions. Not only the result (e.g. SIL 1 for High Temperature Interlock on Methanator) but also the factors applied shall be available. The plant team can deviate if sound justification is available (and properly documented) and might end up with higher or lower requirement (e.g. hard-wired system, which proved to work properly in the past).

Identification of all Safety Functions

Every plant shall compile and maintain a list of all Safety Functions, independently of whether they resulted from Retroactive HAZOP/LOPA exercises, originate from international codes and standards for specific applications (e.g. EN 746-2), or from licensor information. In addition, safety functions without specific SIL shall be added. It is of highest importance that for all of those the same basic principles are applied in the whole Life Cycle (verification, bypassing, modification, testing, etc.).

The worst case, however, would be the avoidance of identifying Safety Functions (e.g. by consciously downplaying the potential consequences – which is, as discussed earlier, the most subjective factor), because of being afraid the outcome cannot be implemented or is too costly.

Conduct multiple LOPAs for the same scenario

Finally, a very painful learning out of an incident described above was, that also internals of control valves can break with a certain probability. Typically, in LOPA evaluations, the initiating event is determined by control loop failure, with a probability of once in 10 years (Initiating Event = 0.1 [1/a]). In our case, the scenario was not triggered by a control loop failure. Neither the probability of the initiating event, nor the consequence (much larger diameter and flow consequently) is the same.

For this specific case, or similar arrangements, additional LOPA evaluations were conducted. The probability for a breakage of internals (as initiating event) was set to once every 100 years (Initiating Event = 0.01 [1/a]). The result was that in most cases, the safety valve is not capable to prevent overpressure in the downstream vessel, ending up in a very high consequence. By installing orifices, the consequences were reduced to a release via the safety valve. Nevertheless, those cases also need to be safeguarded. This leads to the requirement of an additional IPL, which is practically realised by means of instrumented functions, but does not require an RRF > 10. Here a safety function realized in the BPCS is an appropriate way to close the remaining gap.

The basic principles for Independent Protection Layers (IPL) are:

- Effective (against the specific scenario),
- Independent (to the IE but also other IPLs),

- Reliable (the installation must be as simple and robust as possible).

As long as these are met, and the principles respected, and a sound LCM is fully in place, a tremendous improvement in process safety is reached, even without requiring replacement of equipment solely to acquire formal SIL certificates.

References

[1] Willey, R.J. "Layer of Protection Analysis". *Procedia Engineering 84* (2014).

[2] *Guidelines for initiating events and independent protection layers in Layer of Protection Analysis.* Center for Chemical Process Safety. (2015).

[3] Hereña, Peter G. "Validating Multiple Credits from Your DCS… or Not". *Safety in Ammonia Plants & Related Facilities, Vol. 57.* American Institute of Chemical Engineers, (2016).

[4] Preiss R. Struckl M. "Layer of Protection Analysis (LOPA) for Risk-Based Evaluation of Scenarios". Edition TÜV Austria (2014).

[5] Babwah, R., Warzee M., Franceschini, G. "Methanator Vessel Overheating – Why Still a Hot Topic". *Safety in Ammonia Plants & Related Facilities, Vol. 60*, American Institute of Chemical Engineers. (2019).

Corrosion Under Insulation - An Underestimated Damage Mechanism

Corrosion Under Insulation (CUI), despite being incredibly well known, is responsible for significant maintenance costs in the process industry. In year 2016, CUI was responsible for a major leak in Benfield solution piping going to the carbon dioxide absorber in Yara's 2125 MTPD Ammonia plant in Saskatchewan, Canada. A comprehensive site wide CUI inspection program found several other examples, including within the high pressure synthesis gas piping, where significant wall loss was found due to external corrosion.

This paper will share how Yara Belle Plaine administered the CUI inspection program and its learnings from this undertaking. It will highlight how continuous improvement in standard work protocols can help with mitigation. This paper also includes inputs for any future review of API 583.

Arie de Bruijne, Waseem Haider, P.Eng.
Yara Belle Plaine Inc.

Introduction

Yara owns and operates a fertilizer complex in Saskatchewan, Canada consisting of ammonia, urea and nitric acid plants. The facility was commissioned in 1992 and has been in operation since then. The ammonia plant was designed by Uhde (tkIS) with nameplate capacity of 1500 MTPD (1650 STPD) which increased to 2125 MTPD (2340 STPD) as a result of major expansions completed in 1997 and 2009.

Yara Belle Plaine Inc. started late, in 2016, on an inspection program specifically targeting corrosion under insulation (CUI). The program used a risk based phased approach. In the three years that followed the launch of this program, more than 20,000 MT (22,050 ST) of ammonia production was lost due to plant outages taken to address the identified CUI related problems. There was one event involving loss of containment when lean solution piping feeding the carbon dioxide (CO_2) absorber developed a leak releasing Benfield solution. Figure 1 shows another example where significant wall loss was found due to CUI damage to a drain line, in high pressure synthesis loop piping, resulting in a plant outage for repairs.

On top of production losses, maintenance costs and a release to the environment, for a time significant potential existed of physical harm to workers. Luckily, it did not happen in this case.

This is perhaps another example of how industry has underestimated the harm that CUI can potentially bring and it is fair to state that there is a culture that we need to change.

Figure 1. CUI on a drain line in synthesis loop hot gas-gas exchanger, 8E003

The Basics of CUI

General Description

Insulation is used on process piping and equipment for several reasons. Energy conservation, personnel protection, noise abatement, and fire protection are a few of them.

CUI is normally hidden from the naked eye. It initiates when water becomes trapped beneath the insulation with metal surface at the right temperature (0°C – 100°C, or 32°F – 212°F) and is not protected by the proper coating. Water ingress can be due to various reasons but happens mainly due to damaged lagging of insulation or leaking steam tracing.

In carbon and low alloy steels, the mechanism is described as a form of oxygen corrosion where there is general metal loss, cracking, and/or pitting whereas in stainless steels, CUI is generally considered a form of Chloride Stress Corrosion Cracking (Cl-SCC) caused by water containing chlorides. Sources of water containing chlorides can vary; site is in close proximity of seawater, leakage in deluge systems, fire water testing and leaking process drains, to name a few.

API RP 571, 581 and 583 have covered the CUI topic thoroughly and can be consulted for further details.

Factors Affecting CUI

There are a number of factors that affect the onset and extent of damage caused by CUI.

1. Service temperature
2. Nature of service (e.g. continuous or intermittent service)
3. Atmospheric humidity, environmental conditions, climate conditions
4. Localized environmental factors (e.g. proximity to cooling tower), duration of exposure to such conditions, and presence of contaminants such as chlorides
5. Insulation type
6. Mechanical maintenance practices in order to avoid water ingress in insulation
7. Condition of coating and painting on piping and equipment
8. Condition of the insulation barriers and sheeting

The effect of trapped water on the corrosion rate on carbon steel is shown in Table 1 below. It is evident from the table that by covering the exposed surface the corrosion rate increases dramatically. The data for indicative corrosion rates for unpainted carbon steel may be found in ISO 9223 and ISO 9224, and CUI corrosion rates in API 581 Appendix N.

Area	Location	Corrosion Rate Bare CS (mm/yr)	Corrosion Rate CUI (mm/yr)
Europe Mid	Coastal area	0.012	0.50
Europe South	Coastal area	0.005	0.25
Canada	Inland	0.005	0.10
Caribbean	Coastal area	0.012	0.25
Australia NW	Coastal area	0.012	0.25

Table 1. Comparison of corrosion rates

Real Problem with CUI

The damage mechanism of CUI is well understood and documented. However, it is not visible and is spread all over the plant. Consequences are not uniform and can range from minor leaks to major equipment failures. Effects only become noticeable after 20 to 30 years of operation in most cases. Furthermore, European standards compared to ASME give a difference of about 10 years before CUI effects are expected, due to significant difference in applied wall thickness of the equipment and lines.

It is often difficult to justify high costs of inspection programs against un-quantifiable benefits and thus in the competition for capital, CUI is often neglected.

Yara's Approach

Yara has installations kept in operation beyond design life. A number of plants have been in operation since the end of the 1960's with limited or no attention to CUI in the first 25 years.

Problems with CUI started becoming visible in the mid-1990's but were not taken that seriously and were not seen as a major concern. Major incidents with high economic impact drew management's attention, but the cost to manage CUI were a hindrance in going for a full attack on CUI. Many years later, in 2014, Yara developed a best practice based on API recommended practices and Yara plant experiences.

API publications on the subject have been studied during development of Yara's best practice. Based on extended, but not structured experience, from various Yara sites, the best practice deviates from API documents in some ways. The major deviations are summarized below which can be considered in future review of API 583.

The Temperature Range for CUI Susceptible Equipment and Lines

The temperature range for CUI susceptible equipment and lines has been divided in two categories, fully susceptible and limited susceptible, with a different approach per category. These two temperature ranges are:

1. Equipment operating in a temperature range between -5°C and 120°C (23°F and 248 °F) are fully susceptible.
2. All equipment operating at temperatures outside this range are limited susceptible to CUI

This approach means that all equipment and lines are susceptible to CUI, and for that reason, all equipment and lines will be part of a first screening. The second group, outside the standard temperature range for CUI, shall also be reviewed and screened for certain locations where CUI can occur.

Risk Based Approach

CUI is not a trendable degradation mechanism, as locations and corrosion rates cannot be defined with accuracy to determine the remaining life of equipment based on available data. For piping systems and equipment with lower wall thickness, the remaining life can vary between 5 and 50 years for the same type of equipment and lines operating under similar conditions.

The Corrosion Rates

The calculation of the corrosion rate is simplified. Only three categories are used and applied without correction factors for insulation material and condition, complexity, pipe supports, and interfaces. In general, this results in higher but more realistic corrosion rates.

The Yara approach however uses Consequence of Failure (COF) to define the inspection approach but the COF is only defined by the nature of service fluid, without any other consideration.

Non-Destructive Inspection Techniques

Full visual inspection with all insulation removed is the only reliable inspection method to avoid loss of containment. Only for screening the condition of small-bore connections is real time radiography applied. The Non-Destructive Testing (NDT) techniques available without insulation removal only give loss of material over an area, and Yara has experienced very localized corrosion pits resulting in leakages. One example of such damage on a 7.62 mm (0.3 in) wall thickness straight run of pipe is shown in Figure 2.

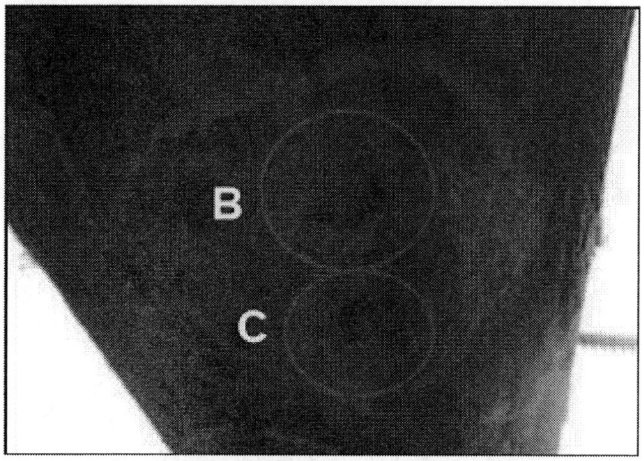

Figure 2. CUI on a horizontal run with leakage due to pitting CUI

Management of CUI Inspection Program at Yara Belle Plaine

Yara Belle Plaine did not have a systematic CUI prevention program in place and the lack of maintenance also led to generally poor condition of insulation. The guidelines as laid out in the Yara Best Practice were followed to the full extent.

The CUI program was handled as a project with following major steps.

1. Organization
2. Planning
3. Preliminary activities:
    a. Documentation
    b. Field verification and review of the documentation
4. Scope of work and definition of priorities
5. Inspection execution
    a. Pre-Inspection
    b. First inspection
    c. Second Inspection
6. Final reporting
7. Setting of future inspection intervals

Organization

Resources put in place for execution of CUI inspection program are listed in Table 2 below.

Discipline	Function	No of resources
Management	Project manager – contractor	1
Maintenance	Static Mechanical MNT engineer -Yara	1
Operations	Operations dayshift supervisor - Yara	1
Inspection	Senior inspector – Yara	1
NDT	Rope access NDT tech's / Inspectors – contractor	3
Support	Rope access insulators - contractor	8
Safety	Safety supervisor – contractor	1
Admin	Admin assistant - contractor	1

Table 2. Organizational structure

Preparation of Required Documentation

Before the team embarked upon CUI work scope definition, a review of available documentation was completed, and where required necessary updates were made after field verification. This was a necessary step before screening for CUI inspection scope could begin.

A summary of documentation reviewed as part of the project included the following:

- Process Flow Diagrams
- P&IDs
- Equipment list
- Updated piping list
- Updated equipment data sheets
- Updated equipment drawings
- Updated piping specifications

- Updated piping isometric drawings
- Updated insulation and painting specification

Some standard templates and new documents developed over the course of the project included the following:

- Local procedure for CUI inspection
- CUI inspection project plan
- Equipment and piping list (for scope definition)
- A template for pre-inspection report
- A template for first inspection report
- A template for second inspection report

The equipment and piping list described above was developed specifically for the project to include following additional details:

- The purpose of insulation on piping and equipment to understand the safety hazards associated with removal of insulation during operation and whether there are any effects on the process.
- Piping and equipment operating conditions such as normal operating temperature, if in cyclic service, if in intermittent service, subject to steam tracing, and are there any large temperature gradients.
- Will the piping and equipment be easy to re-insulate on completion of the inspection.

Scope Definition

Scope definition of CUI inspection program was based on Consequence of Failure (COF). COF was determined considering the type of service fluid without any other associated calculations.

- COF categories for Ammonia:

    A Synthesis Gas
    B Process Gas/Natural Gas
    C Ammonia
    D Other hazardous fluids, like acids
    E HP steam and BFW lines
    O Other

- COF categories for Urea:

    A Carbamate
    B Ammonia
    C Other hazardous fluids, like acids
    D HP steam and BFW lines
    O Other

The simplified workflow shown in Figure 3, together with an example matrix shown in Figure 4, were used to determine inspection coverage, method, priority, and eventually an interval.

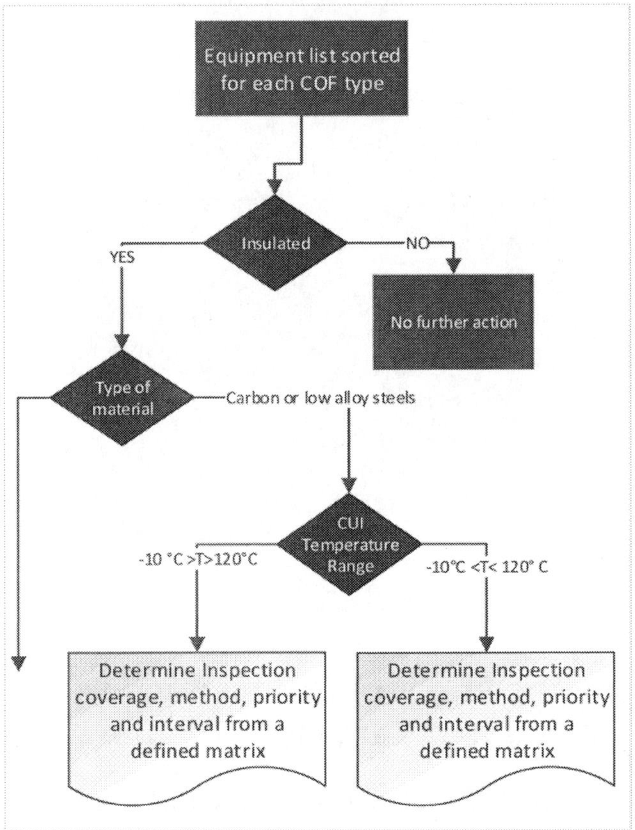

Figure 3. Workflow used for scope definition

Insulated (Y)	COF category (A, B, C, D, E)	Material (CS, SS)	In CUI critical temperature range	Sub-group	Inspection coverage	Inspection Method (Equipment)	Inspection Method (Piping)	Priority
Y	A, B, C, D, E	CS	Yes	1	Equipment and Piping: 100% CUI inspection, including: • 100% main body (if equipment), 100% main line (if piping), 100% supports, 100% small bore connections (DN<50), 100% directly connected dead ends (DN<50)	• Main body: Method A (for large vessels Method D may be applied before Method A) • Equipment connections: Method A	• Main pipe: Method A (Method C, GWUT, may be applied on long run pipes) • Small connections (DN<50, DN<2"): Method A or Method B	1
Y	A, B, C, D, E	CS	No	2	Equipment and Piping: CUI inspection limited to: • 100% supports, 100% small bore connections (DN<50), 100% directly connected dead ends (DN<50) Main body (if equipment) or main line (if piping) not inspected	• Equipment connections: Method A	• Small connections (DN<50, DN<2"): Method A or Method B	2
Y	A, B, C, D, E	SS	Yes	3	Equipment and Piping: 100% CUI inspection, including: • 100% main body (if equipment), 100% main line (if piping), 100% supports, 100% small bore connections (DN<50), 100% directly connected dead ends (DN<50)	• Main body: Method E or F • Equipment connections: Method E	• Main pipe: Method E or F • Small connections (DN<50, DN<2"): Method E	3
Y	A, B, C, D, E	SS	No	4	Equipment and Piping: CUI inspection limited to: • 100% supports, 100% small bore connections (DN<50), 100% directly connected dead ends (DN<50) Main body (if equipment) or main line (if piping) not inspected	• Equipment connections: Method E	• Small connections (DN<50, DN<2"): Method E	4

Method "A" - Visual External inspection with all insulation removed
Method "B" - Profile radiography
Method "E" - Liquid Penetrant (for SS only)
Method "F" - Eddy Current (for SS only)

Figure 4. Example Inspection Planning Matrix

Inspection Execution

Pre-Inspection

The pre-inspection step is required for several reasons; first and most important is to check the general condition of the equipment and lines. Points of attention are:

- Any changes made to the installation which wasn't on the available drawings, like drains and vents, routing of piping, or additional tie-ins.
- Whether the insulation is still installed and the condition of the insulation.

Pre-inspection reports provided the basis for the work management process in Yara's computerized maintenance management system. A pre-inspection report consisted of a list of locations identified for inspection in a given area, with numbered inspection sketches and additional pictures.

An inspection sketch generally consisted of,

- Identified inspection spot including thickness measurement locations
- Reference number of the identified inspection spot
- Location from where insulation has to be removed
- Location where scaffolding is needed

First Inspection: Visual

The inspection locations identified in the pre-inspection report were visually inspected immediately after removal of insulation and before any cleaning of the metal surface. Severity of CUI damage was evaluated in CUI Visual Damage Classes (Table 3) based on condition of the coating and visible corrosion product on surface of the equipment or piping. Depending on the assigned damage class, corrective actions were recommended.

CUI Visual Damage Class	Visual inspection result	Actions
1	Coating still in good condition	Second Inspection (spot check of wall thickness). Reinstall insulation
2	Minor coating damage	Second Inspection (spot check of wall thickness) Repair coating at inspected locations and reinstall insulation
3	Coating completely damaged but no visible wall loss	Second Inspection (spot check of wall thickness) and repair coating at inspected locations Reinstall insulation
4	Corrosion product visible: minor corrosion damage	Clean area for wall thickness measurements before spot check of wall thickness Repair coating at inspected locations and reinstall insulation
5	Huge quantities of corrosion product visible: major corrosion damage	For piping with DN<50 (DN<2") replace the line. All other piping and equipment, remove 100% of the insulation and clean areas for extensive wall thickness measurements

Table 3. Visual damage class and actions

Results of the first inspection were reported on the numbered inspection sketches and supported with pictures before and after cleaning. Results were consolidated in the equipment and piping list discussed earlier.

Second Inspection: UT thickness measurement and corrective actions

Second inspection consisted of wall thickness measurements on the locations identified during the first inspection using manual UT or profile radiography. The number of wall thickness readings were also based on findings of first inspection.

- In case of minor or no visible corrosion damage a localized spot check
- In case of more severe corrosion a scanning of the most affected areas to define the wall thickness profile in order to have a reliable estimate of the remaining life of the equipment or line

In case of severe wall loss, more than the design corrosion allowance, the scope was increased and the cycle repeated. Severity of CUI damage was recorded in CUI UT Damage Class as tabulated in Table 4 below.

UT Damage Class	UT thickness measurement results	Actions
0	No wall loss and no coating damage	No action required
1	No wall loss, coating damage	Repair coating and re-insulate
2	Minor wall loss	Extend inspection locations for further evaluation of the condition the equipment or line. Repair coating and re-insulate
3	Major wall loss (thickness below minimum required thickness according to original design code)	Equipment and main Piping: Extend inspection locations for further evaluation of the condition the equipment or line FFS Level 2 analysis ("Run/Repair" decision) Repair coating and re-insulate Small connections (DN<50, DN<2"): Replace/Repair
4	Severe wall loss	Replace/Repair

Table 4. UT Damage Class and Actions

Future Inspection Intervals

After completion of CUI inspection/repair work, intervals for future CUI inspections are set as per Table 5.

UT Damage Class	UT thickness measurement results	Executed maintenance actions	Inspection interval after maintenance actions
0	No wall loss and no coating damage	None	10 years
1	No wall loss, coating damage	Coating and insulation repaired	10 years
2	Minor wall loss	Coating and insulation repaired	5 years if nominal thickness <= 5mm 10 years if nominal thickness is > 5mm
3	Major wall loss (thickness below minimum required thickness according to original design code)	Equipment and main Piping: FFS Level 2 analysis (Residual Life calculation) for "Run/Repair" decision In case of "Run" decision, coating and insulation repaired; otherwise repaired/replaced Small connections (DN<50, DN<2"): Replaced	If not replaced: minimum between 5 years and Half Life (Residual Life/2) If replaced: 10 years
4	Severe wall loss	Replaced	10 years

Table 5. Intervals for future CUI inspections

Mitigation Measures

A number of changes were recommended to standard work protocols based on the feedback that came from this project. Some of the highlights include:

- Use of thermal insulation for services that required insulation just for personnel protection was stopped
- Improved practices around preparation of metal surfaces, prior to application of coating and the selection of coating for new insulation applications and repairs.
- Use of piping for rigging and support during maintenance activities was abandoned to prevent damage to insulation
- Insulation craftsmanship and caulking methods were reviewed
- Annual insulation surveys by operations team to identify insulation deficiencies to prevent water ingress by repairing insulation damages
- Repositioning of supports and brackets where insulation is replaced and avoiding couplings under insulation lagging
- Begin use of organic coatings on carbon steel equipment and use of aluminum foil for protecting stainless steel

Conclusion

There is no doubt that CUI threatens the safe operation of our installations, but with change in organizational culture that drives continuous improvement using standard work protocols it can certainly be controlled. For sites without an established CUI inspection program a significant initial effort and capital will be required to develop and execute new processes. This case clearly shows that this is an effort worth making.

Yara's approach in dealing with CUI, as detailed in this paper, differs from API recommended practices in some ways. It is Yara's experience that all insulated equipment and lines can be affected by CUI, so having a process that covers both equipment operating within API recommended CUI temperature range and those operating outside should be developed. Furthermore, NDT techniques without removal of insulation give very limited information, hence their use should be avoided especially for the first time when CUI is handled as a project. Finally, CUI is not a trendable degradation mechanism as suggested in API 581. Corrosion rates depend on far too many factors to calculate a reliable remaining life and it is worth mentioning that corrosion rates between 0 and 2 mm/yr (0 – 0.08 in/yr) were observed in this case. A simplified approach for calculation of corrosion rates, as summarized in the paper, will serve good purpose because there are no perfect carrion systems nor coating qualities.

References

1. Yara Guideline 2.06.01 Att 8.1 CUI Management
2. API Recommended Practice 571, Damage Mechanisms affecting fixed equipment in the refining industry
3. API Recommended Practice 580, *Risk-Based Inspection*
4. API Recommended Practice 581, *Risk-Based Inspection Technology*
5. API Recommended Practice 583, *Corrosion under insulation and fireproofing*
6. ISO 9223 Corrosion of metals and alloys — Corrosivity of atmospheres — Classification
7. ISO 9224 Corrosion of metals and alloys — Corrosivity of atmospheres — Guiding values for the corrosivity categories

SAFCO III Ammonia Revamp Project – Plant Start-up and Commissioning

KBR recently executed Basic Engineering for the Reliability Improvement Project (RIP) for Saudi Arabia Fertilizer Company's SAFCO III ammonia plant located in Al-Jubail, Kingdom of Saudi Arabia. Detailed engineering and revamp construction by China Tianchen Engineering Corp (CTEC) was successfully completed in April 2019. The objective of the RIP project was to improve on-stream time and reliability of SAFCO III plant and while also increasing the plant capacity to 1,800 MTPD.

As any other successful revamp project, this endeavor also required interaction with the plant owner's engineering and operation team for defining and implementing the retrofit options with emphasis on process safety and economics. While numerous other small improvements were made to this unit, this paper will share some of the major equipment modifications in detail and note challenges faced during construction, start-up, and commissioning of the modified plant.

Saad Ali Al-Maymoni
SAFCO

Shamik Bhattacharya, Kaushal Bhatt
KBR

Introduction

KBR successfully revamped and commissioned the SAFCO III Ammonia Plant in December 2019. The plant capacity was increased from 1,500 MTPD to 1,800 MTPD (1,650 STPD to 1,995 STPD) accompanied by improvements in plant reliability. This paper provides the roadmap of the project, encompassing revamp features, major equipment modifications, and the execution of commissioning and start-up activities at site.

The SAFCO III Ammonia plant was constructed in 1999 by Tecnimont Arabia Ltd. using the KBR Purifier™.Ammonia Technology. Prior to the Reliability Improvement Project (RIP), the SAFCO III ammonia plant had a nameplate capacity of 1,500 MTPD (1,650 STPD) of 100% cold ammonia product.

KBR performed a revamp study for SAFCO III plant in 2006 which was followed up by a Revalidation Study in 2015. The revamp and reliability improvement features recommended by KBR in the 2006 study remained the same during the course of the revalidation study in 2015. However, some plant operation and maintenance issues were observed during the revalidation study, and improvements for those issues were included in the study. Later in 2015 KBR was awarded the Front End Engineering and Design for the RIP project. After successful execution of the FEED, KBR continued to be involved in post-FEED licensor support to the detail engineering contractor, China Tianchen Engineering Corp (CTEC). The project transitioned to construction in January 2019 which was performed at SAFCO III site under supervision of CTEC in consultation with KBR.

The modified plant was restarted in May 2019 and a 72 hr Performance Guarantee Test Run was performed in December 2019. All the process guarantee parameters were met in the Performance Guarantee Test Run. The process guarantee parameters included Ammonia Production, Ammonia Product Quality, CO_2 Product Quality, CO_2 slippage, Net ISBL Energy Consumption, and NOx content in flue gas at Primary Reformer Stack (summarized in Table 1, later in this paper). The plant has since been operating steadily at ammonia production capacity more than that guaranteed contractually. Also, worth mentioning is the energy consumption of the plant which has been much less than the guaranteed energy number.

SAFCO III Ammonia Process Scheme Features (before RIP)

The SAFCO III ammonia plant in general is based on the KBR Purifier flow sheet, a low energy natural gas reforming process as shown in Figure 1 and utilizes the UOP licensed Benfield solvent for the carbon dioxide (CO_2) removal process.

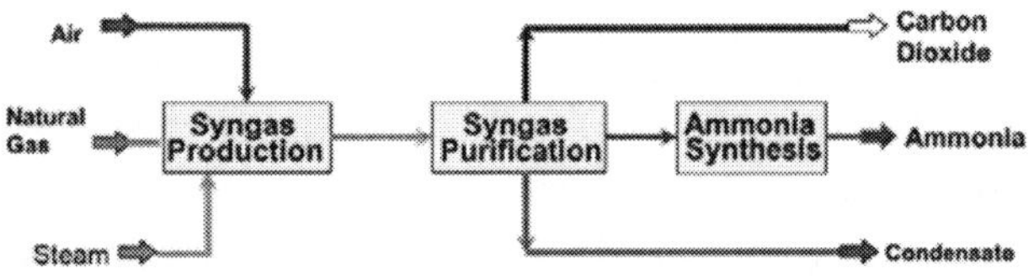

Figure 1. Purifier Process Scheme

The design features of the SAFCO III Purifier plant are:

- Mild conditions for primary reforming via excess air to the secondary reformer.
- Primary reformer with combustion air supplied by the air compressor gas turbine driver and two steam turbine driven induced draft fans.
- UOP's Benfield Lo-Heat solution based process for CO_2 removal
- KBR Purifier for adjustment of synthesis gas hydrogen: nitrogen ratio and removal of all methane and the majority of the argon present.
- Two single bed ammonia synthesis convertors

Description of Revamp and Reliability Improvement Features

Primary Reformer Extension and Convection Section Modification

The radiant section of the reformer had 128 tubes of 155 mm (6.1 in) ID before the RIP. The radiant box was expanded to include an additional row of 32 tubes, making the total number of tubes 160, thus increasing reforming capacity as well as reducing front end pressure drop. The number of burners in the reformer were increased by 40 bringing the total number of burners to 200 thus improving the plant reliability by reducing flame impingement. The convection section was replaced completely. The replaced convection section has nine services as follows:

- Mixed Feed Pre-heat
- Hot Superheated Steam
- Hot Process Air
- Cold Superheated Steam
- Cold Process Air
- Hot Feed Pre-heat
- Hot Boiler Feed Water (BFW)
- Cold Boiler Feed Water (BFW)
- Cold Feed Pre-heat.

During the Revamp Study phase of the project, KBR recommended installation of a new Selective Catalytic Reaction (SCR) bed in between the hot BFW and the cold BFW coils to reduce NOx emissions in the flue gas. However, this feature was not implemented during the project.

The two induced draft fans on top of the convection section were replaced to meet increased flue gas requirements for the higher plant capacity. A forced draft fan was added to supplement the gas turbine exhaust required to burn the increased reformer fuel, primarily during summer months.

Secondary Reformer Replacement

The secondary reformer was a legacy design by C.F. Braun and was in poor condition, leading to its replacement with the KBR design. The secondary reformer is a dual-layer refractory lined vessel with the reformed gas entering from the top of the vessel and the air entering from the side of the vessel via a series of sparger nozzles.

The existing secondary reformer had with a history of refractory failures and repairs due to faulty or improper installation during the initial installation. At least three major repairs were performed before 2015, which made the replacement decision the most reliable solution.

The salient feature of the design is that there is no water jacket or any sort of secondary cooling system. The major method of detecting a hot spot in the secondary reformer is thermal indicating paint applied to the outer surface of the vessel. Metal temperature indicators on the surface of the vessel also provide input to the DCS with an alarm alerting the operator if any issue is observed.

For the new vessel, KBR improved the dual layer design of the secondary reformer by adding a short anchor system for the backup, or insulating, layer. The refractory integrity was improved tremendously by adding more anchors and improving the anchor spacing

During refractory installation continuous and panel casting was used for the backup and hot face. KBR also assisted with the installation of an arched dome with a lower arch angle.

Since startup the new vessel has operated without any major issues. The skin temperature is found to be less than 200°C (392°F) at all locations with design temperature being 300°C (572°F).

Figure 2. View of Primary and Secondary Reformer

Figure 3. Nozzle view of hot and backup refractory

CO_2 Removal System

The existing CO_2 removal system uses the two-stage UOP Low Heat Benfield Process. It consists of a two-stage absorber, a two-stage desorber, a semi-lean solution flash tank, and associated heat exchangers. In consultation with UOP the following changes were implemented in the existing system, for added capacity and reliability improvement.

- UOP Benfield ACT-1 activator was added to increase capacity of the system and to maintain solution circulation rates within capacity limits of existing pumps.
- The existing packing in the CO_2 Absorber's lower section beds was replaced with #3 Raschig Super Rings to increase capacity.
- The existing packing in the CO_2 Desorber's upper section beds was replaced with #3 Raschig Super Rings to increase capacity.
- The existing LP Steam Ejectors were underperforming, so they were replaced to improved system operation and plant reliability.

Addition of 3rd Ammonia Converter in Synthesis Loop

A third ammonia synthesis converter was added to increase ammonia synthesis loop capacity. The new converter increased the final ammonia concentration in the effluent stream, thus increasing conversion per pass and reducing the power consumption for synthesis gas compressor.

The new ammonia converter design follows the C. F. Braun system where an exchanger is not present inside the converter. The basket is a bottle which holds the catalyst inside it. The sweep gas enters through the annulus space between the basket and the pressure shell, and moves to the top of the catalyst bed. After conversion, the gas exits through the bottom nozzle to high pressure steam heat recovery.

This is a true hot wall design as the temperature of the gas is approximately 370°C (700°F). The material of the pressure shell is 2.25Cr-1Mo, ensuring API-941 is satisfied for hydrogen partial pressure and operating temperature. This vessel design has been used in various legacy plants around the world. The catalyst basket is fabricated with 321 stainless steel, ensuring that nitriding is reduced to the highest extent possible.

KBR has approved world renowned fabricators for important vessels like the converter, ensuring fabrication with utmost precision and quality. The selected fabricator also had several KBR inspection and quality hold points with engineering visits to ensure the vessel was built of highest quality.

Replacement of Syngas Compressor Internals

The capacity of the synthesis gas compressor was increased by replacing the bundles in both the low and high pressure cases. This change also helped in improving the reliability of the synthesis loop in the plant.

The speed of the modified Synthesis Gas Compressor was increased to 10,637 rpm from 10,429 rpm.

Increasing the capacity of the existing Synthesis Gas Compressor by changing the low pressure and high pressure cases was evaluated during the process design phase, but not implemented due to prohibitive factors such as high replacement cost and shut down time.

Start up and Commissioning Activities

KBR had the role of consultant and advisor for the commissioning and start-up activities for SAFCO III. This included furnishing advice and recommendations for activities like cleaning the CO_2 removal system, process gas cleaning prior to introduction into the CO_2 removal system and subsequently to the Purifier, dedusting of the

Purifier, etc. The key points for the commissioning and start-up activities are summarized below.

Process Gas Cleaning and Startup of CO_2 Removal Section

KBR recommended the process gas to be clean and free of contaminants during commissioning. The following steps were taken:

- After LTS catalyst reduction and HTS catalyst desulphurization, the LTS bed was de-dusted with process gas passing through the LTS catalyst bed to the flare.
- LTS outlet MOV remained closed during de-dusting to prevent dispersion of catalyst dust downstream.

KBR inspected the process condensate from the raw gas separator bottoms to ensure that the condensate was clear and did not contain any dust or other contaminants. KBR also requested sulfur sample at the HTS outlet to ensure non-detectable sulfur in the process gas before heating up the LTS catalyst.

The plant feed gas rate was increased at intervals of approximately 10% with steam in the system at a ratio between 3:1 and 4:1. The syngas was vented downstream of the Low Temperature Shift Reactor at this time.

SAFCO encountered maintenance issues for the exhaust dampers for the gas turbine driver for the air compressor. The issue was quickly fixed, and the air compressor was put in operation facilitating the start-up activities.

The passivation and vanadation for the CO_2 Removal system was performed in the presence of the licensor, UOP. The Benfield. solution concentration was increased by adding incremental charges of 10 – 12 t/day of ACT-1 activator, until a solution concentration of 26-28% and Vanadium concentration of approximately 0.75% was achieved.

The process gas venting was shifted from downstream of the shift section to upstream of the CO_2 Absorber. The gas was introduced into the CO_2 Removal System in partial increments after completion of the passivation. KBR recommended keeping the slip stream filter and another temporary side stream filter (rented by SAFCO for Start-up) in continuous operation in order to remove any possible dust or other undissolved contaminants in the solution. KBR also recommended performing "Foam Tests" in order to ensure absence of foaming in the system before and after the introduction of process gas.

SAFCO faced some challenges, like mechanical failure of the hydraulic turbine bypass valve and lean solution pump trips. The process gas venting was shifted back to downstream of the shift section. However, the turnover time for the valve and pump repair were quick and solution recirculation was re-established in the CO_2 Removal System. The process gas was charged back after sample analysis for the Benfield solution was checked and verified by UOP and KBR.

Methanator Startup

The methanator catalyst activation was carried out following KBR recommendations, which included:

- Heat up not to exceed 50°C (90°F) per hour.
- Holding the methanator temperature to not more than 330°C (626°F) for activation.
- $CO + CO_2$ slip at the outlet shall be less than 5 ppmv for completion of activation.

The plant load was increased to 70% and was maintained while progressing to starting up the methanation section. The process gas venting was shifted to downstream of the methanator.

Process Gas Cleaning upstream of Purifier

KBR recommended the following in general prior to the gas introduction to the Purifier / Cold Box.

- Air blowing the mole sieve catalyst and piping with high velocity up to Battery limit of the Cold box. This was intended for removal of dust and scale from core exchanger of cold box.
- De-dusting Molecular sieve dryers up to inlet isolation valve of Purifier cold box without having the filter elements in the dryer outlet filters.
- Repeated air blowing to be carried out until the air exhausting from the blow point is visually clean and particle free.

Purifier Cold Box Cleaning

There was no change in the cold box section during revamp. KBR recommended blowing the section from process gas inlet to expander inlet before start-up, in case of contamination due to physical activity done inside the cold box. The following steps/ guidelines were proposed to be considered for air blowing.

- Bursting a temporary rupture disc installed at suitable location in the process feed gas inlet piping.
- Blowing clean, oil free, dry air or nitrogen supplied at 7 kg/cm^2 (g) (100 psig).
- Isolate all other equipment, piping, and instrumentation which were not part of the section considered for blowing at one time.
- Repeat air blows until the air exhausting from the blow point is visually clean and particle free.

The above recommendations were followed in general and inspections of filter elements were made (for the filters at the mole sieve drier outlets). Some catalyst dust were found during the inspections which had accumulated during the catalyst de-dusting following the recommendations mentioned above.

The plant load was increased to 76%. The process gas venting was shifted partially to the suction of the syngas compressor.

The dew point and moisture content of the process / synthesis gas at molecular sieve dryer outlet were monitored. The above-mentioned data were checked and verified before introducing the synthesis gas in the synloop.

The Purifier cool down was initiated by introducing process gas inside the Cold Box and starting the Purifier expander. Some issues were recorded relating to pressure drop increasing in the Purifier core exchangers, and the Purifier had to be isolated and put in deriming mode. KBR recommended modifications to the regeneration cycle (e.g. increasing the heat soak time during heating) for the molecular sieve dryers.

Synthesis Loop Start-up

The synthesis gas compressor had issues with trips based on high vibration during synloop start-up. The issue was resolved in consultation with the compressor vendor, MHI, and recommendations from KBR.

After the compressor start-up, the synloop pressure was increased onwards from 45 kg/cm^2 (g) (640 psig) at minimum governor speed. KBR recommended increasing the pressure at 7-10 kg/cm^2 (100-142 psi) per hour.

The reduction of the ammonia converter catalyst was initiated in the presence of the catalyst vendor, Haldor Topsoe, using the natural gas fired start-up heater. The hydrogen to nitrogen ratio was maintained in the range of 2.6 – 2.8 at the time of reduction of the catalysts. There was some issue with insulation and leaking packing of a synloop valve which had to be fixed.

The first drop of Ammonia was produced towards the end of the first week of May 2019.

SAFCO had some issues with the plant operation during and post plant start-up. The issues included the following:

- Level fluctuation in the high pressure steam drum.

- Sticking of a flow control valve at the 3rd Ammonia Converter outlet.
- Increasing pressure drop in the core exchangers in the Purifier
  - The Purifier had to be taken out of operation a few times based on the increasing trend of moisture content and decreasing dew point readings in the Molecular Sieve Dryer outlet stream.
  - The moisture carryover was observed to be significantly higher for one of the dryers (B) while it was in operation with respect to the other one (A).
  - The plant capacity was restricted to 70% of the nameplate capacity while operating in Purifier bypass mode.

SAFCO requested that KBR diagnose the above-mentioned issue. KBR investigated the issue closely and prepared a list of recommendations and checking action points to resolve the issue of moisture ingress to the Purifier:

- KBR performed detailed checking of the plant equipment and operation (upstream of the Purifier), including analyzing the plant data to check if the problem originated from any equipment / area upstream. This included the following equipment:
  - Methanator Effluent Exchanger and Ammonia Refrigerant Chiller
  - Methanator Effluent Cooling Water Exchanger
  - Methanator Effluent Separator
  - Piping inlet and outlet of the Molecular Sieve Dryers

After performing the above checking and analysis the following issues were diagnosed:

- The internals for the Methanator Effluent Separator were not meeting the water removal specification at the separator outlet.
- The inlet piping at dryer inlet was sloped towards dryer B. Thus, the liquid water carryover from the methanator effluent separator showed up in the desiccant of dryer B both during operation and regeneration modes. This caused the moisture carryover to the Purifier to increase significantly when Dryer B was in operation.

KBR recommended replacing the internals for the Methanator Effluent Separator. SAFCO followed the recommendation and executed the procurement and installation of the new internals on a high priority basis. This required the plant to undergo a shutdown intermittently in order to implement the repairs. However, the issue was resolved completely after the installation of the new separator internals and the nameplate capacity of the plant was achieved shortly thereafter.

Performance Guarantee Test Run

The plant performance was closely monitored for few weeks by SAFCO and KBR post start-up. It was observed that the plant consistently continued achieving capacity more than the nameplate. KBR requested the plant data from SAFCO, and the analysis of the data provided confirmed the increased plant capacity achieved and low energy consumption. SAFCO and KBR mutually agreed to schedule the Performance Guarantee Test Run (PGTR) for the plant.

KBR arrived at site during agreed time frame and the PGTR procedure was discussed and agreed between SAFCO and KBR. The PGTR was executed for Three (3) days / seventy-two (72) hours. The required plant data was collected during the above-mentioned time frame and analyzed by SAFCO and KBR on site. The results were mutually agreed upon, accepted and signed by SAFCO and KBR. Table 1 summarizes the results from PGTR:

Items	Guarantee Values	Measured 3-Day Average
Ammonia Production*	1800 MTPD (1984 STPD)	1809 MTPD (1994 STPD)
NH_3 Product Quality Ammonia Water Oil	 99.9 wt.% min 0.1 wt.% max 5 ppmw max	 99.96 0.04 < 1 ppmw
CO_2 Product Quality CO_2 mol% (dry basis) (H_2+Inerts mol%) (dry basis)	 99.0 min 1 max	 99.19 0.81
CO_2 slippage, ppmv	1000	1000
Energy, LHV basis	7.50 Gcal/MT (27.0 MMBtu/ST)	7.31 Gcal/MT (26.3 MMBtu/ST)
NOx in Flue gas, ppmv	75	69

Table 1. Guarantee and PGTR Results

Lessons Learned

1. Close teamwork between Owner, Contractor, and Licensor team is a success key for smooth and safe Commissioning and Start-up.
2. Resources, tools, or facilities for commissioning should be identified and prepared during engineering stage as much as possible. This helps in avoiding extra downtime and schedule delay.

Conclusions

The SAFCO III plant start-up experience highlights the importance of thorough plant pre-start-up checkout, and effective commissioning practices.

Despite the challenges faced, the plant successfully met all its energy consumption, production, and product quality guarantees.

References

KRIBHCO, India & KBR: "Conventional Ammonia Plant Revamp with Purifier", AIChE 2013.

Project Records, including daily, and weekly minutes of meetings and status reports

Safe Operations of Low Temperature Shift Reactors Under Harsh Conditions

Selecting a highly active and stable catalyst for the Low Temperature Shift (LTS) converter is of high importance when considering the potential harsh operating conditions it may work under. Under the risk of condensation (wetting), sulfur poisoning, and high temperatures (due to poor temperature control), a high-quality LTS catalyst is essential to ensure proper catalyst operation without any loss of production.

Clariant ShiftMax® 217 was developed with enhanced activity to allow higher conversion at low temperatures, high stability to mitigate the risk under upset conditions and finally an excellent crush strength in the oxide and reduced state for overall robustness.

This paper summarizes several case studies where ShiftMax® 217 achieved excellent performance under harsh conditions. This paper discusses how OCI Nitrogen and OCI Iowa Fertilizer Company have benefited from the stability and high activity of Shiftmax® 217.

Keegan Dunn
OCI Iowa Fertilizer Company

Andy Vluggen
OCI Nitrogen

Christoph Krinninger
Clariant Corporation

Introduction

OCI is a major ammonia producer operating eleven ammonia production sites around the world. This paper will discuss two plants in particular; OCI Nitrogen in Geleen, the Netherlands, and OCI Iowa Fertilizer Company in the USA. Both of these ammonia plants use Clariant ShiftMax® 217 in their Low Temperature Shift (LTS) reactors. Both plants experienced water exposure events with their LTS catalysts, with the Iowa unit also experiencing high inlet temperatures and sulfur ingress. What follows are the details of those events and the safe recovery at both facilities.

OCI Nitrogen

Ammonia plant 3 (AFA-3) of OCI Nitrogen in Geleen, the Netherlands, had been operating

ShiftMax® 217 since 2012. In November 2015 an incident occurred that required the immediate shutdown of the plant. The focus of operations was stopping the plant as soon and safely as possible. Therefore, the normal shut-down procedure could not be used and the High Temperature Shift (HTS) and LTS reactors were taken out of operation in a non-standard manner (no circulation with nitrogen and direct depressurization of the gas make up section). Operations was able to minimize the ingress of oxygen, but condensation of steam could not be prevented, resulting in severe wetting and partial oxidation of the HTS and LTS catalysts.

During the shutdown period, condensate was drained that had accumulated in the inlet and outlet lines of the HTS and LTS. However, it was not possible to drain all the water, because of the piping layout. This meant that any remaining water in the lines would end up going into the reactors upon restart, thereby wetting the catalysts. Wetting of the catalyst can lead to a thermal shock on the catalyst surface, which can mechanically damage the catalyst and lead to an increase in pressure drop.

OCI Nitrogen Safe Catalyst Recovery

To recover the wetted HTS and LTS catalysts, Clariant advised OCI Nitrogen to use a slow and well-controlled dry-out procedure for both shift reactors to minimize the water vaporization within the pores of the catalyst pellets which can cause physical damage. The following dry-out procedure has been used (Figure 1):

- Before startup, the inlet and outlet lines of the HTS and LTS were drained.
- Nitrogen circulation was started at low pressure to slowly remove moisture from the catalyst pores.
- The drain ports were left partially open during the nitrogen circulation until water was no longer present.
- The catalyst beds were heated up in nitrogen circulation with a rate not greater than 30°C/hr (54°F/hr). The inlet temperature was held at the point where the catalyst bed temperatures did not follow the inlet temperature anymore. This was an indication that the dew point was reached. The inlet temperature was maintained 10°C-15°C (18°F-27°F) above the estimated steam dew point for approximately 12 hours to ensure the slow vaporization of the liquid from the catalyst pores.
- After this period, the heat up continued at a rate of 30°C/hr (54°F/hr) up to 225°C (437°F), at which point the reactor was ready for the normal start-up procedure. When the recommended temperature was reached, the LTS reactor was bypassed until process gas was introduced.

As can be seen in Figure 2 there is a slight decrease in the conversion of the LTS catalyst after the incident, which is caused mainly by the lower inlet temperature. As seen in Figures 3 and 4, CO slip and pressure drop before and after the incident are stable and low as desired.

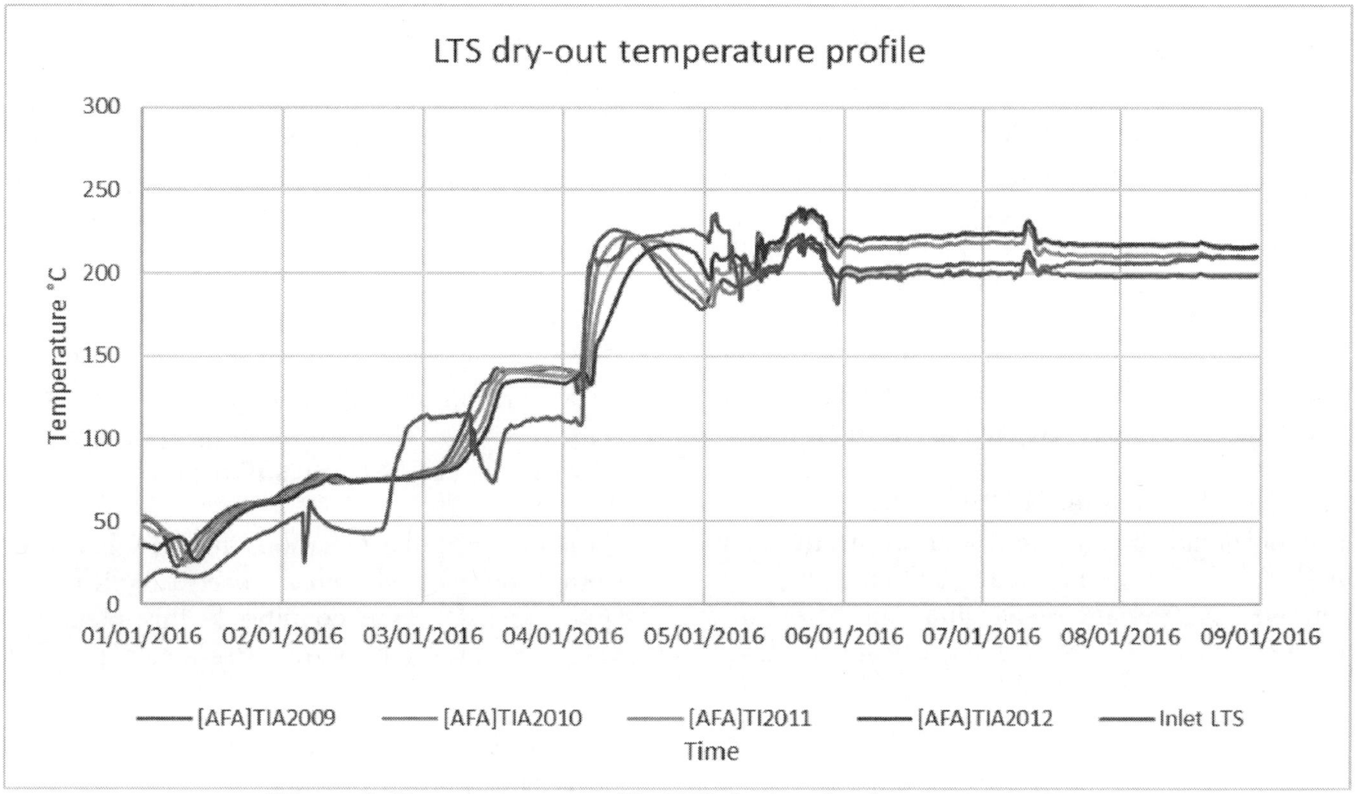

Figure 1. Temperature profile LTS during the dry-out procedure

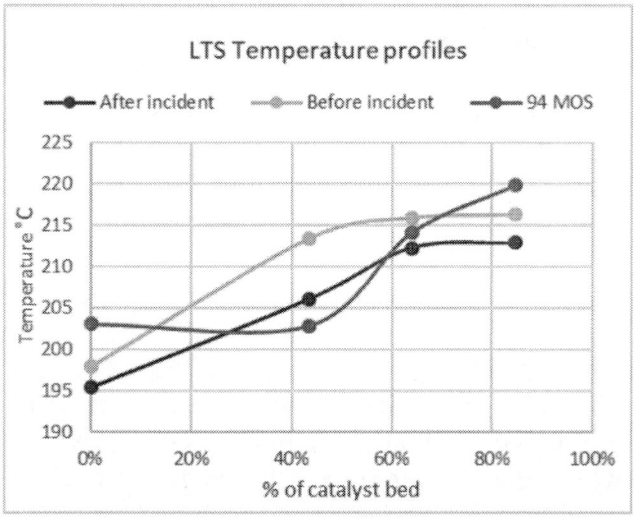

Figure 2. LTS temperature profiles AFA3

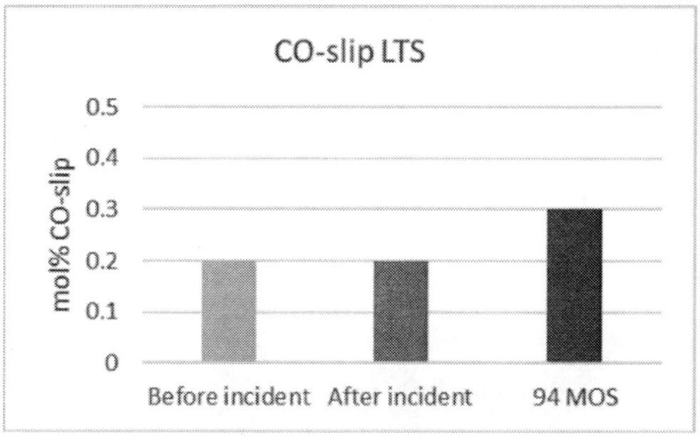

Figure 3. CO slip outlet LTS AFA3

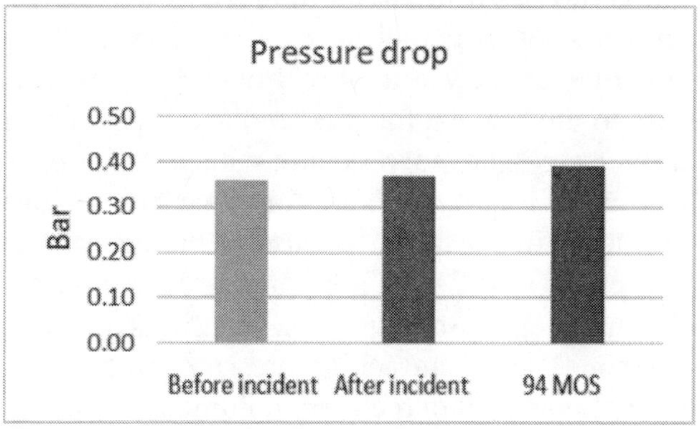

Figure 4. Pressure drop over LTS reactor AFA3

OCI Iowa Fertilizer Company

The OCI Iowa Fertilizer Company (IFCo) operates a world-scale ammonia plant, based on the KBR purifier design, that was commissioned in the first quarter of 2017. During commissioning, the isolation valves for the LTS did not function properly, and often either the inlet or the outlet isolation valve would not close when directed to do so by the control system. This caused more than a few instances where the LTS catalyst (Shiftmax® 217) was exposed to steam or liquid.

Due to manufacturing and preservation issues, the upstream boiler feedwater preheater and boiler (103-Cs) were fouled and did not properly remove heat from the HTS effluent gas before introduction to the LTS. Because of this, the LTS was operated with an inlet temperature at or above that of typical end of run conditions from the start of operation. The inlet temperature exceeded 232°C (450°F) for 7 to 8 months.

The lack of heat removal also meant that during plant startups heat to the LTS could not be effectively controlled and the LTS would experience several temperature swings above 260°C (500°F) until the plant steadied out and ran routinely with an inlet temperature of 235°C (455°F).

Once repairs were made to the 103-Cs, the inlet temperature was lowered to 209°C (408°F) near the end of run conditions.

Complicating the issue was the performance of the Hydrodesulfurization unit (HDS). The HDS converts organic sulfur in the incoming gas stream to H_2S by adding hydrogen from the process to the raw natural gas feed and then passing it over a cobalt molybdenum catalyst. The plant sources of hydrogen to use for this purpse: a low-pressure stream to the suction of the natural gas compressor (102-J), a high pressure stream from the first-stage discharge of the synthesis gas compressor (103-J), and a high pressure stream from the main ammonia recovery column (124-D).

During the commissioning phase, the 103-J discharge stream was damaged in a transmitter fire, and the 102-J stream could not be used as the 102-J was out of operation. The ammonia recovery unit was the only stream that could be used, but due to design issues, the stream was unreliable at best.

Due to these issues, the hydrogen flow to the HDS was sporadic. Although the concentration of organic sulfur in the natural gas the plant receives is low, over time sulfur poisoning of downstream catalysts can still be a concern.

To summarize the situation, the LTS had been wetted, overheated, and exposed to sulfur. Despite these extreme conditions, this plant was able to operate at 115% of nameplate capacity.

OCI Iowa Fertilizer Company Safe Handling of Repairs

The repairs at IFCo were not straight forward but over time they were corrected to the best of the facilities ability, without a major outage. The LTS isolation valves were removed, rebuilt, and their gearboxes were replaced. This repair corrected the issues with steaming at shutdowns or startups. A redesign project for the ammonia recovery system was undertaken which allowed hydrogen to be consistently supplied to the HDS reactor.

The issue of the 103-Cs was one that could not be solved without a major outage. However, working with Clariant and others, boiler feed water in the plant was balanced such that the inlet to the LTS was at an acceptable temperature during stable operations. Clariant was able to work with in-house technical resources and ensure that even with the damage that occurred to the catalyst at IFCo would be able to run until the scheduled 2019 turnaround.

Figure 5 shows the history of the LTS performance. The dark yellow section indicates the period of highest potential sulfur poisoning while the light yellow shows the highest temperature operation, sometimes over 238°C (460°F). It can be seen that the catalyst still operated for 20 months before it began deviating from equilibrium. When the catalyst was replaced the CO slip was low enough that the methanator temperature was acceptable.

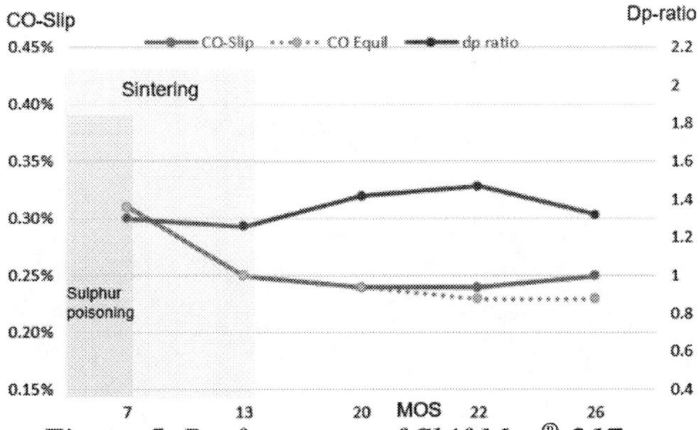

Figure 5. Performance of ShiftMax® 217 at OCI Iowa Fertilizer Company

At the same time, the pressure drop of the ShiftMax® 217 remained stable and within acceptable parameters. Figure A1, in the Appendix, shows the rate of deterioration based upon the thermocouple readings from the LTS.

Figure A1 also indicates that the repairs to the system slowed the rate of degradation. Based on initial estimates, the predicted end of run performance would be reached by November 2018. After the modifications and repairs, this was re-evaluated and extended to beyond November 2019 which allowed the plant to operate until the scheduled turnaround date during the summer of 2019.

OCI Iowa Fertilizer LTS Turnaround Findings

The LTS catalyst in the IFCo plant was replaced during 2019 turnaround, and replaced with another charge of ShiftMax® 217.
When the LTS vessel was inspected it was discovered that the catalyst had settled over 4 ft (1.2 m), which caused a collapse of the hold-down grid. A typical amount of settling would have been about 1.5 ft to 2 ft (0.4 m-0.6 m) for this reactor. The amount of settlement is shown in Figure A2 and the collapsed grid is shown in Figure A4.

The hold-down grid was installed above the thermowell support, as the thermowell support was installed lower than the installation drawings called for, as seen in Figure A3. Even so, this should have been corrected during the initial loading, but was not. This discovery highlights the importance of having a quality catalyst services company as well as the support of your catalyst vendor during installation, to help avoid instances such as these.

There was a minimal amount of dusting found for the number of cycles that the LTS catalyst had been subjected to, and only the very bottom catalyst layer showed signs of high temperature damage. The top hold-down grid screen, also showed signs of charring.

Conclusion

The lessons learned for the Geleen plant show that damage to wetted LTS catalyst can be minimized by using a proper dry out procedure before restarting production. The ShiftMax® 217 catalyst has shown itself to be robust, even in case of a severe wetting incident, hence no early changeout was necessary.

The IFCo plant, by partnering with Clariant, was able to operate its ShiftMax® 217 LTS catalyst under even more severe conditions for 2.5 years and operate until a scheduled turnaround. The Iowa plant was able to operate wetted catalyst above typical temperature ranges at rates exceeding 115% of the nameplate capacity for longer than anyone expected.

These incidents demonstrate that the combination of a robust and highly active catalyst, such as ShiftMax® 217, and close cooperation between operating plant and the catalyst vendor leads to safe plant operation and significant savings due to avoided downtime.

Appendix

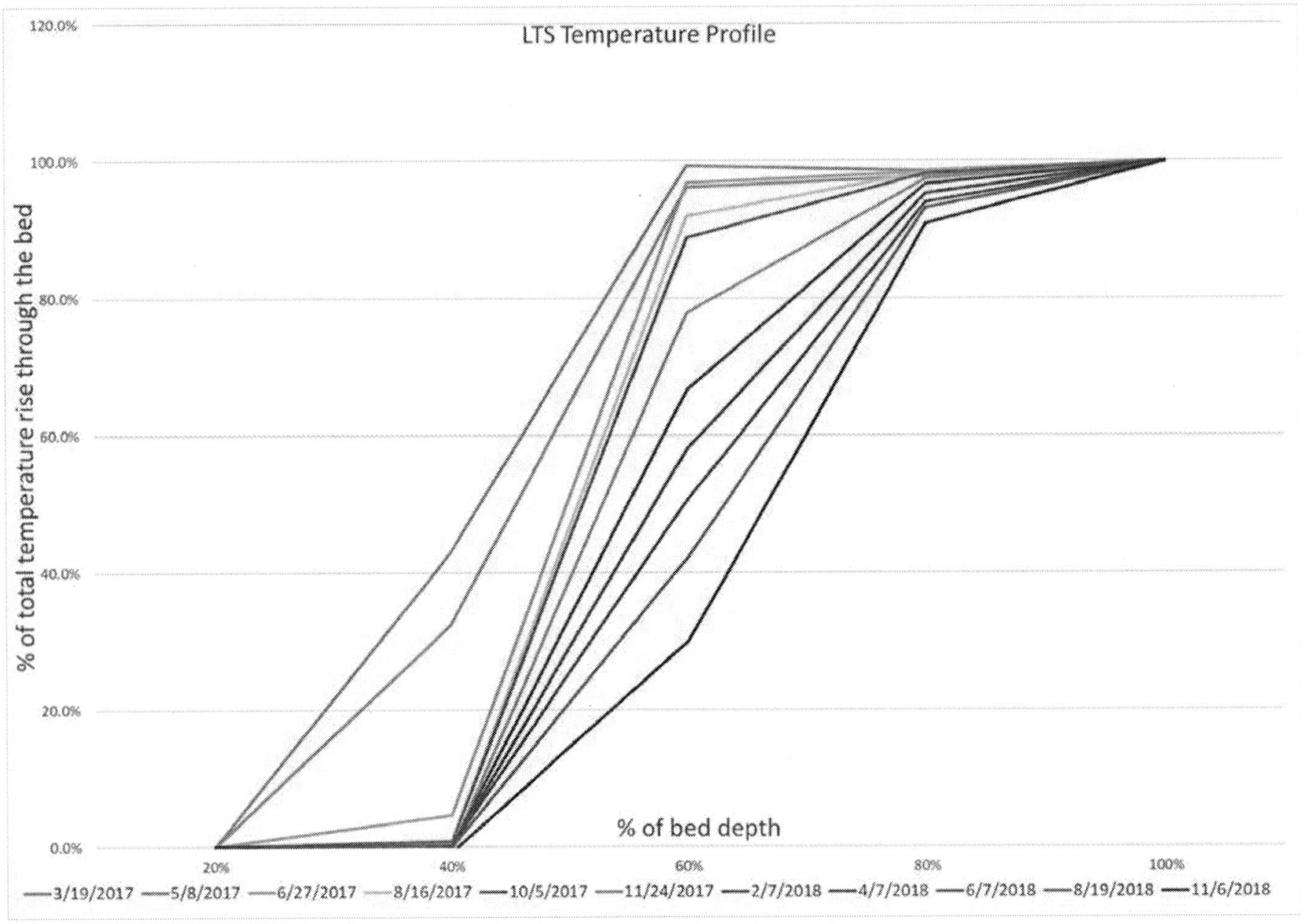

Figure A1. LTS Temperature Profile Shift Over Time

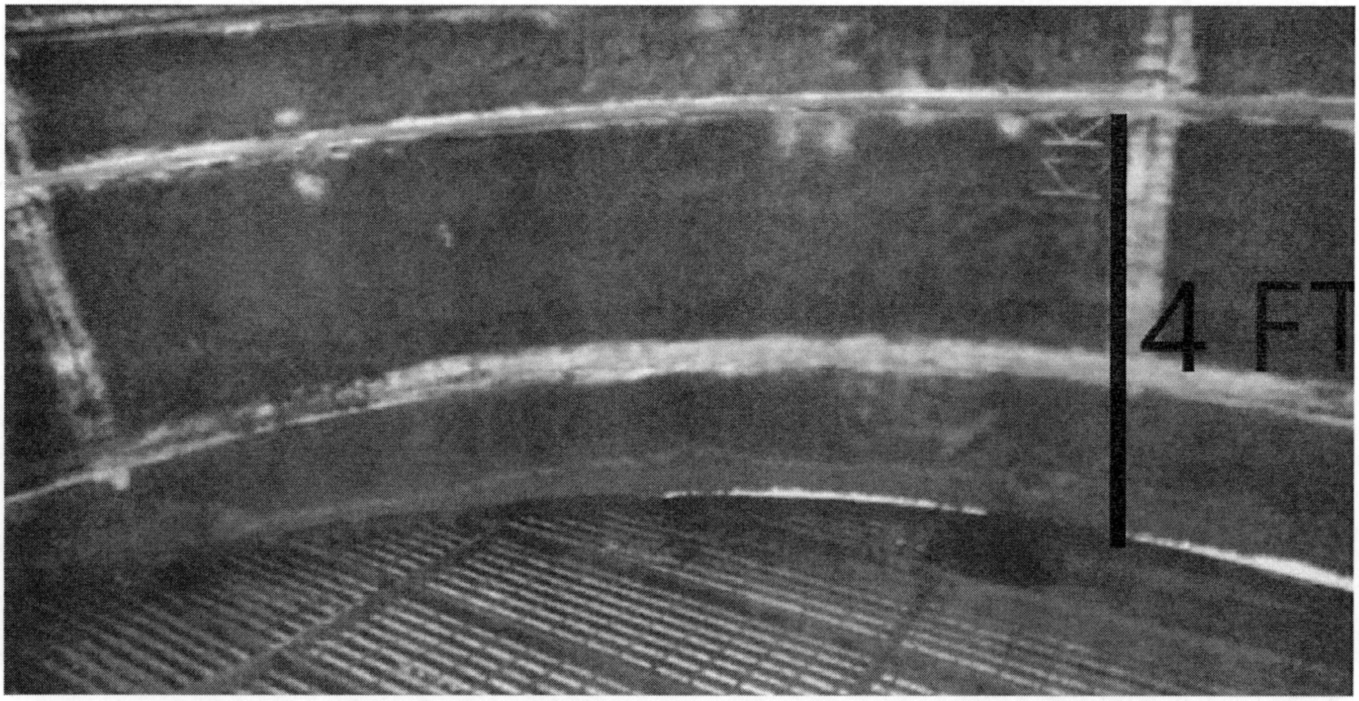

Figure A2. Original outage markings showing original hold down grid location

Figure A3. Thermowell support holding grid up

Figure A4. Breakover point of hold down grid

Experience with Repair and Replacement of the Second Ammonia Converter (R-0502) in the Ammonia Plant-1A of Pupuk Kaltim Fertilizer

Plant-1A is one of the plants in Pupuk Kaltim Fertilizer Company. It was built in 1999 and has 2000 MTPD ammonia production capacity, using Topsoe technology with two Ammonia Converters. The Second Ammonia Converter, R-0502, experienced two failures in November 2017 and February 2019, in the form of cracks on its bottom head. In July 2020 the Second Ammonia Converter unit was replaced. This paper explains the causes of failure and methods employed for repair, and the experience replacing the converter. Furthermore, the performance of the plant afterwards will also be elaborated on in this paper.

Anindito Priyambudi, Yudhistira Perdana Putra, Majus Luther Sirait, Akhmad Sajidin

PT Pupuk Kalimantan Timur

Introduction

In November 2017 the Second Ammonia Converter (R-0502) in Ammonia Plant 1A of Pupuk Kaltim suffered a crack in the bottom head. The plant (and converter) had been in service since 2000. The decision was made to replace the converter with a new unit, and repairs were made to the existing reactor. The vessel repairs began on January 11, 2018 and were completed on March 13, 2018, 62 days later.

After about 8.5 months, on February 21, 2019, a similar problem occurred. The second repair program took 83 days, starting from March 4, 2019, to May 25, 2019, with some improvements from the first repair. After that, the plant was able to continue running until the replacement converter was installed over 43 days beginning June 15, 2020.

This paper describes the method for the second repair program, the vessel replacement experience, and plant operating conditions afterward.

Plant and Equipment Description

Pupuk Kaltim is a subsidiary of PT Pupuk Indonesia Holding Company and currently has a production capacity of 2.98 million tons of Urea per year, 1.85 million tons of ammonia per year, and 350 thousand tons of NPK per year. Pupuk Kaltim also produces organic fertilizer with a capacity of 45 thousand tons per year.

Ammonia Plant-1A is one of the five Ammonia Production Units owned by PT Pupuk Kalimantan Timur. Ammonia Plant 1A began commercial production in February 2000 and has been in operation for over 17 years with ammonia production capacity of 660,000 t/yr (730,000 ST/yr) and urea production capacity of 570,000 t/yr (630,000 ST/yr) urea.

The Haldor Topsoe ammonia technology is used with two Ammonia Converters in series, namely the Primary Ammonia Converter (R-0501) and the Secondary Ammonia Converter (R-0502). A diagram for the Synthesis Loop showing both Reactors is shown in Figure 1.

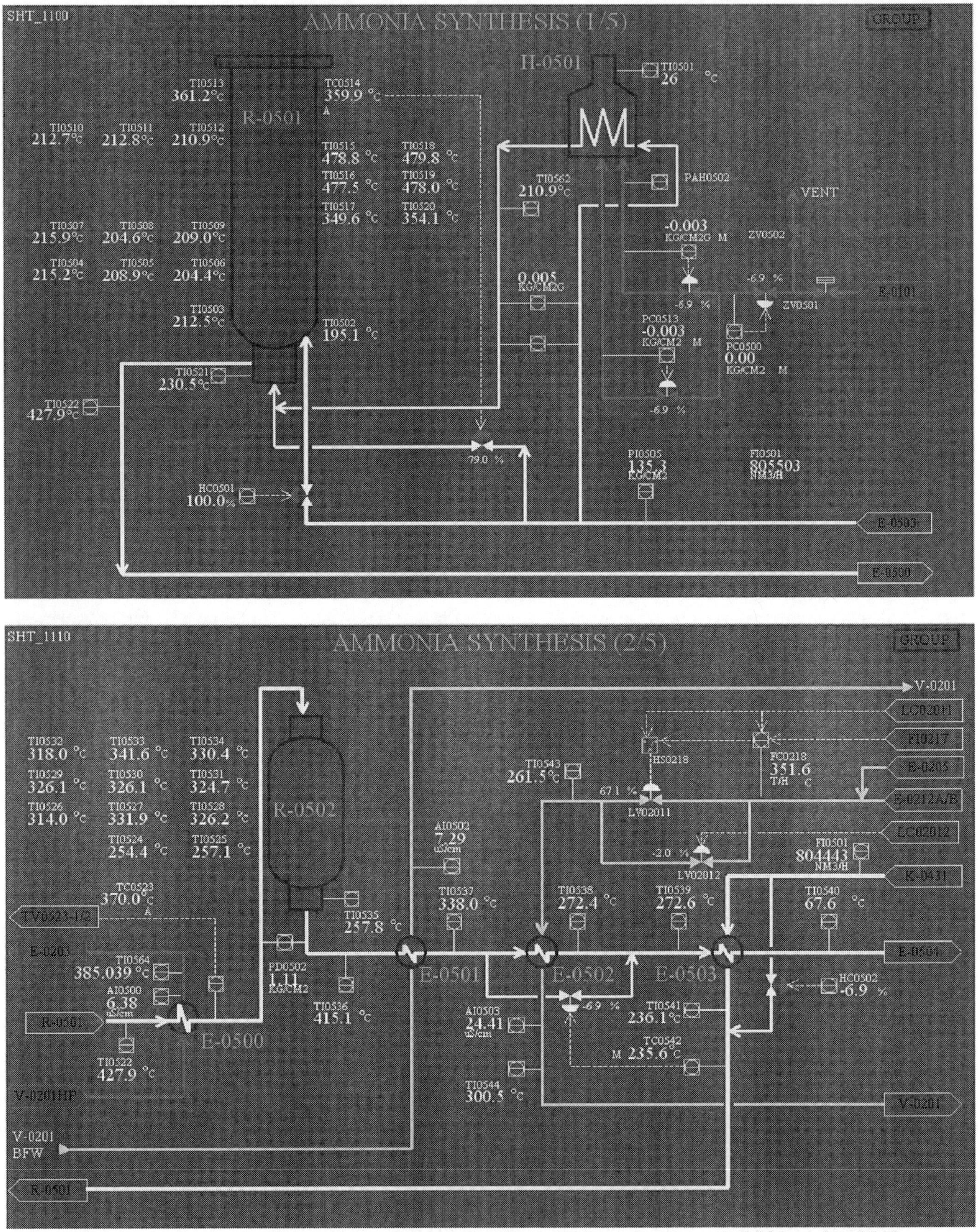

Figure 1. Ammonia Plant-1A Synthesis Loop System Diagram shows R-0501 and R-0502.

The technical datasheet for the Secondary Converter R-0502 is shown in Table 1.

Design Code	AD-Merkblatter
Service Fluid	Synthesis gas
Operating / Design Pressure	135 barg (1957 psig) / 152 barg (2204 psig)
Operating / Design Temperature	376-420°C (708-788°F)/ 450°C (842°F)
Catalyst Volume / Type	137.6 m^3 (4900 ft^3) / Topsoe KM1
Material of Vessel	SA-542 Type B CL4 (see comment)
Material of Center Screen	INCONEL
Material of Outer Screen	321 SS
Diameter	3000 mm (9.8 ft)
Inner Length	21,930 mm (71.9 ft)
Shell Thickness	119 mm (4.7 inches)
Bottom Head Thickness	85 mm (3.3 inches)

Table 1. R-0502 Technical Data Sheet

First Bottom Head Crack and Repair in 2017

In 2017 a crack developed in the bottom head of the R-502. A repair was performed from outside of the vessel, so the catalyst was not removed. Before crack removal, hydrogen removal at the bottom head was performed at 400°C (752°F) for 24 hours. The crack was then removed by hot gouging and hot grinding at 200°C (392°F). The repair area was then beveled and root pass welding using the gas tungsten arc welding (GTAW) process with ER-90S B3 filler metal, with 200°C (392°F) preheat temperature. The weld was then filled until capped using the shielded metal arc welding (SMAW) process with electrode E-9018 B3 and post heat treatment at 400°C (752°F) for two hours.

The vessel material was 2.25 Cr-1Mo-V, so post weld heat treatment (PWHT) process is required at 670°C (1238°F) for 3.5 hr, per the initial weld procedure specification (WPS) from the manufacturer. Local PWHT was performed from outside the vessel. The implementation of PWHT is divided into three areas, due to a finite element analysis (FEA) which mentioned that the equipment could collapse if the PWHT was done simultaneously.

Second Bottom Head Crack in 2019

R-0502 is a radial flow converter use to convert synthesis gas to ammonia at high pressure and temperature using a catalyst. The Converter was built with SA 542 Type B Class 4, a 2¼Cr-1Mo enhanced material. This type of material has high strength, so the vessel thickness and weight can be reduced, but it makes the material more prone to cracking. [1]

The R-0502 developed a crack in the bottom head which was repaired in 2018. The Converter had a second crack failure in February 2019 in the same location. A crack failure was found that penetrated from the internal catalyst support welds to the outer surface of the bottom head. The reactor was in operation approximately 15 yr before the first failure and 8.5 months after the first repair.

After the plant was shut down, an inspection was immediately performed. From the results of the visual inspection and penetrant test, an open crack was discovered on the bottom head in the area of the first external repair weld from March 2018. The crack length was about 150 mm (5.9 in), as shown in Figure 2.

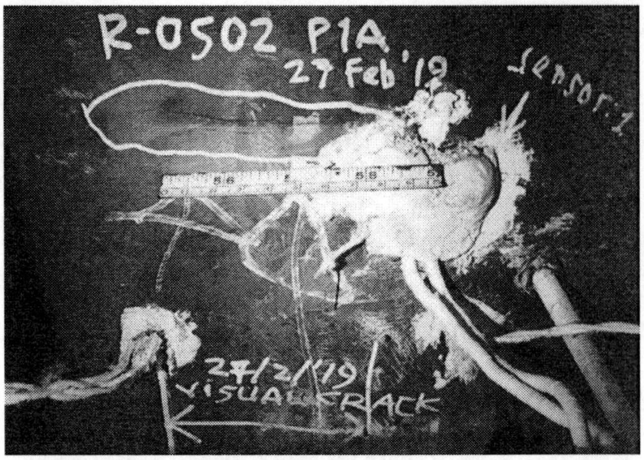

Figure 2. Leak at bottom header due to cracks with the length of crack about 150 mm (5.9 in)

Further inspection using phased array ultrasonic testing (PAUT) determined that the cracks occurred in weld metal from the March 2018 repair, and a new crack location was found. Crack location details can be seen in Figure 3.

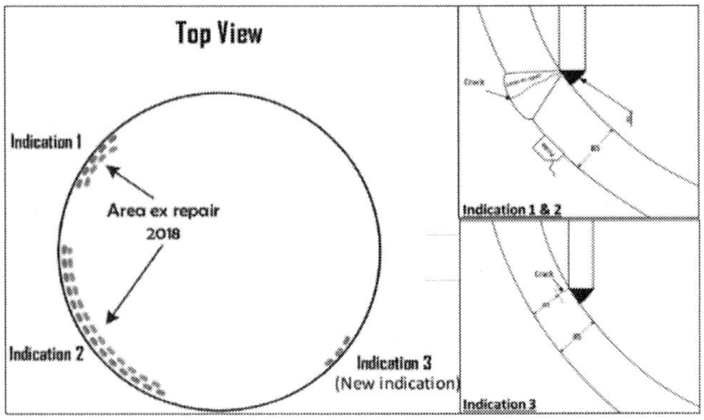

Figure 3. Sketch of location of cracks on the bottom head

The cause of failure is the initiation of cracks in the catalyst support weld to the bottom head. Hydrogen diffuses and accumulates critically at the initiation of the crack and causes embrittlement of the material due to hydrogen embrittlement (HE). Embrittlement causes a decrease in fracture toughness up to 50%. [2]

Since the replacement equipment ordered after the initial failure had not yet arrived, the vessel was repaired again. There were many challenges during the repair, including repairing the shell under an inert atmosphere, because the ammonia synthesis catalyst was still in place and the vessel was under nitrogen blanket. This presented significant complications to the welding process. Repairs could only be performed from outside areas and staged PWHT techniques were required to maintain the mechanical integrity of the equipment to prevent it from collapsing and improve the repair process to keep the equipment from failing again after repairs.

In the course of the inspection and repair after the second failure, an interesting phenomenon was discovered. Cracks were identified in all zones of the first repair, and new cracks appeared in zones that had not been repaired. During the crack removal process, a transversal crack was found and was very difficult to eliminate. Delayed cracks were also observed during a repair. Consequently, the repair took 83 days, compared with 62 days for the first repair. Transverse cracks are difficult to eliminate by ordinary grinders, so pen grinders were used. Photographs of the transverse cracks can be seen in Figure 4 and Figure 5.

Figure 4. Circular and transverse cracks on the bottom head after grinding.

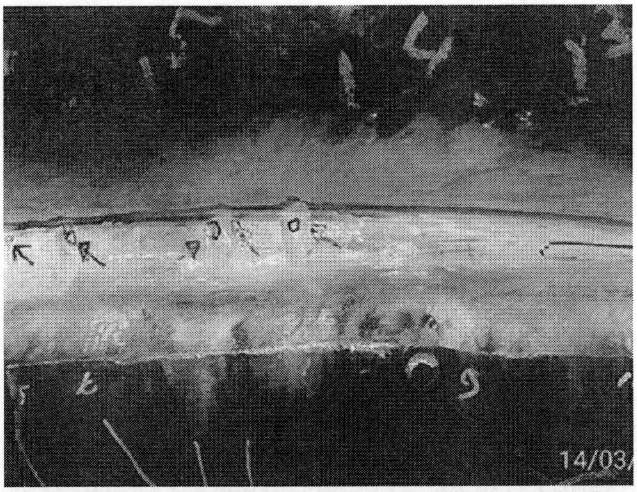

Figure 5. Transverse cracks were removed after using the pen grinder.

Second Bottom Head Crack Repair

Since the first repair was effective for only about eight months, there a need to improve the repair procedures. The method of the first repair (2018) to the second repair (2019) is generally similar, with the exception of several changes to the method of repair, including:

1. **The method of welding.** The 2019 repair method used welding by a tempered bead welding (TBW) technique, while in 2018, only conventional welding was used.
2. **The selection of the welding electrode.** During the 2018 repair, a standard type of E-9018 B3 welding electrode was selected. At the beginning of the second repair, standard type E-9016 B3 was used, but cracks appeared on the weld metal. Then a special electrode E-9016 B3 was used. The special type of E-9016 B3 is an electrode with low hydrogen flux, and the chemical composition of the weld metal is limited by the embrittlement temper factor (X-factor), $X \leqslant 15$, $Cu \leqslant 0.20\%$ and $Ni \leqslant 0.30\%$, which can reduce the risk of cracking.
3. In the 2019 repair, the preheating temperature was raised to 250°C (482°F), compared to only 200°C (392°F) in the 2018 repair.
4. Use non-destructive testing (NDT), PAUT and penetrant testing (PT) to monitor for delayed cracking. In 2019, delayed crack monitoring was carried out for a minimum of 48 hours, each after weld repair and after PWHT. During the 2018 repairs it was not done at all.
5. In 2019, local PWHT was carried out twice, when the weld metal thickness reached 50% (43 mm, or 1.7 in) and when after the weld metal thickness was 100% (85 mm, or 3.3 in). Whereas in 2018, PWHT was carried out only once, when the thickness was achieved at 100% (85 mm or 3.3 in).
6. The PWHT temperature was increased from 670°C ± 5°C (1238°F ± 9°F) to 695°C ± 5°C (1283°F ± 9°F) after conducting a vessel mechanical integrity study in the FEA modelling simulation.

The repair with two stages of PWHT, once at 50% thickness and once at full-thickness (with the preferred welding rod), was successful as indicated by good hardness (max 222 HB). However, problems were again encountered when cracks were discovered in 19 new locations after PWHT due to the use of standard E-9016 B3 welding wire. After using the special type E-9016 B3, the crack did not appear again even after preheating or after the PWHT.

Finally, after 14 months of operation, no failures occurred. Even during the examination by PAUT before the new equipment was installed, no defects were found. So it can be concluded that the repair method was successful.

Discussion

The PWHT: The PWHT process during the first repair in 2018 was likely not sufficient because the PWHT temperature was only 670ºC (1238ºF). When compared to the standard, ASME Section VIII Division 1, the minimum temperature is 675ºC (1247ºF). In addition, the local PWHT was applied from the outside only, so the metal temperature on the inner side did not reach the minimum required temperature.

Hydrogen Embrittlement: When pressure vessels operate in hydrogen service with high pressure and temperature, hydrogen can diffuse into the weld metal repair if the PWHT is not adequate. In addition, the bottom head supports the vessel's skirt, with a high load of 120 MT (132 ST). This resulted in cracks appearing as both circular and transverse cracks that we know as hydrogen embrittlement.

Temper Embrittlement: The second repair was completed with some improvements, but with an ordinary welding wire SMAW E-9016 B3, although while the TBW technique was used, cracks still appeared after PWHT at a temperature of 695°C (1247°F). The material of 2.25Cr-1Mo-V could develop cracks due to heat treatment, which was initially marked by the decrease of toughness and shifting the temperature of the ductile fracture transition to brittle fracture when exposed to a temperature range of 371°C-593°C (700°F - 1100°F). This is commonly known as temper embrittlement. So, it is natural to have cracks occurring when PWHT is conducted, although it was not found before PWHT.

The Special Electrode Welding Wire: The use of E-9016 B3 special welding wire, which in addition to a weld control with low hydrogen flux, can reduce the risk of cracking. Chemical composition of the weld metal is limited by the tempering factor of embrittlement $X \leq 15$, $Ni \leq 0.30\%$ and $Cu \leq 0.20\%$ to reduce this risk. The X factor can be calculated with the following formula, where P, Sb, Sn and As are in ppm.

$$X = \frac{10P + 5Sb + 4Sn + As}{100} \leq 15$$

R-0502 Replacement

The 2019 repair to R-0502 remained in service until June 15, 2020 when Ammonia Plant 1A was shutdown to install the replacement R-0502. Major stages of this program included plant shutdown (including system cooling down and depressurization), catalyst replacement, vessel replacement, and finally catalyst activation and plant start-up.

Plant Shut Down & Cooling Down

Ammonia Plant 1A shutdown began with the intent to utilize synthesis gas, circulated by the syngas compressor and cooled with the plant refrigeration system, to cool the synthesis loop reactors and catalyst to 130°C (266°F). The nitrogen blower would then be used, with the refrigeration system in service, until the temperature was reduced to 40°C (104°F). Unfortunately, when the syngas compressor was being shutdown it tripped, leading to a trip of the refrigeration compressor.

The skin temperatures of R-0502 at the time ranged from 197°C to 226°C (387°F to 439°F). Because there is no bed temperature indicator in R-0502, the skin temperatures were used as the primary indicators to monitor the cooling process.

The refrigeration compressor could not be restarted, so cooling had to proceed using nitrogen circulated by the nitrogen blower and cooled with cooling water. The synthesis loop was also pressurized, purged, and depressurized to remove flammable gas and to assist with cooling the equipment and catalyst. To speed up the cooling rate, on June 24, 2021, efforts were made to remove the outer insulation of the Reactor and spray water onto the sides of the reactor surface using a water ring attached to the top of the Reactor. This method proved to be effective, as the cooling rate was improved significantly, as can be seen, in Figure 6 and Figure 7.

After the average skin temperatures on the Reactor were below 40°C (104°F) the vessel replacement began.

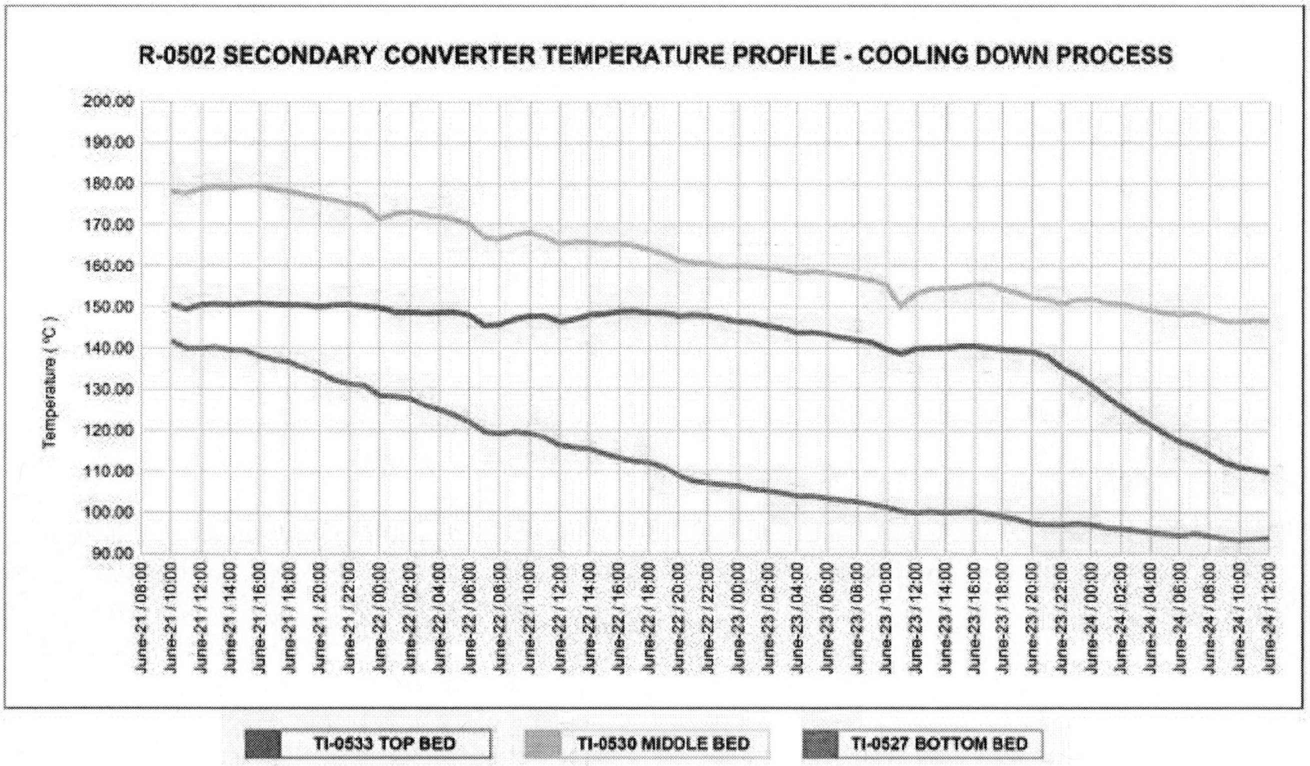

Figure 6. R-0502 Cooling profile by nitrogen circulation along with pressurization-depressurization.

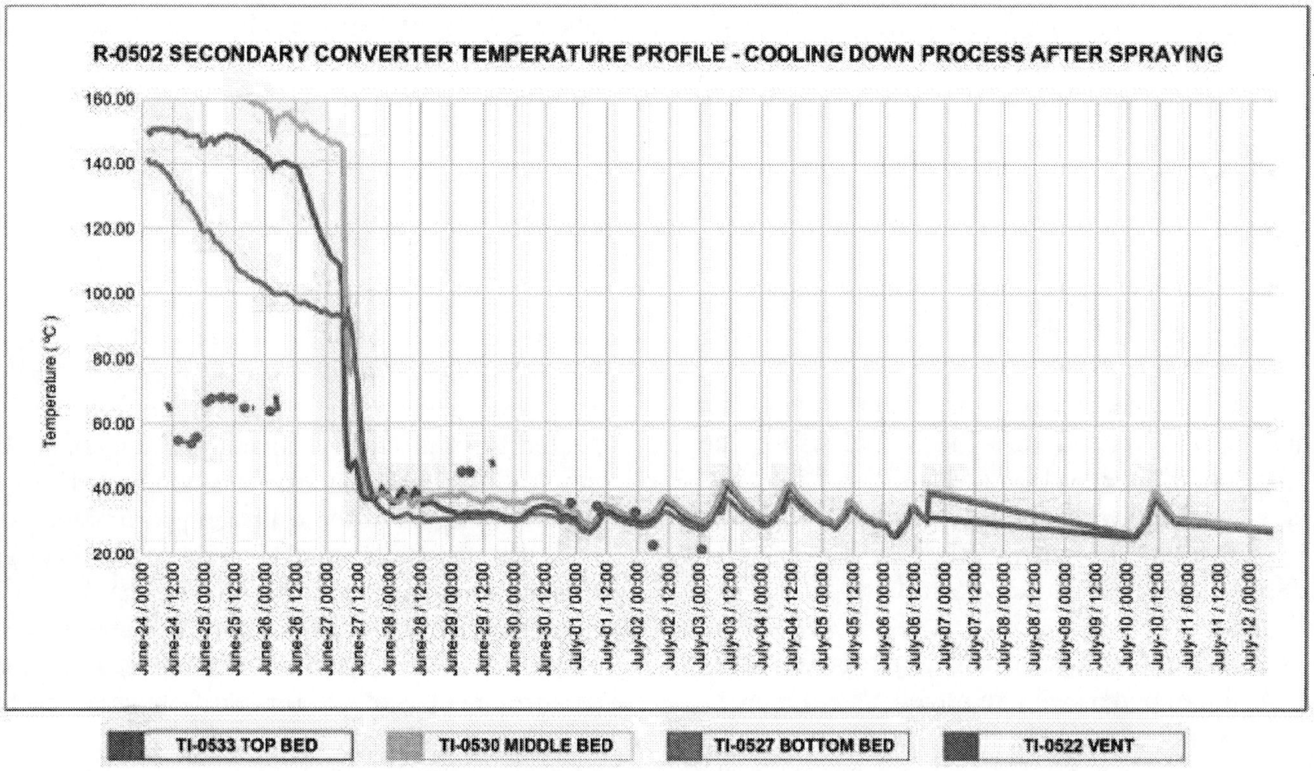

Figure 7. R-0502 Bed cooling down profile after using spraying method

R-0502 Vessel Replacement

The replacement pressure vessel shell material was changed from 2¼Cr-1Mo-V (F-22 enhanced) to 2¼Cr-1Mo (conventional F-22) to reduce the risk of cracking. This resulted in a thicker and heavier pressure shell, with a weight of about 82 MT (90 ST). Because of this, the support structure had to be reinforced by thickening the concrete column and adding 2 piles. The process of strengthening the concrete foundation column was carried out before the turnaround.

The R-0502 replacement was planned for 26 days which consisted of:

1. Unload catalyst and demolish existing vessel - 13 days.
2. Install new vessel and load catalyst - 12 days.
3. Reinstall piping system - 5 days (some of this work was performed in parallel with catalyst loading).

Catalyst replacement (unloading and loading) was carried out by a CATTECH vendor from the United Kingdom, in collaboration with a local agent PT PRO FLUID. The process of unloading the catalyst was carried out using the vacuum method while the system was purged and blanketed with nitrogen. The unloading system used is closed-loop nitrogen circulation. The nitrogen used for system security and unloading was supplied via the Nitrogen Tank / Lorry.

After the new reactor was installed the synthesis catalyst was loaded into the vessel using the dense loading method. 382 MT (421 ST) of catalyst was loaded with a density range in each layer of loading ranging from 2210 $kg/m^3$to 4785 kg/m^3 (138 lb/ft^3 to 299 lb/ft^3).

After the catalyst was loaded, all temporary seals on the top of panels were removed, and the screen cover installed. All damaged graphite gaskets were also replaced. After cover segment plate nuts were tightened the nuts were tack welded. Finally, the top cover manhole to the nozzle and bottom cover were installed, followed by welding the lip seal gasket at the top cover and diaphragm gasket at the bottom cover, using ERNiCr3 (Inconel 82) welding rods. Penetrant testing for the root passes and fill passes were performed until no defect was found.

Catalyst Activation Obstacles

The activation of ammonia synthesis catalyst at R-0502 was completed in 17 days, beginning on August 27, 2020, and was completed on September 13, 2020.

There was a problem with the syngas compressor due to dry gas seal leakage, so the synloop pressure had to be lowered, and the activation process restarted. At that time, the reduction had been ongoing for approximately 37 hours. The dry gas seal repair was completed in two days. During the repair process, the catalyst temperature dropped to 140°C (286°F).

The H_2 / N_2 ratio in the inlet R-0501 during the activation process was difficult to maintain at an optimum ratio of 2.8 – 3.0 due to problems in the Hydrogen Recovery Unit.

R-0502 Performance After Start-Up

Because the purpose of the vessel replacement was due to mechanical damage of the vessel rather than a decrease in the performance of the catalyst, the performance of the catalyst was not expected to improve significantly. Analysis of the new catalyst performance was conducted, which were comprised of the following parameters: operational, physical, and activity of the catalyst. Evaluation results for the physical and activity indicators above proved that the catalyst is in good condition and able to achieve warrantied performance, except for the physical factor, which indicates that the pressure drop is higher than guaranteed.

As for the mechanical performance regarding the previous failures at the bottom head of the Converter, the vessel is currently being monitored periodically, by sampling for flammable gases near bottom area of the Converter, as well as maintain the operating conditions, especially reactor pressure and temperatures according to Key Operating Parameters. The vessel is also regularly checked visually to monitor for any abnormalities. The vessel has been operating for over seven months without any failure or problem.

Conclusions

The 2018 repair to the bottom head of the R-0502 reactor failed after 8.5 months. This failure was due to a combination of factors caused by operating in hydrogen service, high temperature, and high pressure. The enhanced F-22 material used in the original R-0502 was more brittle and susceptible to cracking than standard F-22 material. These factors were exacerbated by improper PWHT during the 2018 repair.

The repair method with two phases of PWHT – at 50% thickness and 100% thickness – was successful because it allowed a good hardness value (maximum 222 HB). The problems with the initial repair, indicated by cracking following PWHT, was due to the use of a standard E-9016 B3 welding wire. After using the special E-9016 B3 welding wire, which is more resistant to hydrogen embrittlement, cracks no longer appeared after PWHT, and the vessel was able to withstand operation for another 14 months before vessel replacement.

Overall performance of the R-0502 Reactor after vessel replacement is generally good, both for the catalyst performance and mechanical integrity aspects. Until now, the vessel has been operating for over seven months, according to guaranteed performance, without any failure or problem.

References

1. Michael B. Rugh and Stanley Reid, "Repair and Failure Analysis of a Crack in an Ammonia Converter" AIChE, 2002.
2. Achmad Rois and Jaka Kirwanto, "Experience with Second Ammonia Converter (R-0502) Bypass at Ammonia Plant of Kaltim-1A", AIChE, 2018.
3. "Temper Bead Welding", Welding Technology Institute of Australia, Paper No. TGN-PE-02, 2006.
4. Rian Wishnu R, "S-50 NH3 Converter Performance Analysis Report", Pupuk Kaltim, 2020.
5. Rian Wishnu R, "R-0502 Plant-1A Catalyst Activation Report", Pupuk Kaltim, 2020.

Advanced Process Control (APC): Striving for Real Time Optimization at 2194 MTPD Ammonia Complex

This paper details the various steps involved in APC implementation, starting from pre-testing, challenges faced during DCS modifications done on stream without hindering plant operation and commissioning experience of APC without a single process upset. It also highlights the aspects that need to be taken into account to ensure successful installation and operation of APC after commissioning.

Safi Ullah Khan, Muhammad Waqas Quraishi, Muhammad Majid Latif
Engro Fertilizers

Introduction

Due to the complex and dynamic nature of plant control, key process parameters tend to deviate from their set points regularly. The operators tend to operate the process efficaciously, yet conditions move a little bit away from the true limitations as human constraints come into play. Because an operator, being a human being, wants to get the maximum amount of time to respond to a disturbance, they try to operate plant in an acceptable region away from any constraint. This provides the need for an automated process controller to get maximum profit while pushing plant to its limit.

Advanced process control (APC) is a process control and optimization technology which uses the multi-variable interactive nature of process units to reduce variability. It drives the process to an optimum point on a minute by minute basis through its Real time optimizer (RTO), an internal dynamic model of the process.

APC acts as an automated "operator" with the capability to carry out optimization every minute by changing set points of base layer controllers. APC is designed for controlling highly interactive, multivariable processes at several constraints simultaneously and operates at supervisory level. Figure 1 depicts the hierarchy in typical industrial control.

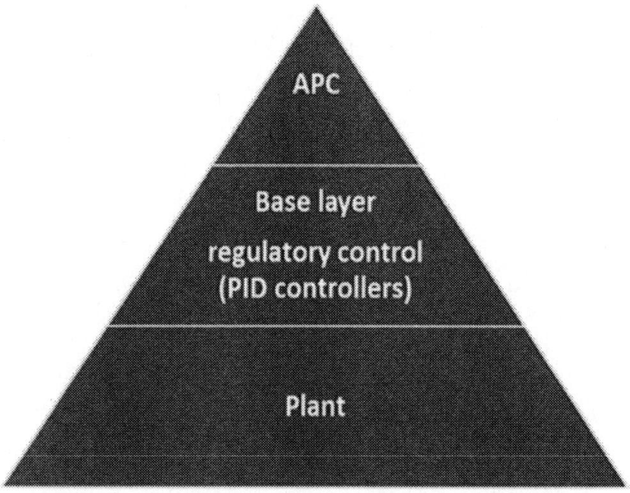

Figure 1. Hierarchies in typical industrial control

APC system is designed on three types of variables listed below:

a) **Manipulated Variables (MVs):** These are the independent variables whose conditions are manipulated by APC to maintain control Variables in their limits e.g. feed gas flow, air flow, steam flow, etc.

b) **Controlled Variables (CVs):** These are the process conditions to be controlled e.g. Steam /Gas ratio, plant load, Gas/Air ratio etc.

c) **Disturbance Variables (DVs):** These are measured disturbances (changes) in the process that influence the CVs, but that are not under APC control. For e.g.: Feed composition, ambient temperature, cooling water temperature, etc.

APC Implementation Steps

The implementation of an APC system involves several project phases listed below:

1. Scouting Study
2. Base Control Loops assessment
3. Functional design of controller (FDS)
4. Step testing
5. Detail Design Specification
6. Model Development
7. DCS Switch interface programming
8. APC commissioning and tuning

Scouting Study

In order to see readiness of plant for APC and realize margins available for production gain, a scouting study is conducted as a first step. For Engro's Ammonia plant, this was conducted in 2017 by IPCOS, a market leader for Advance process control. This study is based on detailed data analysis of all critical PID controllers' behavior on a minute by minute basis and identifying margins available for production gain. This study also includes a critical check that required instrumentation and infrastructure is available, a pre-requisite for APC implementation. APC project timeline and cost can be significantly altered by any hardware modification not identified in scouting study at later stages of the project.

After doing this Preliminary cost benefit analysis based on scouting study, Engro decided to move ahead for APC implementation on its state of the art ammonia unit (Amm-3)

Assessment of Base Layer Controllers

Since APC works above the foundation layer as a supervisory control, the regulatory PID control loops form the base layer of the control system. Before start of this phase, the APC server was installed and linked with the DCS server via an OPC server. Before installation of the APC server, a detailed Failure Mode and Effects Analysis (FMEA) of the system was conducted for risk assessment. Also, a Management of Change (MOC) evaluation was conducted per site protocol. APC data collection software was installed for the sake of recording the controllers' data on a one-second basis.

As a first step, all controllers were analyzed to identify problems like hysteresis, stickiness, noise, range of instruments etc., in critical control loops via DCS trend analysis and APC Data collection software. Various techniques were then applied, like controller tuning through software, maintenance of control valves, and calibration of instruments, to achieve more stability in plant operation. This is shown in Figures 2a and 2b.

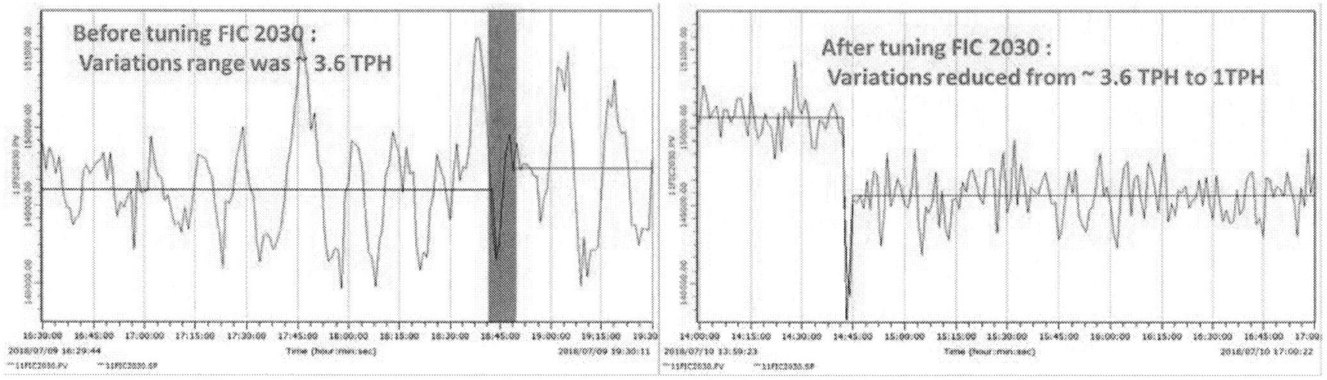

Figure 2a. Variation reduction achieved after base layer controllers tuning of Main Steam Valve (FIC2030)

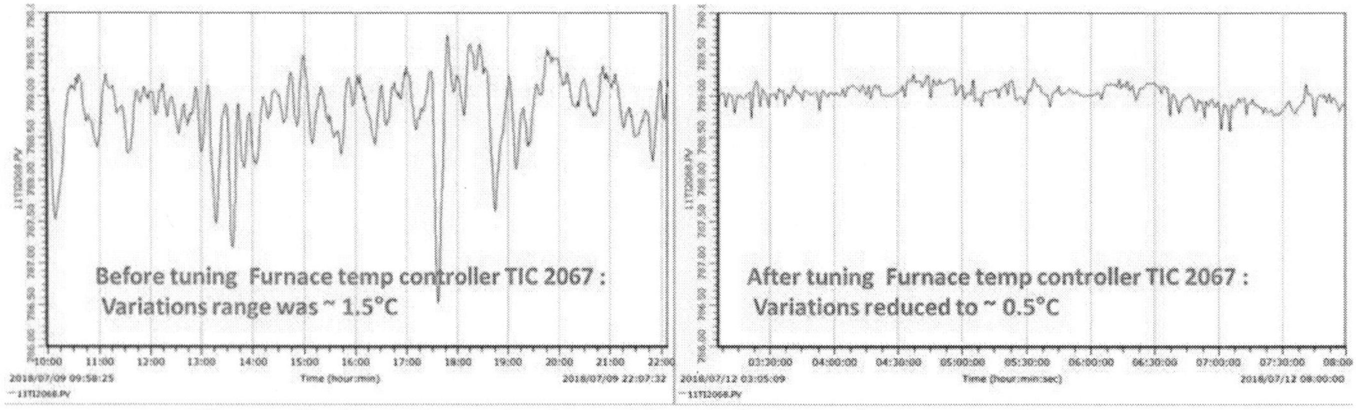

Figure 2b. Variation reduction achieved after base layer controllers tuning of Primary Reformer Outlet Temperature (TIC2067)

Tuning of critical controllers on a running plant is a precarious task as it has the potential of tripping the plant. Risk assessment for each controller, for which tuning activity was to be performed, was carried out in Team PHA and a controlled strategy was adopted to ensure safe plant operation during this activity. It is to be noted that this activity was to be performed on controllers in DCS while Emergency shut down system (ESD) and its block valves were intact to ensure safe plant shutdown in case of any process upset resulting from this activity.

Controllers that have hand jack or bypass available were utilized in order to take main controller out of process control before doing any tuning. Controllers without any bypass or hand jacks poses serious threat so soft limiters were put on controller output in DCS in order to avoid any inadvertent behavior during its tuning. This due diligence between the team ensured that task is done smoothly without any process upset.

Functional Design of Controller

In this step, a preliminary list of MVs, CVs and DV's was created after explaining the limits and constraints of plant to the APC engineer. The control philosophy of the process was explained to signify the various safety and process constraints together with equipment limitations, especially the control on the duty of the waste heat boiler to keep all operating parameters within limits defined by boiler vendor to ensure safe plant operation.

In order to highlight obstacles for the next step in implementation (Step testing), preliminary step changes were given in the MV's in series and data was collected to indicate problems like step sizes being too small to make a substantial change in relevant CVs above its noise level. System response time to a change in MV is also noted to save time in next phase.

Since the success of the APC controller greatly depends on how the functional designs are formulated to tap potential margins available in the process, this step requires great synergy between the APC engineer and plant process engineers.

Step Testing

Once base regulatory controller performance was evaluated and improved with all issues identified in FDS, step testing began and continued for two weeks. A series of step moves were made in each identified MV to see its impact on CVs. This was done to establish precise correlation between MVs and CVs thus enabling accurate modeling of APC. In order to ensure smooth operation, DVs were also given step change to enhance model ability to predict changes and proactively take action. The only DV identified was gas composition which significantly alters plant operating parameters. A preliminary process model was also developed in parallel during step testing to ensure good quality data.

Since APC is data driven, those models are as good or bad as the quality of step test data. Hence this step in APC implementation is absolutely **the most critical and most time-consuming step** in an APC project.

Detail Design Specification

After Step testing, detail design specification of the APC was developed enlisting the final list of Manipulated variables, Controlled variables and Disturbance variables. In addition, the related optimization strategy and constraints for each variable were concluded based on discussions between Engro and the IPCOS engineer. The list of main MV's and CV's are given in Table 1.

No.	Manipulated Variable	Controlled Variable
1	Feed Gas Flow	Plant load
2	Process Steam Flow	Steam to Gas ratio
3	Process Air Flow	H/N in synthesis loop
4	Furnace Bottom Temperature	Waste heat boiler inlet temperature

Table 1. List of Main MVs and CVs

Process Modeling

In the next step, the model was developed by an APC engineer using the data collected during step testing and model building software process. MV and CV relationships coupled with DVs were developed through simulation of different conditions and employing different model equations and techniques available in model predictive control (MPC) software to build process model.

DCS Switch Interface Programming

This step involves making APC switching logic in DCS so that the APC can read the base layer controller set points and write new set points after running its model every minute, once the APC model is commissioned and APC is to be put in operation. This was a cumbersome task as any mistake in scripting while logic building can trigger failure of APC application. Another important aspect of this was to make sure there is smooth transfer of control from APC to DCS, when required.

For this purpose, a dummy logic was initially developed for a controller that was out of service during normal plant operation and testing was done on it. Once the logic was successfully executed on test controller, it was deployed for real MVs identified for the APC controller.

APC Commissioning and Tuning

In this phase, APC model is linked with the DCS through the switch logic described above. However before turning APC on for the first time, switching logic and the fall back mode for each MV was tested rigorously so that in a case of any APC hardware / software failure the plant control is reliably shifted back to the DCS, ensuring no safety risk arising from APC system failure. Each MV was tested to ensure smooth transfer of control from APC to DCS, ensuring the DCS is tracking the set point given by the APC to the PID controllers. Subsequently, the APC controller was taken in operation step by step by adding one MV at a time under APC control. During this phase, fine tuning of the APC model was also done after careful evaluation of its performance with respect to plant control and optimization. Since APC doesn't compromise on plant safety, parameters related to waste heat boiler (WHB) mechanical integrity were given high priority so the APC takes fast action whenever a controlled variable tends to move beyond the operating limit defined in APC.

Benefit Assessment of APC

Increase in Plant Stability

As described above in base layer controller assessment, implementation of APC significantly improved plant stability. This resulted in a decrease in variation of critical controllers as depicted in Table 2.

Sr. #	Controller Description	Gains after tuning
1	Furnace outlet temperature	Variations reduced from **1.5 to 0.5 °C**
2	Process steam flow	Variations reduced from **3.6 to 1 mt/hr**
3	Process Gas flow	Variations reduced from **900 to 400 Nm3/hr**
4	Process Air flow controller	Variations reduced from **1000 to 200 Nm3/hr**

Table 2. Variation reduction of Main Controllers

Increase in Yield

Since APC always pushes the CVs to their upper/Lower limit depending on the types of variable, the yield of the plant increases. To assess this, an in-house evaluation of the production gain versus the plant load was analyzed and it was observed that production has increased by 0.58% on same load, resulting in improvement of ~0.1MMBTU/ton of plant efficiency.

Enhanced Reliable Plant Operation

Since the duty of the WHB was limited owing to its mechanical issue, it hampered plant operation by not being pushed to its limit to increase yield. Hence it was considered as a major controlling parameter during APC modeling as the WHB stable operation was of utmost importance to ensure safe plant operation. Data analysis in Figures 3a and 3b show that the APC effectively controlled the boiler inlet temperature to its maximum allowable limit defined by the boiler vendor. The variability of the WHB inlet temperature reduced from 7°C to 4°C (13°F to 7°F), thus ensuring safe plant operation while reaping maximum yield.

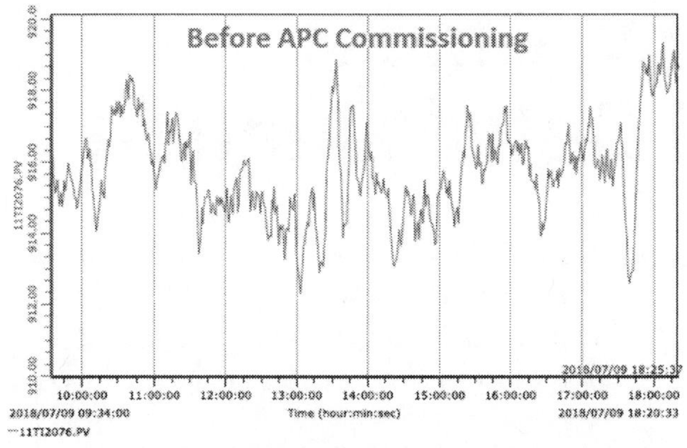

Figure 3a. WHB Temperature Variation Before APC

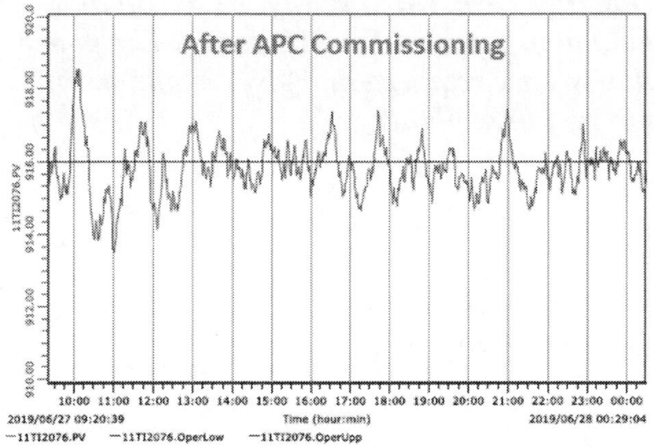

Figure 3b. WHB Temperature Variation After APC

Conclusion

Advancement in technology has enabled plant operations to run more smoothly and efficiently. This is evident for Engro as after commissioning of the system, the preliminary data analysis indicates stable plant operation giving consistent ammonia production while breaking the highest ever ammonia production record in the history of Engro's plant operation. Moreover, as the duty of the waste heat boiler was limited, APC greatly managed to keep all operating parameters within limits defined by boiler vendor to ensure safe plant operation.

References

[1] *Multivariable Predictive Control: Applications in Industry*, First Edition. Sandip Kumar Lahiri.John Wiley and Sons Ltd. Published 2017 by John Wiley and Sons Ltd. Pp55-60

Process Safety Times: Their Meaning and Determination

It is essential that a process plant goes into a safe state in case a failure occurs in the process or its control system. To ensure this, the plant is equipped with a safety instrumented system (SIS). As part of the demonstration that this safety instrumented system is adequately designed it is required to confirm that for each of its functions its response time is shorter than the so-called process safety time.

The potential of any failure to give rise to a hazardous event is assessed in the Hazard and Operability (HAZOP) study and evaluated by means of a Safety Integrity Level (SIL) rating. Based on that, process safety times must be determined for all instrumentation loops with a SIL rating. The paper presents examples of where process safety time calculations are required and gives examples of critical applications in ammonia and urea plants.

Klaus Noelker, Carina Will, Christoph Meissner
thyssenkrupp Industrial Solutions AG

Introduction

For the design of new process plants, as well as for major modifications in existing ones, it is standard to carry out a safety assessment. Usually this assessment contains the elements of:

1. identifying the hazards
2. identifying the safeguards
3. designing the safeguards adequately to mitigate the hazards.

Steps 1 and 2 are usually done in a HAZOP (hazard and operability) study where one systematically searches for hazards created by abnormal operating conditions and identifies protections or safeguards against them. Examples for such safeguards are:

- instrument functions (e.g. alarms);
- physical elements (e.g. rupture disks and pressure safety valves);
- instrument functions (e.g automatic shutdown functions acting on trip valves).

For step 3, a layer of protection analysis (LOPA) or safety instrumentation level (SIL) assessment must be done to adequately select and design the proper safeguards. In these, the risk of the unmitigated hazard is evaluated, and the requirements as to what level of risk reduction the safeguard must provide are defined.

Risk in this context is the product of likelihood and severity of consequences and might be separately evaluated for risks to humans, to the environment, or to the assets of the plant.

Essential for the effective risk reduction by a safeguard is its reliability. Taking the example safeguards from above, it is clear that an alarm annunciated to the operator via DCS screen has a higher potential for being overlooked and not triggering the desired counteraction compared to a pressure safety valve which opens when its set-point is reached, independently from any operator intervention. For instrumented safeguards, their function can be subdivided into the three consecutive steps of:

- sensing the deviation;
- solving the logic function for initiating the counteraction;
- execution of the counteraction (e.g. changing a valve position).

A function is only as reliable as the weakest element in this chain. For this reason, multiple sensors or actors come into play to increase the reliability.

The above is a simplified description of the procedure and is meant only to provide the context in which the process safety time is needed.

Process Safety Time

For a safety instrumented function not only the reliability of its sensing, logic and acting elements is essential, but also the speed in which they act. The safety instrumented function, which is usually a kind of shutdown function, must be fast enough to avoid the unwanted hazard.

This is expressed by the concept of process safety time and instrument response time.

- Process safety time (PST) is defined by International Standard IEC 61511-2 [1] as "the time period between a failure occurring in the process or the basic process control system (with the potential to give rise to a hazardous event) and the occurrence of the hazardous event if the safety instrumented function is not performed". This definition is consistent with those in other industrial standards such as IEC 61508 [2] and API 556 [3]. The process safety time is a function of the process and its equipment in a certain unmitigated hazard scenario.
- Response time (RT) is the time "to detect the out-of-limit condition and complete the actions necessary to stop the progression of the process away from a safe state" [4], [5]. If there are several independent layers of protection, response times can be different for all of them. Contributions to the response time are as shown in Table 1 (similar to [6]).

Element and type		Speed	Example
Sensor	pressure	fast	50 ms
	temperature	slow	a few seconds
	Level	fast	50 ms
Analog input module		fast	10 ms
Logic solver	DCS or ESD	fast	ESD: 100 ms
Digital input / output		fast	10 ms
Actor	pilot and control valve	slow, depending on type and size	one to several seconds
	circuit breaker for motor	fast	a few milliseconds

Table 1. Contributions to response times for instrumented functions and qualitative classification of speed.

In many cases, the unwanted change of the process variable PV is not immediately stopped when the instrument response is finished. This is expressed by the process lag time, which is the time from completion of the safety system's action until the process moves away from the hazardous condition.

A typical example for lag time is when temperature or pressure continue to rise for a while after the closure of a shutdown valve has been completed.

The situation is illustrated in Figure 1 for the timeline of an incident in a simple configuration.

- PV: The process variable under consideration, e.g. pressure, temperature, level
- PST: Process safety time: time from leaving the normal range until reaching the design limit
- RT: Response time: Time from detecting a deviation until completion of the counteraction.
- PLT: Process lag time as defined above.

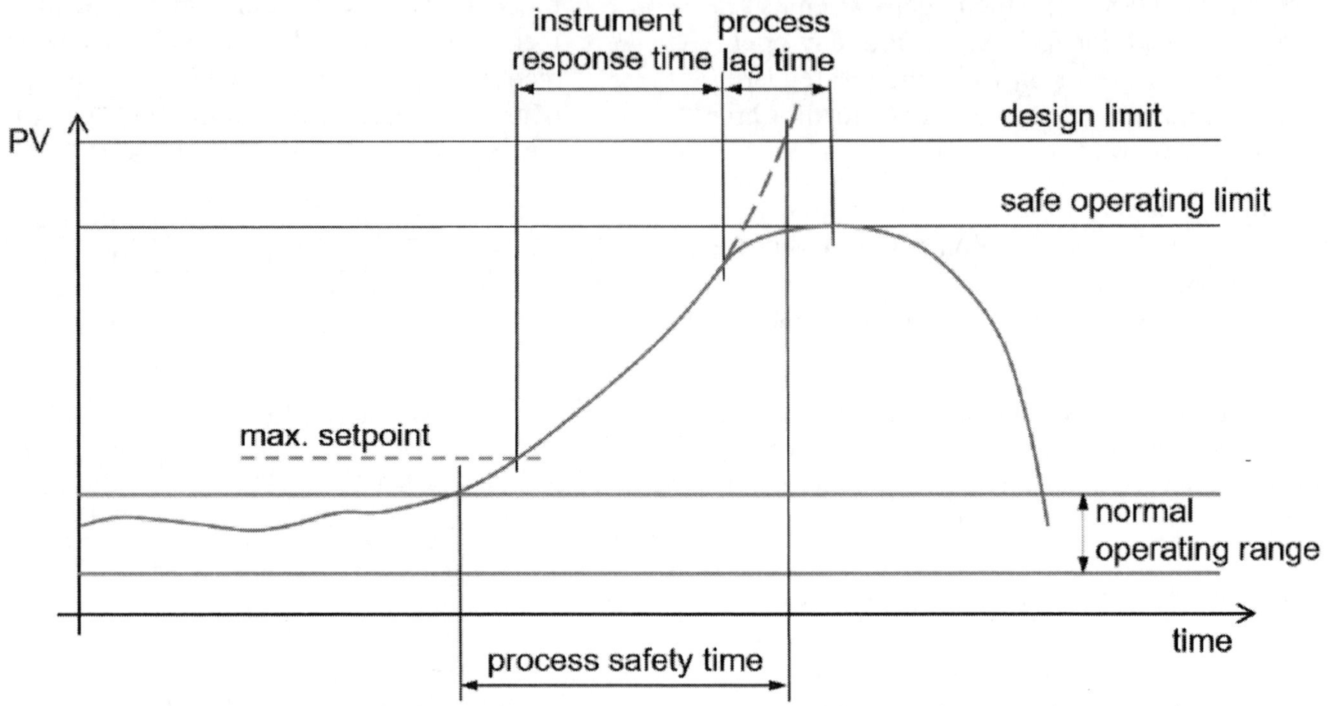

Figure 1. Timeline of a hazardous event: Process variable (PV) leaving the normal range, exceeding the design limit (unmitigated, dashed line) or staying below limit (mitigated by instrument response, solid line).

It is evident that a safeguard can only be effective when the response time plus the process lag time are shorter than the process safety time:

PST > RT + PLT

The determination of the process safety time is the starting point to determine the maximum available response time for the safeguards. If the above equation is not fulfilled adjustments can be made on both of its sides.

The response time of the safeguards cannot be arbitrarily shortened. Table 1 illustrates that travel times of large valves can take some seconds. They cannot be shortened at will, and usually requires extensive hardware modifications. In such a case, one must consider increasing the process safety time. A look at Figure 1 shows that for a given process this is only possible if either the safe operating limit is increased, or the normal operating range is narrowed, i.e. that the threshold value triggering the safety function is modified.

The task of determining the process safety time and the response time of a safety automated function is comparable to sizing a pressure safety valve. Having a safety valve alone is not sufficient, it also must be selected with sufficient capacity and the correct set pressure. Correspondingly, the safety automated function must be provided with parameters which allow for correct action with respect to detection of deviation and timely reaction to handle the incident.

Case Study 1: Dry Run Protection of the High Pressure Ammonia Pump in a Urea Plant

Process

In a urea plant a critical piece of equipment, whose failure can cause harm to humans, the environment, and the plant's asset, is the high pressure (HP) ammonia pump. This pump feeds liquid ammonia to the high pressure synthesis loop of the urea plant. Typically, a vessel is installed on the pump's suction side for removal of inert gases, providing sufficient net positive suction head (NPSH) and hold-up volume. In order to maintain a specified liquid level, the vessel is equipped with a level control realized by differential pressure transmitters, acting on a control valve on the inlet.

Abnormal operating conditions can occur if the ammonia feed from battery limit is cut or the level control fails. In this case dry running of the pump would be the next consequence, resulting in damage of the pump and potential discharge of toxic ammonia. To avoid this scenario, the running HP ammonia pump must be stopped before the suction vessel is emptied completely. This instrument loop has a SIL rating of 1.

Process Safety Time

In this scenario, the process safety time (PST) is the time period between the process value (level) falling below the low level trip value and the suction vessel being completely empty. In other words, it is the time the pump needs to empty the vessel when no feed from the battery limit is added to the vessel.

In order to calculate the process safety time, the following information is required:

1. Geometry and dimensions of the vessel (general arrangement drawing)
2. Elevation of the level transmitter's lower and upper tapping points at the vessel
3. Low level trip value (percentage of level transmitter indication)
4. Maximum capacity of HP ammonia pump

In the first step, the liquid volume at the low level trip value has to be calculated. A typical trip value to start with is 5% of the level transmitter range.

Using the vessel geometry and the range of the level transmitter, this 5% value is converted into a liquid volume. By dividing it by the pump flow rate the time the pump needs to empty the vessel completely is calculated.

If assumptions or simplifications have to be made for the calculation, the alternative with a shorter process safety time should always be selected, in order to stay on the safe side. In this example this means the maximum pump capacity is used (not the normal one) and that zero incoming flow to the vessel is assumed. (This could be the cause for the low level in the vessel.)

Table 2 shows the values for the calculation. The liquid volume in the vessel at low level trip amounts to 0.8 m^3 (211 gal). At a rated capacity of 165 m^3/h (726 gpm) it takes the pump 17.5 s to empty the vessel. After falling below the low level trip value, a time of 2.5 s is needed to stop the pump. This is the response time, consisting of the response time of the sensor, the time for processing the information in the emergency shut down system (ESD) and in the motor control cabinet (MCC) for stopping the motor. There is no process lag time (PLT) as no more ammonia leaves the vessel after the pump is stopped.

Instrument Response Time

The data in Table 2 show that the response time of the instrumentation system to stop the pump is much shorter than the process safety time. This means, the time is sufficient for the desired safeguard to stop the pump to prevent a hazard. Furthermore, also the initially selected trip value of 5% level indication can be maintained.

Process safety time (PST)	
Liquid volume in vessel at low level trip (5% of transmitter)	0.8 m^3
Maximum pump capacity	165.0 m^3/h
Process safety time (PST)	17.5 s
Response time (RT)	
Response time level transmitter	1.0 s
Response time ESD	0.5 s
Response time MCC	1.0 s
Response Time total (RT)	2.5 s
Process lag time (PLT)	
Process lag time	0.0 s
Conclusion: PST > RT + PLT: Sufficient safeguard	

Table 2. Exemplary values for the process safety time calculation of the HP ammonia pump dry run protection.

Potential Improvements

If the analysis had shown the level threshold of 5% was not sufficient one would have to modify the system in order to fulfil the equation:

PST > RT + PLT

Modifications are possible on both sides of the equation:

One could increase the level threshold to a percentage which would provide a longer process safety time (PST). The disadvantage is that the threshold comes closer to the normal operating point and narrows the range for allowed fluctuations of the process. This would not change the response time.

Alternatively or in addition to that, one could shorten the response time. By a look at Tables 1 and 2 it becomes clear that there is usually not much room for that. The largest contribution is usually the valve travel time. One might shorten the travel time by selection of a different valve type or increasing instrument air pressure.

Unwanted Corruption of Process Safety Times

A process safety time can be changed inadvertently as these two examples illustrate.

Modification of a Plant

A safeguard might turn ineffective when an operating condition of a plant is modified, for example, if throughput is increased. Even if a HAZOP study for this change is carried out one might think the old hazards and their safeguards remain applicable, without change, because no additional risk is introduced as everything except flowrates remains unchanged.

Higher velocity in the process means faster rate of changes, and thus the slope of the red curve in Figure 1 becomes steeper. This shortens the process safety time while the response time remains the same.

The analogy to this is the sizing of a safety valve. The set pressure and location of the valve remain suitable parameters but its relieving capacity might be too small when an upstream pump or compressor is upgraded in capacity.

Therefore, also in plant modifications which “only” increase rates but don’t introduce new equipment, the process safety times should be rechecked in order to verify adequate design of the safety functions.

Modification of Trip Initiators

A plant might suffer from unwanted trips because a process parameter frequently reaches a trip value during operation, without being out of control. For convenience, an operator might in good faith want to increase the plant’s availability by moving the trip value away from the normal operating range in order to allow for a larger fluctuation of the process parameter.

He may not be aware that such a practice can compromise the plant’s safety, and might think the plant is always safe due to the presence of the trip function which shuts down the unit once an unsafe condition is recognized. He might further think that the selection of trip values is alright as long as they are below the design limit (e.g. for high pressure or high temperature).

Here is a risk that should not be underestimated. As can be seen from Figure 1, a higher threshold value mean shortening the process safety time as its starting point is moving to the right. The shorter process safety time is not necessarily still longer than the instrument response time which remains unchanged.

A similar situation arises when a time delay is added to a trigger value, meaning the safeguard is only activated after the process value is above the threshold value for certain time. Operators sometimes do this in good faith in order to avoid spurious trips.

It is a management task to ensure such changes are only made after a thorough evaluation of the impact on safety.

Case Study 2: Methanator Protection in an Ammonia Plant

General Risk

In most ammonia plants, a methanation step is required for final synthesis gas purification. The methanation reaction converts unwanted carbon monoxide and carbon dioxide to methane and water, while hydrogen is consumed.

Both reactions are highly exothermic and produce a noticeable temperature rise across the methanation reactor even at fairly low impurity inlet concentrations. As a rule of thumb, for every percent CO methanated, the gas temperature will increase by approximately 74°C (133°F) and for every percent of CO_2 it will increase by approximately 60°C (108°F).

In case of upstream process upsets causing higher CO or CO_2 inlet concentrations the temperature rise due to the methanation reaction will be very prompt and severe. The rate of reaction is not limited by kinetics and will proceed to equilibrium, and the temperature rise might affect the reactor shell which can lead to exceeding the design temperature and potential loss of containment.

It is common practice to implement four to six temperature transmitters at various depths of the catalyst bed acting on a trip of the methanator, for example by a 2 out of 4 or similar logic. In the following example, it is shown that this arrangement is of limited utility to detect a rapid temperature increase on the top layer of the catalyst and is not fast enough to safely shut down the system. Therefore, it is crucial to implement other measures to prevent high CO and CO_2 inlet concentrations as the cause of the temperature increase. Only when the CO or CO_2 concentration increases slowly, the temperature sensor arrangement can safely shut down the methanation section.

Therefore, two elements are part of an evaluation: An estimate for the rate of CO or CO_2 increase can determine whether the trip acts fast enough. The final concentration of CO or CO_2 can be used to evaluate the severity of an incident: If the equilibrium temperature of the final concentration is below the design temperature of the equipment, the scenario is safe by design. Anyway, it is necessary to identify and evaluate all causes for higher CO and CO_2 concentrations at the inlet of the methanation section.

Potential Causes

The safety time for the increase in temperature depends on the cause which leads to it. If it is shown that a safeguard is applicable for a certain rate of temperature increase it is logical that the same safeguard is also suitable for a lower rate of increase.

Group	No.	Cause	Safe-guard
A: high CO_2	1	Wrong feed gas temperature to absorber	S1
	2	Lean solution poorly regenerated	S1
	3	Wrong temperature of lean solution	S1
	4	Low flow or no flow of lean solution to the absorber top	S1 / S2
	5	Semilean solution poorly regenerated	S1
	6	Low flow or no flow of semilean solution to the absorber	S2
B: high CO	1	High inlet temperature to HT shift	S1
	2	Low inlet temperature to HT shift	S1
	3	Deactivated catalyst in the HT shift	S1
	4	High inlet temperature to LT shift	S1
	5	Low inlet temperature to LT shift	S1
	6	Deactivated catalyst in the LT shift	S1
	7	LT shift bypassed	S2

Table 3. Causes for temperature excursion in the methanation. Safeguards are:
S1: Methanator temperature measurement suitable as trip initiator.
S2: Methanator temperature measurement providing little process safety time only and other trip initiator needed.

Plant configuration	2,200 mtpd ammonia
CO shift unit	2 stage (HT, LT)
CO_2 removal unit	two-stage amine based
Methanation normal inlet condition	CO: 0.3 % mole (dry) CO_2: 500 ppm v (dry) 300°C / 572°F
Methanation design condition	450°C / 842°F
Temperature trip	410°C / 770°F by 2 out of 4 logic with 4 transmitters at different elevations

Table 4. Key data for methanation case study

Potential causes are listed in Table 3, grouped into those related to high CO_2 (A) and those related to high CO (B), for a plant configuration as per Table 4.

Evaluation of scenarios and assumptions

To evaluate the risks, each scenario has to be investigated. When assumptions have to be made, the approach shall be conservative as discussed in case study 1. Key process data are listed in Table 4.

Due to the limited space available, only some of the causes from Table 3 are discussed in the following paragraphs. Low flow / no flow of semilean solution – A6

When the semilean solution flow is disturbed, either by a malfunctioning control valve or an issue with the semilean solution pumps, the CO_2 slip will increase. As a worst case scenario the CO_2 concentration on the outlet of the absorber will increase significantly (approximately 15%).

The rise in CO_2 concentration and the corresponding temperature rise in the methanation are severe and very rapid as illustrated in Figure 2. The increase in CO_2 concentration at vessel inlet starts at 60 s (dashed purple line). The temperature at adiabatic conversion to equilibrium starts to rise sharply with a rate of approximately 2.8°C/s (5°F/s). The actual temperature in the vessel depends not only on time but also on location and is between the inlet temperature and the equilibrium temperature. It mostly depends on:

- Catalyst activity: The higher the catalyst activity at the top layer is, the faster is the temperature rise there and the higher is the place where equilibrium is reached and the highest temperature is reached.
- Cooling effect by catalyst and vessel walls: During the event the gas, heated up by reaction, releases part of its heat to the surrounding colder material during the transient phase. By this time-dependent effect the gas gets colder than the equilibrium temperature.

However, if the event is not mitigated the surrounding material will reach the gas temperature and a new stationary temperature profile for gas and vessel will be established.

For the calculation some simplifying assumptions have been made. The actual temperature in the reactor and at the four transmitters #1 to #4, located in the catalyst at different elevations from top to bottom in direction of the gas flow, is between the dark blue line (inlet) and the red line (equilibrium) of Figure 2.

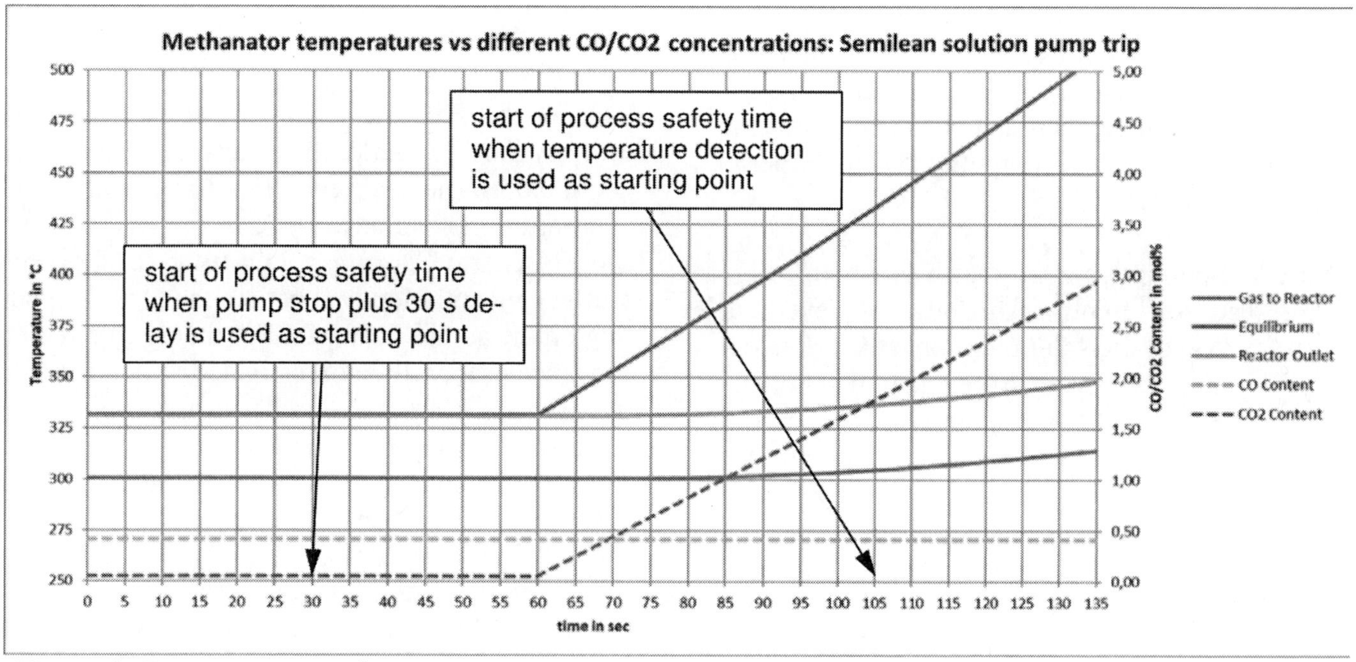

Figure 2. Temperature and concentration profile for increase in CO2 concentration by trip of semilean solution pumps (simplified). Dashed lines: concentrations, right axis. Solid lines: temperatures, left axis.

Figure 2 shows that the equilibrium temperature crosses the reactor design temperature (450°C or 842°F) 52 s after the start of the temperature rise. At that time, the four trip transmitters read as per Table 5, and two of them have crossed the trip threshold temperature of 410°C (770°F).

Transmitter	Calculated temperature (°C)	Time since trip value (s)
#1	439	12
#2	427	7
#3	393	0
#4	365	0

Table 5: Transmitter readings in unmitigated scenario when equilibrium temperature reaches design temperature of 450°C. At 2 out of 4 logic, this is 7 seconds after triggering the trip.

It may be worth noting that these figures are taken from a simulation which does not account for the hysteresis in the temperature output of the thermocouple, caused by the insulating effect of the thermowell. Therefore, in reality the transmitter reading is lower than the actual temperature.

The theoretical safety time for this event is 7 sec. This means, the instrumentation response time must be shorter than that. This is almost impossible to realize, since this time must accommodate transmitter and ESD response time and the travel time of the reactor inlet valve to close, plus a safety margin. For this reason, the safety of the system has to be ensured in a different way.

The recommended solution is to look for an earlier event which is easily detectable and which leads to the CO_2 concentration rise. In this case, this is the stop of the semilean solution flow to the absorber.

The standard approach by thyssenkrupp Industrial Solutions to provide a suitable safeguard for this scenario is to install a second logic (in addition to the temperature trip) which already trips the methanation when both semilean pumps are not running because relying of the high temperature trip of the methanation alone cannot prevent local temperature exceeding the design temperature of the vessel.

When the pump trip is used as a safeguard, Figure 2 reads differently. The initiating event is now the stop of the solution pump at 0 s. It takes approximately 60 s from that event until a noticeable increase in CO_2 concentration reaches the methanator, due to remaining solution in the CO_2 absorber packing and time for gas flow to reach the reactor. The process safety time is now 60 s longer than calculated above, i.e. 67 s.

A part of this increase is consumed for other purposes: Even during normal operation of the plant, switch over between pumps is done frequently, and when pump A trips, pump B starts by automatic takeover. Therefore, a time delay to processing the lost running signal has to be implemented to avoid unwanted and unnecessary trip of the methanation section.

When the methanation trip is initiated by a pump trip plus 30 s delay, the process safety time is 37 s (7 s as above + 60 s since pump stop – 30 s delay)

As an alternative to this trip logic one might think of increasing the available process safety time by lowering the trip temperature (410°C or 770°F, in this example). This is a valid approach and from Figure 2 it is evident that this would increase the process safety time. However, it would also lead to a trip where the temperature increases above this figure in a safe scenario (e.g. in A3, see below) where a temperature of 400°C (752°F) is reached in a safe condition.

Low flow / no flow of lean solution – A4

When the lean solution flow is disturbed, either by a malfunctioning control valve or an issue with the lean solution pumps, the CO_2 slip will increase. As a worst case scenario the CO_2 concentration at the outlet of the absorber is similar to the inlet concentration of the lean absorber (approximately 5%).

During the time the absorber packing needs to run dry the CO_2 outlet concentration will change from 0.05% (normal operation) to 5% (inlet concentration of lean absorber), in the case of a no flow scenario. During a low-flow scenario the outlet concentration will be less than 5% depending on the actual flow rate. For that reason the no flow scenario will be investigated in detail, since it is more severe and a proper protection for that case would also cover the low flow scenario. As per the previously given rule of thumb, 5% of CO_2 will cause a temperature increase of approximately 300°C (540°F), which leads to a theoretical methanator temperature of 600°C (1112°F). This is well above the usual design temperatures of methanator reactors. This scenario is very critical.

The dry-run scenario is a complex dynamic process with many parameters influencing the absorption performance. As a conservative approach, a linear increase of the CO_2 slip over time is assumed.

The behavior is similar to the scenario described above, but the temperature rise is slower. Approximately 20 s after the pump trip, the gas composition starts to change at the inlet of the methanator due to the hold-up time between the absorber outlet and the methanator inlet.

By the same way as done for scenario A6 above, a process safety time of 54 s is calculated. This could be sufficient to initiate a trip but since there are so many uncertainties in the calculation and because with solution flow measurement or pump running signals there are alternative trip triggers available these should be used to improve the safety function response.

Wrong temperature of lean solution – A3

The absorption performance of the solution depends on the temperature of the lean solution. In many plants the inlet temperature of the lean solution is controlled by a bypass valve around a final water cooler for the solution. This temperature may be disturbed by a control loop failure of this system. In this case, the temperature is somewhere between the cooling water supply temperature and the solution temperature upstream of the water cooler. In both cases CO_2 absorption is affected, but the CO_2 slip only increases to approximately 10,000 ppm (1%) for the higher temperature (66°C or 150°F). This corresponds to a maximum methanator temperature of 400°C (752°F), at equilibrium.

A second effect might be that the gas-gas heat exchanger which is used to preheat the methanator feed by heat exchange with the methanator effluent causes an additional temperature increase at the methanator inlet. This may occur if the inlet temperature control fails. This temperature rise will be fairly slow, since the whole catalyst bed needs to be heated. More importantly, the catalyst will evenly heat up over the whole bed height, which allows the temperature transmitters to initiate a safe shut down in case they should exceed the temperature threshold of 410°C (770°F).

Incorrect temperature or deactivated catalyst in one CO shift reactor – B1 to B6

When the catalyst of a shift converter is poisoned, the CO content at the outlet will be the same as the inlet concentration. This scenario is quite theoretical since catalyst deactivation, even following poor operation, is a matter of hours or days. Even if this is taken as fast, the consequential temperature rise for the methanation is slow compared to some of the above scenarios and can be detected and stopped by use of the temperature transmitters in the reactor, if the plant is not shutdown due to poor performance of the CO shift.

Conclusion

As described above, some events are mitigated by temperature transmitters while for others it is necessary to use other signals in order to increase the available process safety time for a safe shutdown. This has been marked by safeguards S1 and S2 in Table 3.

Summary

Knowledge of process safety times is needed to validate the safety functions of a plant. Their calculation is not only a legal requirement (e.g. under the CE certification in Europe) but it is also a useful means to determine a maximum permissible threshold value (triggering a safety function like a shutdown) which still allows for operating fluctuations.

Such threshold values must not be changed for convenience by an operator without studying the impact. They are part of a safety instrumented system and have an impact on the process safety time which must be longer than the response time of the safety function they trigger.

The case studies show that there are deviations from normal process conditions that happen so fast their detection cannot serve to shut down the process quickly enough to avoid a hazard. In this case, an alternate trigger which gives an earlier warning must be found (e.g. case labelled "S2" in case study 2).

Determination of process safety times is a complicated task because it involves dynamic, transient processes, like heating up of vessel volumes, for which knowledge of the process is a precondition. Care must be taken to ensure the results are accurate when simplifications are made.

References

[1] International Standard IEC 61511:2016 Part 2

[2] International Standard IEC 61508:2010 Part 2

[3] American Petroleum Institute, API Standard 556, second edition, 2011

[4] G. Barnard, W. Creel, Impacts of Process Safety Time on Layer of Protection Analysis, American Institute of Chemical Engineers 2015 Spring Meeting, 11th Global Congress on Process Safety, Austin, Texas, USA, 2015

[5] CCPS: Guidelines for Initiating Events and Independent Protection Layers in Layer of Protection Analysis, Center for Chemical Process Safety, American Institute of Chemical Engineers, New York, USA, 2015

[6] Dil Whetherill, IEC 61508 and Functional Safety System Selection, Measurement Technology, Ltd., England, 2005

Experience and Lessons Learned in Management of Change at Yara

Despite huge improvements in company's procedures and attention to control the management of change in the various Yara plants and sites, we still face incidents and near misses that are in essence due to failures in managing the risks generated by changes to either installations or procedures. This paper will highlight the areas where further improvements are needed and how training, sharing of lessons learnt, continuous improvement, and the help of a new common tool are being used to reduce the probability of failure due to improperly managed MOC.

Vincent Duponchel
Yara Belgium SA/NV

Introduction

The implementation of a Process Safety Management system using the Center for Chemical Process Safety's (CCPS) twenty elements has enabled us to highlight weaknesses in our process safety. Through a review of the incident management, we have highlighted the root causes of our accidents and we found out that Management of Change (MOC) was not at an adequate level in our company. The review of our different failures caused by improperly managed MOC has allowed us to establish the way forward and to implement a step improvement in our way of managing changes.

From where we started

A perfect MOC Standard

In Yara, our MOC standard was inspired by a position paper developed by insurance brokers [1]. In this paper, we can find clear guidance on **keys steps** and **keys roles/responsibilities** for a solid MOC process.

The interpretation of the standard

At our 25 sites, this standard has been interpreted differently which led to several deviations discovered during audits:

- ✓ Concepts invented by site to reduce the "bureaucracy" of MOC:
  - *Simple/trivial MOC* to by-pass the analysis (based on a person's feeling)
  - *Emergency change* to use the fast track (even if there is no emergency)
  - "Changes with same specification" to by-pass the system
  - Temporary MOC are (almost) never properly used
- ✓ The entire MOC process was seen as a succession of signature/approval with high bureaucracy

✓ At many sites, Risk Analysis requirements are based on MOC owner/project manager decision

Discussions held during plant visits about obvious changes showed a misunderstanding of the essence of MOC:

"If I do a MOC for this type of modification, I will need to do a MOC for every modification"

The results

In 2017, 10% of the 1000 Process Safety (PS) incidents (accidents, near misses, and hazardous conditions) recorded had the root cause of "improper MOC". But even worse is that 3 out of 5 of our Tier 1 accidents (as defined in the API recommended Practice 754 [2]) were clearly linked with lack of MOC.

The analysis

A root cause analysis of our MOC failures was performed and the simplified results are presented in the Ishikawa fishbone diagram shown in the Appendix, Figure A1.

Based on these results, an action plan was initiated to improve the execution of MOC in three main aspects:

- The competence of people
- The MOC process understanding
- Removing the time constraint

Competence of People

Sharing the failure of MOC

Sharing our experience is key in our PSM system. Every month, more than 100 PS incidents are analyzed in which some are presented in the PS monthly report, showing direct causes and root cause of the accident. The selection of the presented incidents is based on their potential severity and some of them are linked to an improperly managed MOC.

Shared learning incidents that have a MOC root cause are recorded in the Shared Learning / Major Risk (SL/MR) database for use in future toolbox meetings or safety committee meetings. An example of a Shared Learning is presented in the Appendix, Figure A2.

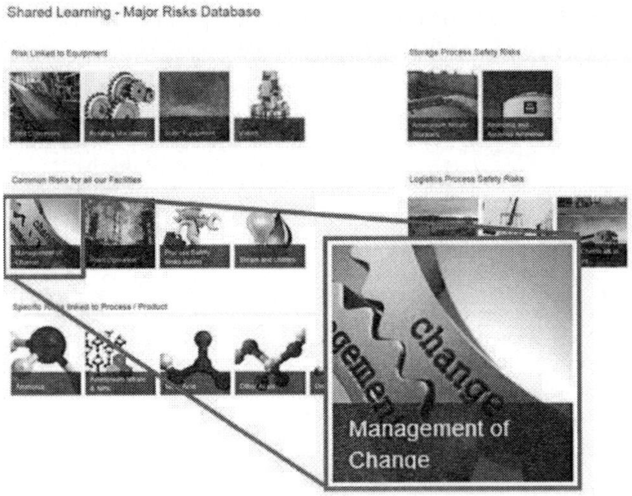

Figure 1. Shared Learning Database

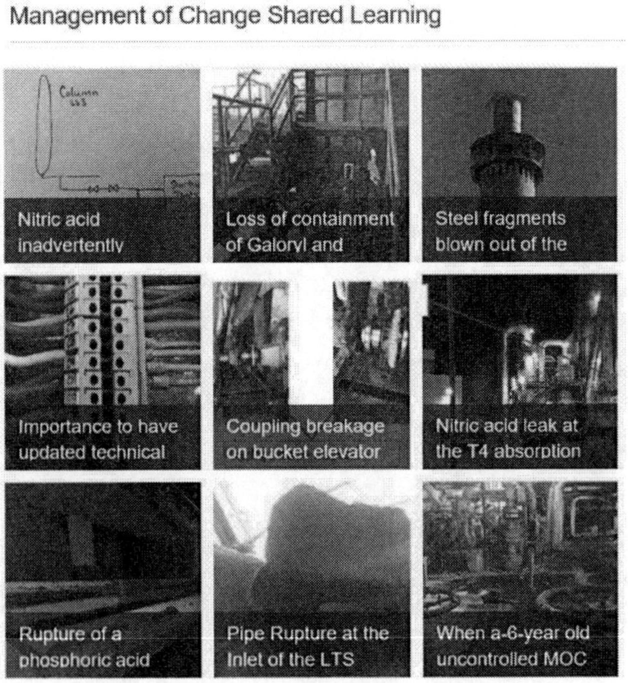

Figure 2. MOC Shared Learnings

Ensuring we have the correct discipline

In the new Yara MOC tool, every MOC starts with an impact assessment to analyze the potential consequence of the change, and the part of the PSM that might be impacted. This is including the four risk recipients People, Environment, Assets, and Reputation. Based on this impact analysis, the different discipline engineer required to assess the MOC are selected.

Correct use of the Risk Assessment Authority

The Risk Assessment Authority (RAA) Role was present in Yara MOC standard for several years but it was not implemented well at the sites.

In fact, the MOC process requires risk assessment processes appropriate to the size and nature of the change to be selected and executed. Sites must have people who are capable of taking this view and supporting the execution the MOC process.

MOC Process Knowledge

Defining a clear MOC process

Yara's previous MOC standard gave direction and main steps without going in depth. The first action taken was to elaborate on the different steps of the MOC process workflow (Figure 3), and then define the different roles involved in each step and how to ensure that they will perform adequately in each step.

Figure 3. Yara Process Steps for Handling MOCs

The MOC training part

To be fully efficient, our training is based on case studies and high involvement of the trainees.

All Yara employees in production plants shall follow a MOC awareness training via a web based platform called "Together We Learn" The key steps in the MOC process are presented, and the training ends with a test of "MOC or not MOC".

People involved in the MOC process shall follow the standard MOC training which consists of 3 modules:

- Session 1: MOC as key element of PSM
- Session 2: How to improve our MOC process
- Session 3: Roles and responsibilities in MOC and Process workflows

In between modules 1 and 2, homework is required to analyze 15 MOC failures at different steps in the process. These are based on actual Yara events.

People with the role of Risk Assessment Authority shall receive specific Process Safety Management training, in addition to the training above. This includes:

- Yara PSM boot camp, which consists of a review of the Yara PSM system and contains more than 50 cases. The training ends with a test in which the future RAA needs to score above 90%.
- Follow the specific MOC training for RAA.

- Follow and validate a retraining with a blank case of MOC (at least twice a year) based on a failure scenario from a site (case study to be analyzed individually and prepared before the feedback session).

The Time Constraints

One of the main reasons for MOC failure is that properly analyzing the change takes time. However, repairing mistakes is also time consuming, so how can we ensure the correct management of change analysis with a limited time?

Different process workflow for different uses

One of Yara's main advantages is that we operate more than 25 plants. With identical workflows for all sites, we can build our experience based on the feedback from all sites, which will make the improvement curve faster than anyone else. Starting from this statement, we created several types of workflows using our experience to ensure the quality and the consistency in every type of change.

The standard work process flow

The standard work process flow uses checklists to ensure the quality of the analysis is maintained. These checklists allow us to define the complexity of the change and the different discipline engineers needed for a proper analysis. They also allow us to dynamically adapt the work flow to make the best use of our resources based on the selections. The selections made by the people involved generate reminders to ensure all required training, documentation updates, DCS/ESD backup, etc. in the different steps of the work process flow are completed.

Simplified process work flow

A list of process workflows for some of the most common changes implemented at our sites was developed. In the past, these changes were not always analyzed, due to time constraints.

These typical changes are:

- Change of catalyst type
- Alarm/trip modification
- Change of raw material
- Change of valve
- Change of Maximum Demonstrated Rate
- Management Of Organizational Change
- Hot spots management
- Leak Management (including Clamp installation)
- Long term override of safety functions
- Modifications during maintenance

The last six items were often not registered as MOC at our sites, despite being direct cause or contributing factors of PS accidents. "Modifications during maintenance" was added following an incident caused by a normal maintenance operation that required modifications due to unavailability of spare parts, so replacement parts from a different manufacturer that were "almost" replacement in kind were used. In these cases, MOC is required to ensure a correct recording, analysis, and approval of these changes (Appendix, Figure A3).

Performance follow-up of the MOC system

The performance of the MOC process must be regularly monitored and assessed using both a routine review of KPIs and also periodic audits. These steps help assure the site management team that the system is actually being used in the way it is designed and intended. A specific MOC process owner must be nominated to take a lead role in operating and monitoring the system.

Indicators and dashboard

Standard MOC dashboards are created using Microsoft Power BI, a business data analysis tool. Information is imported from different databases, such as incident reporting tools to monitor the performance of our MOC system. The information presented includes:

- Number of MOCs in the system (in different steps) with a limitation of what each plant can handle,
- Number of urgent and temporary MOCs,
- Number of completed MOCs with open/overdue close-out actions,
- Number of MOCs with documentation not finalized,
- Numbers of incidents/near miss due to MOC process failure.

Audit

The MOC process shall be reviewed by 2 levels of audit:

- Regular "Systematic Inspection" which are checklists focusing on an individual case. From the case review, several questions can be set either to "safe" or "at risk" based on the analysis of the assessor. RAAs are entitled to perform these reviews.
- Integrated to the site audit plan (internal and corporate), with a focus on the system itself and integrating systematic inspection outputs. The audit material provides clear guidance on what should be checked to ensure a good efficiency of the local MOC process.

The better understanding of the MOC process and the consequences if the system is not well managed have led to detection of inadequately managed changes that would have not been reported before. This is illustrated by the results of systematic inspection (Graph 1). Many more items are detected at risk during the course of the MOC process than before.

Conclusions

Improving our MOC routine has not prevented us from the risk of incidents from changes incorrectly made in the past, but we saw an immediate gain in the competency of people to detect and manage new MOCs.

A plan to involve the workforce is key for a successful implementation.

- Start with an adapted training for all employees based on their involvement in the MOC process. The awareness training is ensuring the detection of modifications, the standard training the good management of the MOC and the RAA training the adapted level of risk assessment and scrutiny of the MOC
- Foster change detection by the mean of shared learning, showing how shortcuts have led to major event in Yara.
- Give more responsibility to shift supervisors to initiate and risk assess Temporary MOC they might performed in the normal life of the plant (leak & hotspot management for example).

Also, creating different specific process workflows for changes that are commonly done in plants, by involving several sites to develop them, is improving the quality and the consistency of the MOC process:

- For temporary changes that were not analyzed due to lack of time, simplified workflows provide adapted analysis to increase the sites' safety.
- For changes using the standard workflow, the use of check-lists leads to a gain of time (by involving only the required competence) and quality (by ensuring that nothing is forgotten in the different steps of the MOC).
- For typical changes, simplified workflows bring the analysis to the point using an adapted workflow.

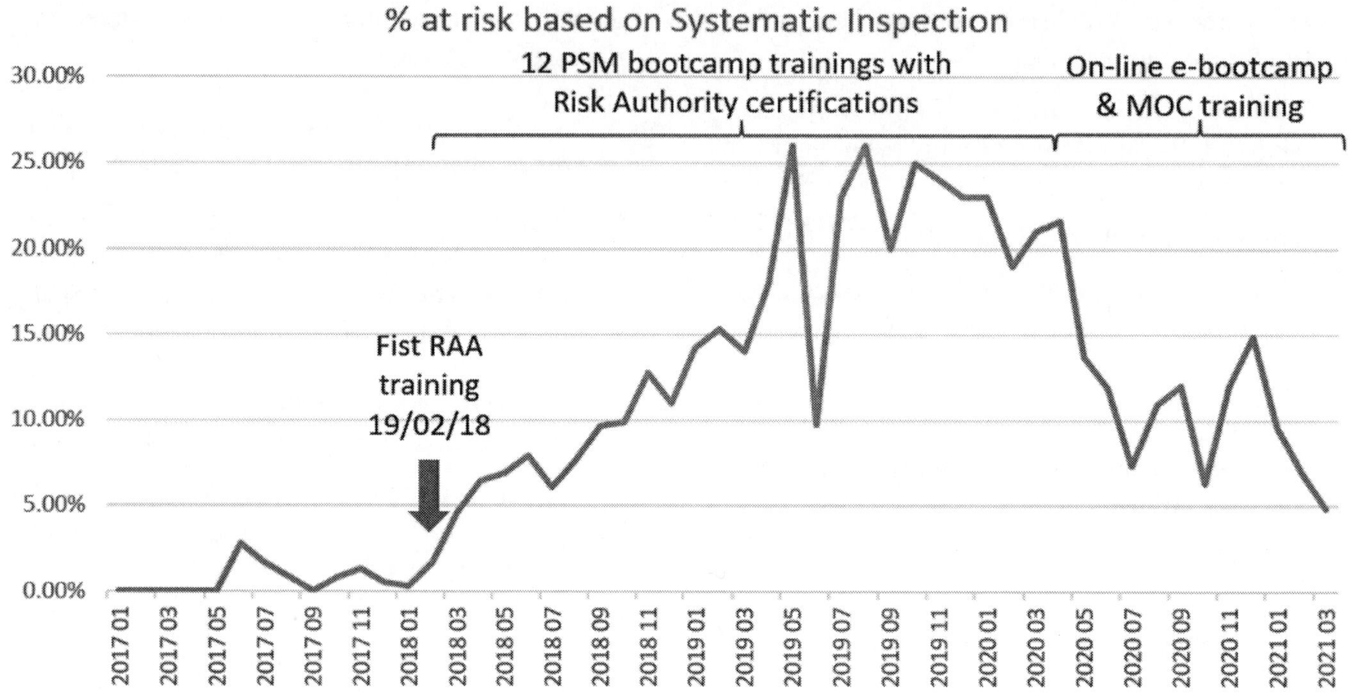

Graph 1. At risk % from systematic inspections of MOC

References

[1] Risk Engineering Position Paper – Marsh: https://www.marsh.com/uk/insights/research/risk-engineering-position-paper-management-of-change-moc.html

[2] API Recommended Practice 754: Process Safety Performance Indicators for the Refining and Petrochemical Industries

Appendix

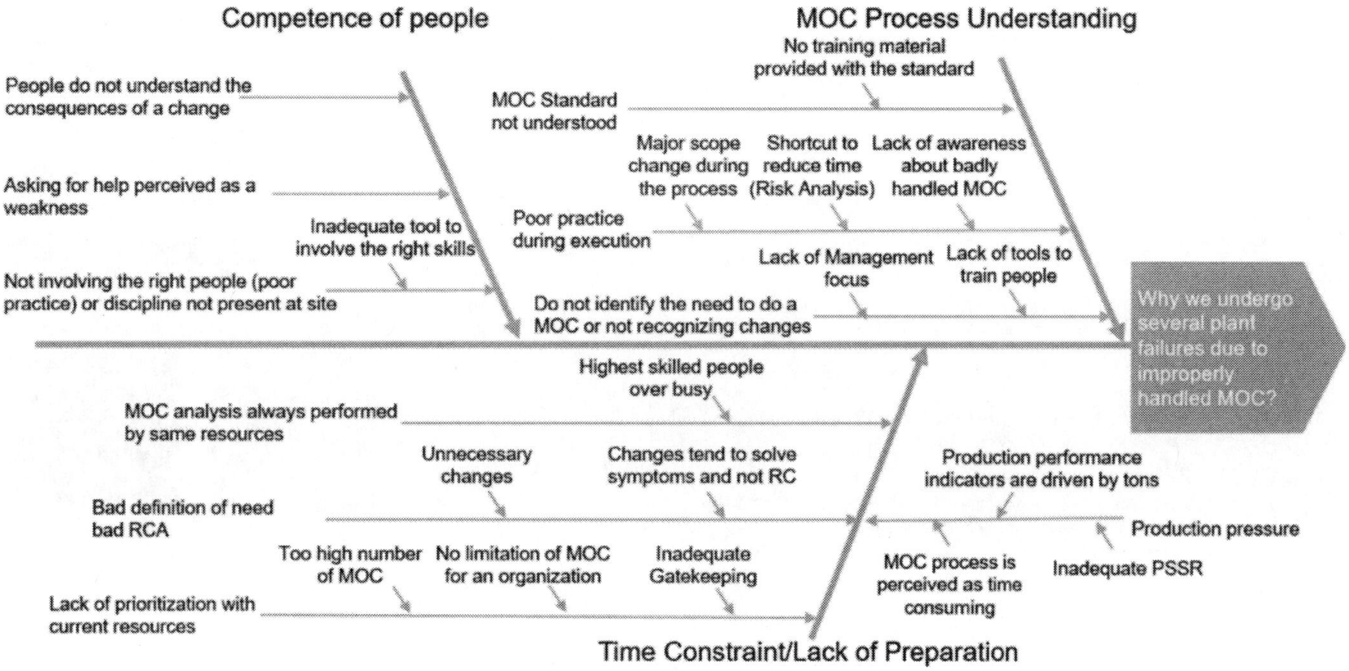

Figure A1. Root Cause Problem solving

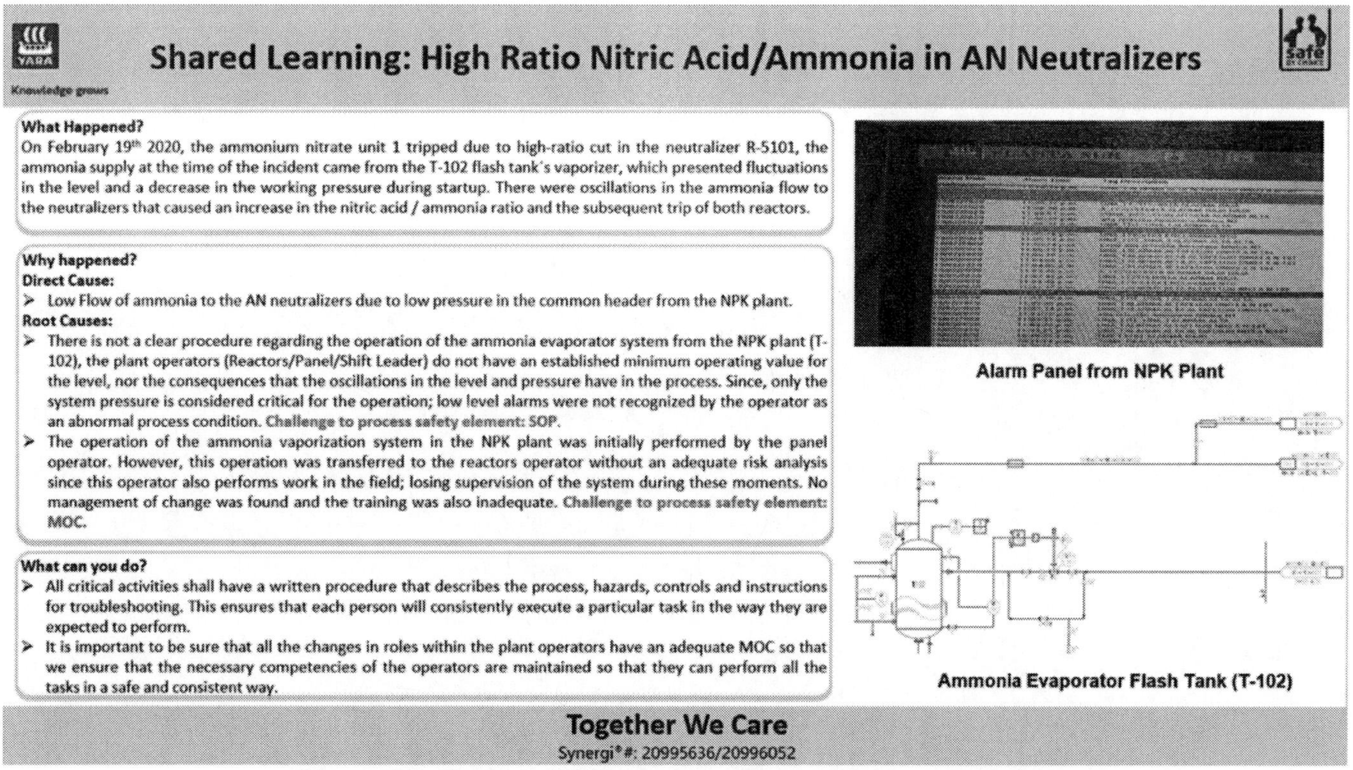

Figure A2. Typical PS Shared Learning with Yara Template

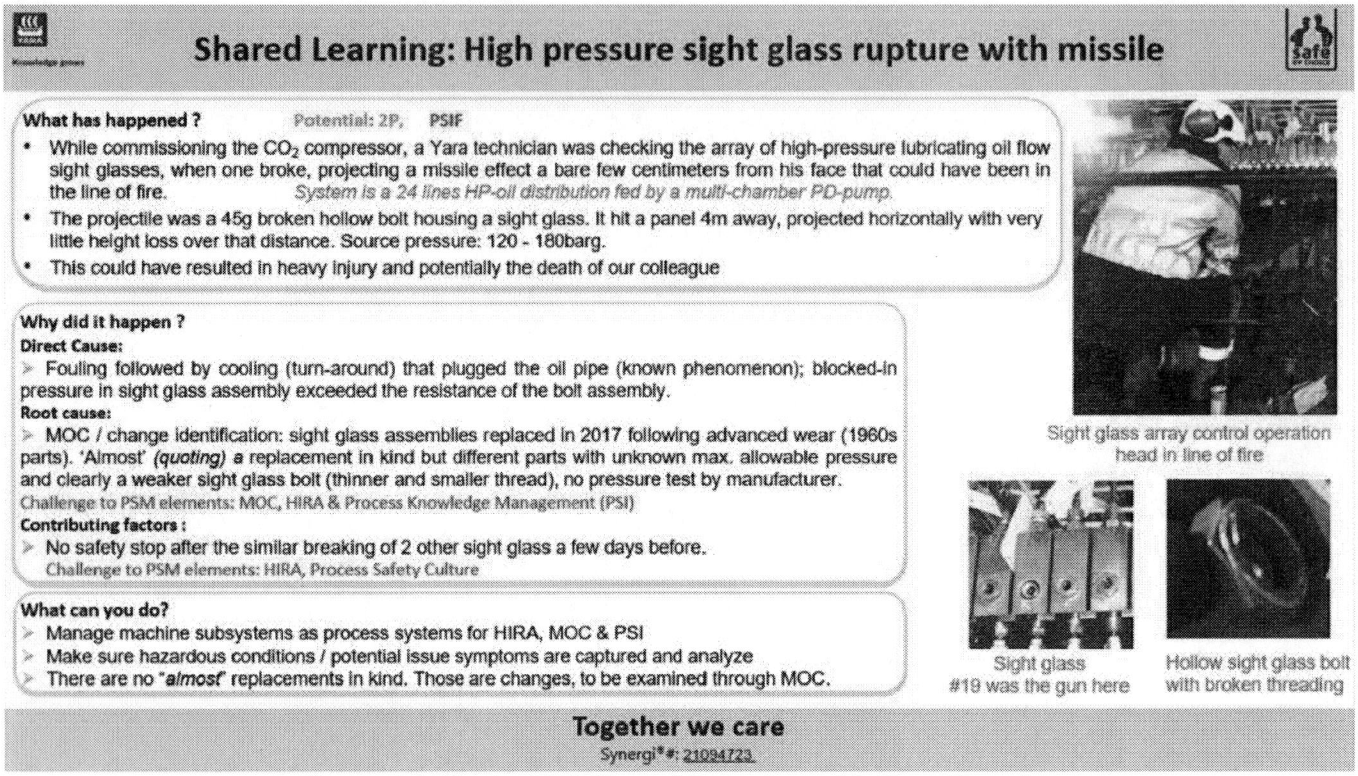

Figure A3. Typical PS Shared Learning with Yara Template

Operation, Inspection and Maintenance of Ammonia Tanks: A Kavala Fertilizers Case Study

This paper describes the case study of Kavala Fertilizers Ltd (KF) and the measures taken over the last four decades to ensure the safe and efficient operation of their ammonia storage tanks. KF has operated a refrigerated ammonia sphere since 1965 and an atmospheric, single-wall ammonia tank since 1982. Several non-destructive tests (NDT) took place within the framework of preventive maintenance of the equipment, with a focus on improving the safety and operation of both tanks. After each NDT inspection, a root-cause analysis recognized the main cause of the findings and based on their evaluation several measures took place to avoid similar defects in the future.

Based on the experience of the NDT's executed in the past, the inspection team generated and followed a methodology to decide the frequency and to select the most appropriate NDT methods for any specific ammonia tank in the future.

Dr Theodora Kouloura
COO, Kavala Fertilizers Ltd

Introduction

Kavala Fertilizers Ltd (KF) is the only ammonia and nitrogenous fertilizer producer in Greece. Production facilities are located in North-East Greece and include several production units, including an ammonia plant, a nitric acid and two sulphuric acid units as well as an AN/CAN fertilizer production unit. Additionally, various utilities, such as production of demineralized water and cooling water, power generation from waste heat, ammonia storage tanks, and off-site facilities (bagging and palletising units, storage tanks for acids, and bulk storages for sulphur and fertilizers) are involved.

More specifically, KF has operated a semi-refrigerated ammonia sphere since 1965 and a refrigerated, single wall, atmospheric ammonia tank since 1982. Both tanks are subject to certification and are inspected, maintained, and operated following the rules and codes of well-known practices.

In this framework, several methods of non-destructive testing (NDT) were carried out in the past during the preventive inspection of the tanks, in order to improve their safety and operation. The inspection team, based on its experience previous testing, follows a methodology and determines the frequency and the most appropriate NDT method for each tank.

Design Conditions of the Tanks

The ammonia sphere, TK-901 (Figure 1), was constructed by Nuovo Pignone in 1965 for storage of the final product of the ammonia plant. TK-901 has a capacity of 1,800 MT (1,980 ST) of ammonia.

Figure 1. KF's semi-refrigerated ammonia tank TK-901

The sphere has a volume of 3,000 m^3 (790,000 gal), an internal diameter of 17.9 m (58.7 ft), and thickness of 20-25 mm (0.8-1.0 in). It is supported by 12 legs, partially insulated up to 800 mm (31.5 in) length, and the center of the sphere is 10.5 m (34.4 ft) above ground level. The material of construction is carbon steel (ASTM A 299 FBQ). The tank was initially insulated by fiber cement, composed of 30 mm (1.2 in) of FENOLITE, two-10 mm (0.4 in) layers of asbestos cement, and one layer of impermeable glass fiber. In 1985, after approximately 20 years of operation, the original insulation was replaced with polyurethane foam, along with two layers of vapor barrier material and protective aluminum sheets.

The design pressure of TK-901 is 4.5 bar_g (65 psig) and it operates at a pressure of 2.7 – 3.0 bar_g (39 – 44 psig) and a temperature of approximately -2°C to 2°C (-28 to 36°F).

The atmospheric pressure ammonia tank, TK-1201 (Figure 2), was constructed by UHDE in 1982. It is a single wall storage tank capable of storing liquid ammonia unloaded from ships. The capacity of the tank is 15,000 MT (16,500 ST) of ammonia. It has a volume of 22,000 m^3 (5.8 MMgal), 36.6 m (120 ft) internal diameter, and 22 m (72 ft) height. The tank was constructed according to API 620. The material of construction is the cryogenic steel TTstE29/TTstE36 (ASTM.A516-Gr60/Gr80). The shell and roof of the tank are insulated by polyurethane foam in situ flat aluminum sheets (110 mm (4.3 in) and 170 mm (6.7 in) respectively). The bottom is insulated by 50 mm (2 in) sand, 150 mm (5.9 in) foam glass, and 200 mm (7.9 in) foam concrete along the outer zone.

Figure 2. KF's ammonia tank TK-1201

The design pressure of TK-1201 is 1100 mm H_2O/ -50 mm H_2O (43.3 in H_2O / -2.0 in H_2O). The normal temperature of the liquid ammonia is -33°C (-27°F).

Common NDT's Used for Ammonia Tank Inspection

Several series of NDT have been conducted on TK-1201 and TK-901, to ensure their safe operation. NDT is a testing and analysis technique applied to evaluate the properties of a material, component, structure or system for characteristic differences, or welding defects and discontinuities, without causing damage to the original part.

There are two categories of these tests. The first one includes the most common techniques like Visual Inspection, Magnetic Particle (MP) inspection, Ultrasonic Test (UT) and Hydraulic Test. The second one is the "new generation" of NDT, and includes Alternative Current Field Measurement (ACFM) and Acoustic Emission

(AE). The following are summaries of these common techniques.

Visual Inspection (VT) is one of the oldest and most widely used NDT method. This method includes the visual observation of the surface of the object under investigation in order to evaluate any discontinuities (i.e. corrosion, misalignment of parts, physical damage, and cracks). VT can be utilized by looking at the test object directly or by making use of optical instruments such as mirrors, borescopes, and computer viewing systems. It is a cost effective method that can provide immediate results based on a minimal preparation and without requirement of special skills. However, it is a suitable method only for surfaces that can be easily accessed, and only larger defects can be detected.

Magnetic Particle (MP) is the NDT process which uses magnetic fields to find discontinuities at or near the surface of ferromagnetic materials. The magnetic field can be created with a permanent magnet or an electromagnet, which requires a current to be applied. The magnetic field will highlight any discontinuities as the magnetic flux lines produce leakage, which can be seen by using magnetic particles that are drawn into the discontinuity.

Ultrasonic Testing (UT) is one of the more common non-destructive testing methods performed on materials. This technique utilizes high frequency mechanical energy to conduct examinations and measurements on a test area. Typically the UT inspection system consists of an ultrasonic transducer, pulser/receiver, and display unit. A pulser/receiver is an electronic device that can produce high voltage electrical pulses to the transducer. When driven by the pulser, the transducer generates high frequency ultrasonic sound energy and delivers it into the material in the form of sound waves. When there are discontinuities, such as inclusions, porosity, or cracks in the sound path, part of the mechanical energy will be reflected from the discontinuities' surface. The reflected sound waves signal received by the transducer is then transformed back into an electrical signal and its intensity is shown on the display unit. The sound waves travel time can be directly related to the distance that the signal has travelled. From the signal, information about reflector location, size, orientation, and other features can be determined.

Hydraulic Testing (HT) involves the pressurization of equipment with water in order to examine the ability of withstanding the applied stress and testing for any leakages. That pressure typically is ten to fifty per cent over the design operating pressure. Hydraulic test assures standards and durability of a vessel over time. The benefits of an assessment are that it greatly reduces the cost and provides a quick turnaround. However, this method requires to be performed with caution by competent personnel.

Alternative Current Field Measurement (ACFM) is an electromagnetic technique used for the detection and sizing of surface breaking cracks in metallic components and welds. It combines the advantages of the alternating current potential drop technique and Eddy Current Testing in terms of defect sizing without calibration and the ability to work without electrical contact respectively. The ACFM probe introduces an electric current locally into the part and measures the associated electromagnetic fields close to the surface.

Acoustic Emission (AE) detects and monitors the release of ultrasonic stress waves from localized sources when a material deforms under stress. AE testing works by mounting small sensors onto a component being tested. The sensors convert the stress waves into electrical signals, which are relayed to an acquisition computer for processing. As the damage grows in the component, there is a greater release of energy.

AE and ACFM are categorized as "new generation" NDTs and have advantages and disadvantages when compared with the classic NDTs. First of all, AE is an affordable test that can be executed while the equipment is in service, is very fast, and can provide immediate results through a personal computer. The most important of these is the ability to inspect a whole tank without decommissioning the tank. Additionally, the data are collected and stored in a manner that makes it easy to follow up on findings of previous inspections. In the initial steps of the AE method the inspection of the floor on cylindrical tanks when no room for installation of sensors on the bottom was a weak point. But the last decades a specific methodology is implemented with sensors established on the periphery of the tank.

On one hand, it is necessary to take into consideration some drawbacks of AE tests:

- In case of serious findings the storage tank must be shut down for more thorough inspection and repairs, removing the cost saving benefit of avoiding the shut-down of the tank. This cost can be high and in some cases (e.g. plants which operate only one ammonia tank) not permitted.
- The execution of AE requires removal of some of the tank's insulation, in order to attach the AE probes. The partial removal of the insulation is a disadvantage since it creates a risk for icing in the area of removal and weak point that could lead to insulation deterioration.

On the other hand, ACFM is a very fast and accurate inspection method, but it has the disadvantages of internal inspection methods analyzed in the next section.

Disadvantages of Internal Inspections

There are serious disadvantages involved in opening of an ammonia tank in order to execute internal inspections:

- Loss of operability of the tanks during the preparation of the tank, the execution of the NDT's, and the startup of the tank
- One has to warm up the tank and this can give rise to extra stresses because the tank slightly increases in size.
- Completely emptying the tank is not easy. Emptying the tank with pumps will leave some liquid ammonia. This remaining ammonia will have to be evaporated. Processing of this gaseous ammonia vapor can be difficult. Depending on available facilities, the ammonia can be absorbed in water, it can be condensed and transferred for use, or it can be flared.
- Before one can enter the tank all remaining ammonia vapor present in the tank has to be removed. This requires purging with nitrogen followed by air.
- When entering the tank residual oil from compressors (with dissolved ammonia in it), may be present which has to be physically removed by vacuum and washing.
- When the internal inspection is completed, it is important to remove the air before putting ammonia in the tank. Residual oxygen in the tank increases the risk for stress corrosion cracking (SCC).
- Mixing of air and ammonia vapor must be controlled to stay out of the explosive mixture regime.
- Sometimes not all facilities are available at a plant to perform the necessary steps in a safe and environmental friendly manner. This can complicate the safe decommissioning and recommissioning of the tank, by requiring special rental equipment.
- The above-mentioned procedures that have to be applied when entering a tank take a lot of time (several weeks) and air could be used in the decommissioning step to remove ammonia gas from the tank, and ammonia gas could be used to remove the air at commissioning. In this way time could be saved by not having to perform the intermediate nitrogen purge step (to prevent the formation of an explosive atmosphere) [2]. However the contents of the

tank would go through the explosion limits of the ammonia/air mixture. This is not a preferred method, even if flame arrestors are used. It is clear that the long time needed for an internal inspection is a big disadvantage compared to a non-intrusive inspection [3].

- In any case the oxygen in the ammonia tank with or without the presence of ammonia could cause several problems and it is towards the safety of the tank to be avoided
- A few ammonia plants need a tank to store the product to keep the plant running. Having to stop the ammonia plant for a long period due to a tank inspection can be very costly.

Rules of thumb for Inspection Intervals decision

There is no standard or protocol for the determination of the inspection intervals [1] specifically for the atmospheric pressure ammonia tanks. The findings from previous inspection, the experience of the user as well as the general inspection of the equipment by a third party certification body could indicate the date and the appropriate method of the next inspection.

A Risk Based Inspection (RBI) which involves the planning of an inspection on the basis of the information obtained from the above noted analysis/parameters should be executed. As primary elements of RBI estimation of probability of failure and consequences are determined. As secondary elements of RBI, inspection requirements and effectiveness as well as its impact on the ammonia tanks were considered.

Based on the experience from the findings, an inspection of the tank under operation could be executed in the five years intervals. On the other hand, an internal inspection is decided when several critical indications are detected as well as indications of external corrosion, failure of insulation and new Regulations/Technical standards are issued. Considering that ammonia tanks suffer from oxygen inside the tank as well as from fluctuation of the temperature above and below the dew point of the ambient air, the inspectors are focused to avoid an internal inspection. There are references of atmospheric ammonia tanks in operation for more than twenty years that were never internally inspected.

Inspection History of Cylindrical Tank, TK-1201

The tank had been internally inspected once in 1991, according to the designer's instructions. The inspection plan included MP Test of the shell, floor, and all attachment welds, along with shell thickness measurements (due to external corrosion from icing underneath of the insulation). A leak test using vacuum box of all roof and floor welds was also performed.

Some weld defects were discovered, especially at the bottom welds, and some broken anchors were found. All findings were repaired, and tested with radiography testing and MP.

The atmospheric ammonia tank does not follow a specific Code or Standard for inspection intervals and the recommendation based on bibliography was to avoid empty them. According to the experience with AE it was decided to proceed with AE testing from time to time.

During the period of 2003-2004 the tank was inspected using AE. The AE data collection was conducted by Euro Physical Acoustics (MISTRAS GROUP) and included a loading and unloading cycle of the tank, as shown in Figure 3. The data was evaluated using the MONPAC method, developed and licensed by Monsanto.

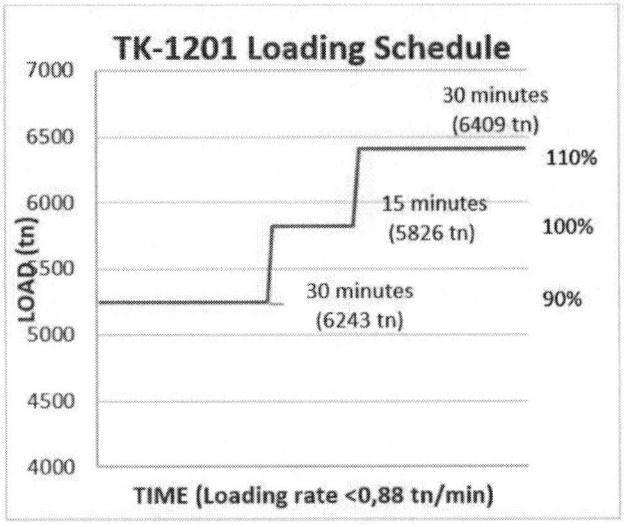

Figure 5. First loading cycle in 2016 inspection.

According to MONPAC analysis and grades (Figures 4 and 5) during both cycles, the global behavior of the tank was stable and no active leaks were observed on the tank floor.

The MONPAC procedure, applicable to both new and in-service vessels, provides a method of AE testing and data analysis based on laboratory results and field tests of over 1000 metal tanks, pressure vessels, and piping systems. The related specific software is used to grade the intensity of any source of acoustic emission recorded during the test.

The MONPAC method is an experience-based method and is able to identify and grade AE activity, resulting, for example, from cracking, corrosion damage, overstress, in order to determine if there is a need for detailed local inspection. It does not give information on defect size or orientation but grades the AE activity within the prespecified zones on the vessel (from "insignificant" to "intense") and gives recommendations for the extent of any follow-up inspection. Location is primarily "zonal" and indicates the "active" area/zone. Planar location is, sometimes, possible when a source is detected by at least 3 sensors. This defines a smaller search area. The method does not identify manufacturing defects if they have not propagated during subsequent service.

GRADE	COLOUR	INTERPRETATION	RECOMMENDATIONS
A	Green	Very Minor Source	*No action* Results recorded for further test
B	Cyan	Minor Source	*Visual Inspection* Surface defect such as corrosion, pitting, gouges and cracked attachment welds etc.
C	Magenta	Source	*Indications-Flaws and/or Defects requiring* immediate or short term follow-up evaluation. Evaluations may be based on further AE data analysis, rete, or complimentary NDE by other NDT methods or AE re-examination.
D	Yellow	Intensive Source	*Significant defects requiring* immediate follow-up inspection using complimentary NDE methods.
E	Red	Critical Source	*Major defects Requiring* immediate shut-down and follow-up NDE.

Figure 4. MONPAC grades.

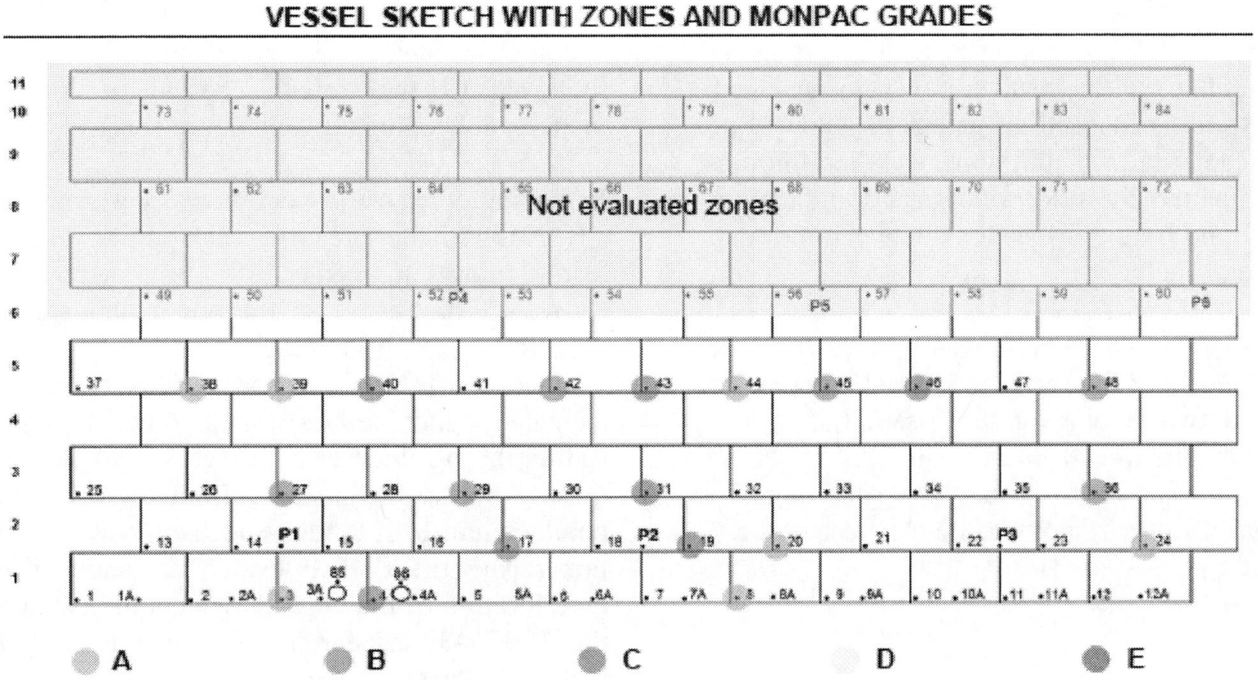

Figure 5. Vessel sketch with zones and MONPAC grades

During 2016, AE data collection was again conducted by Euro Physical Acoustics, and included two loading/unloading cycles. The MONPAC analysis of this data showed that TK-1201 did not appear to have any locations with leakage characteristic.

In conclusion, after the first internal inspection of the tank, which indicated a few defects caused either during the decommissioning or from the initial installation, KF follows a program of AE inspection of the tank under operation.

Inspection History of Ammonia Sphere, TK-901

Since 1965, the ammonia sphere TK-901 has been globally inspected eight times. Three of the inspections were internal and the remaining five used AE while the sphere was in operation. After each inspection, findings were recognized and remedial actions taken. As a result of this, not only were the findings repaired, but many measures were taken to prevent the same occurrences in the future.

In 1985, the first internal inspection of TK-901 was performed. The duration of the inspection was several months and a total replacement of the initial insulation was executed. Inner welding inspections were conducted using MP. The inspection showed critical findings on the equator and upper tropical zone. The findings were repaired by welding and a thermal process (annealing) was executed. Due to extended damage of the insulation and icing/humidity, it was decided to proceed on exterior welding inspections using MP but no findings were detected. There were also no findings for at the welds between the support legs and the shell. Although no thickness reduction was found in the measurements of pipelines. In order to facilitate the operation of emptying the tank in the future, a modification of the tank was decided and a new bottom outlet pipeline had been installed. Finally, a hydraulic test of TK-901 was executed at a pressure of 4.8 bar_g (70 psig), with satisfactory results, following the requirements by the Standard and the Third party Certification rules.

In 1991, a second internal inspection of TK-901 was executed. The duration of the inspection was two months, but the insulation was not replaced. Inner welding inspections using MP showed no significant findings, but some smaller findings were repaired by mild grinding. No indications were found by the inspections between the foot-holds and shell.

In 1995 an AE inspection was performed while the sphere was in service. The AE inspection required two weeks for the installation of the detectors, one week for test while TK-901 was in operation, and two weeks for repairing the insulation (after removing the AE test probes). (Figure 6).

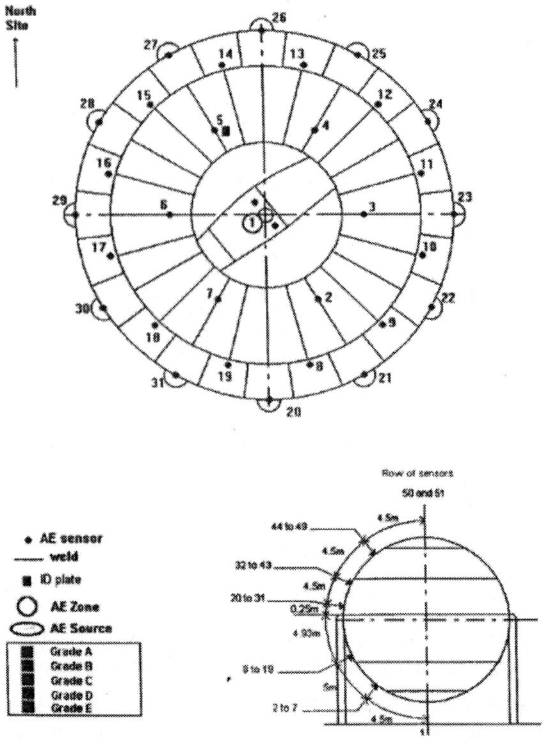

Figure 6. AE inspection – Position of sensors

This was the first time that a "new generation" NDT like AE was carried out at KF. Although there were two findings, no actions were taken. The findings, which were at the equator and at the bottom of the sphere, were categorized as class A, and no action was needed according to the MONPAC procedure. The AE inspection lasted five weeks (including the period of installation of the probes and removal of them) keeping the tank in service all this time against total almost two months period of internal inspection keeping the tank out of service.

In 2001 another AE inspection was performed on TK-901, with the same five week preparation and testing duration as the 1995 test. AE results showed a finding C on the upper hemisphere. According to MONPAC procedure, further evaluation was required. For this reason, UT was executed. and showed small cracks, 2 mm (0.08 in), and undercuts. It was speculated that these cracks originated from the repairs conducted in 1985, and the decision was made to not repair them. The evaluation showed no deterioration of the class A defects indicated by the 1995 AE inspection. Finally, the foothold welding inspection, with the use of MP, revealed two new findings. These two non-continuities were repaired by grinding.

In 2005-2006 TK-901 was inspected again, using AE. The preparation and testing were completed in five weeks, as were the earlier AE tests. The findings were similar those found in 2001, but this time three class C indications were found. After UT inspection, the cracks and undercuts (possibly originated from repairs of 1985) were not repaired. Finally, the leg to shell welds were inspected using MP, and revealed two findings. These two non-continuities were repaired by grinding.

In 2010 TK-901 was inspected again internally. The decision was based on the necessity to fully replace the insulation which was destroyed. The duration of the inspection was several months and TK-901 was sandblasted and its insulation totally replaced (Figure 7). Moreover, according to the new seismic factors, the tank support tie-rods were reinforced (Figure 8) according to the national law.

Figure 7. Setting up the new insulation

Figure 8. Installation of new tie rods.

After MP inspection of the interior welds there were difficulties evaluating the findings due to the weakness of MP test (many very small indications appeared which were impossible to be evaluated), and ACFM, was chosen to inspect the inner welds of TK-901. This was the first time ACFM, another "new generation" NDT, was used for TK-901. There were no notable findings (not even in the area where C findings were detected from the previous A test), and minor findings were removed by mild grinding. The exterior welds were also inspected with ACFM and non-notable findings were repaired by mild grinding. Finally, a hydraulic test was successfully performed on the sphere. The difficulties on evaluation of the findings from the classical NDTs and the decision to proceed on implementing modern inspection method (ACFM), as well as the upgrade of the tie rods due to the new seismic regulation, were the basic reasons to have an extended period of this inspection of the tank.

In 2016 TK-901 was inspected again, using AE. There were no significant findings. Based on these test results, TK-901 was certified for the next five years. In Figures 9 and 10 the sensors' position as well as the AE loading cycle are illustrated.

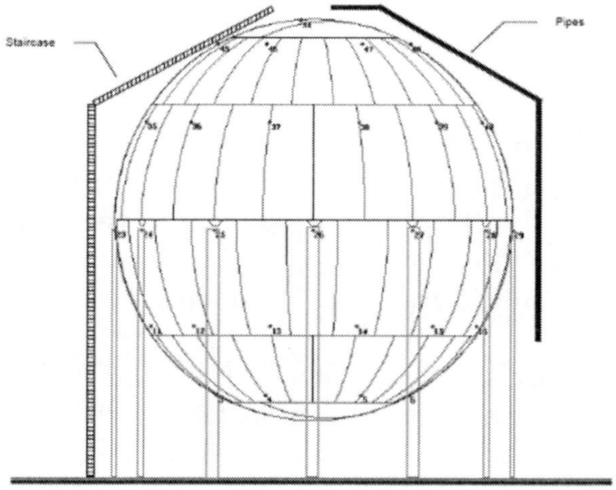

Figure 9. Vessel sketch with dimensions and sensors positions during AE tests in 2016.

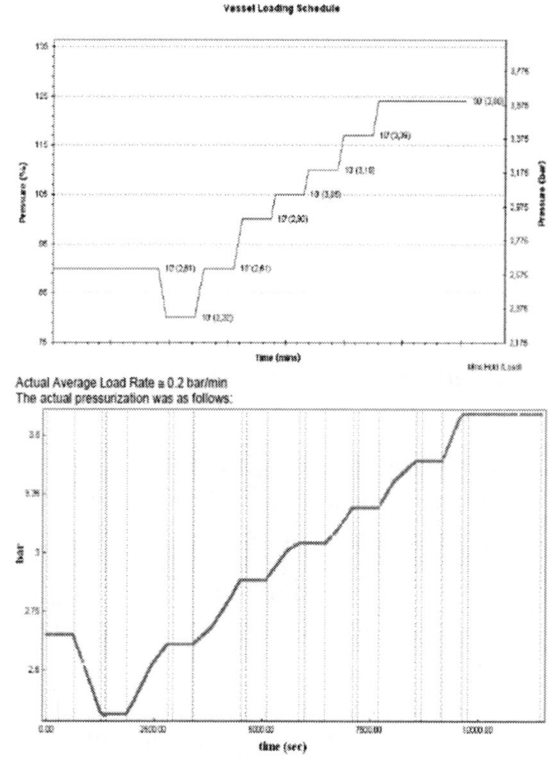

Figure 10. AE loading cycle in 2016.

Conclusion

Ammonia tanks, atmospheric or pressurized, are equipment that need to be inspected properly and on a regular basis. This paper describes the importance of the NDT inspections for ammonia storage tanks.

There are several inspection techniques, and one of the most important differences between them is the operational condition under which the NDTs are effective. Some NDT must be executed internally after the decommissioning the tank and others are capable of recognizing possible defects while keeping the tank in operation.

A basic disadvantage of the AE method is the external inspection. The insulation might be affected and if not repaired properly corrosion problems due to humidity and moisture intrusion could develop.

The essential advantage of the "new" NDT generation, like the AE test, is found no defects are suspected. The more commonly used NDTs are more preferable when there are indications that defects will be found, and subsequent repairs will be necessary.

Based on Root Cause Analysis (RCA):

1. All the replacement of the insulation of the tank is kept under very precise rules in order to avoid icing on the external surface of the tank
2. Any welding in the internal surface of the tank is followed by thermal treating to avoid initiation of SCC

Finally, we have learned that the long shut down and other disadvantages of internal inspections due to requirements of decommissioning and commissioning of the ammonia tank can be avoided. If AE examinations are used, immediate results will be available while the tank is still in operation. AE is a faster method (no need to empty the tank or to refill it properly) to find and evaluate possible defects but fully incapable to maintain them.

Based on the experience of the previous inspection the current inspection plan is to execute AE while keeping the tank in operation. Even more importantly, when insulation must be repaired or replaced, it is recommended to avoid exposing the internal parts of the tank to atmospheric conditions (oxygen). During these situations the tank could be filled with nitrogen.

Acknowledgements

We would like to express our gratitude to all those who gave us the possibility to complete this paper. We want to thank:

1. Kavala Fertilizers Ltd President and the Board of Directors, who give us the opportunity to attend the conference
2. AIChE, who invited us to attend the 65th Annual Safety in Ammonia Plants and Related Facilities Symposium.

References

1. Recommendations for the safe and reliable inspection of atmospheric, refrigerated ammonia storage tanks, EFMA, 2008
2. Guidance for inspection of and leak detection in liquid ammonia pipelines, Fertilizers Europe Association, Issue 2014
3. "A Non-intrusive Inspection of a 25000 m^3 Ammonia Storage at DSM Agro". Ammonia Technical Manual, 2009.

Damage Mechanisms and Lifetime Estimations in Steam Reformer Tubes

Creep damage is the primary cause of reformer tube failure in steam methane reformers (SMR). However, other damage mechanisms related directly or indirectly with creep material deterioration might also have a significant role in reformer tube failures. Additionally, remaining life assessment of steam reformer tubes is a critical point in ensuring steam reformer reliability. The article will describe different case studies where the end users benefited from the experience of a collaboration in addressing potential risks and improve plant reliability.

Dr. Pablo Cardín, Pedro Imízcoz, Dr. Dietlinde Jakobi
Schmidt+Clemens Group

Introduction

Reformer tubes are one the most critical and costly components in steam methane reformers (SMR). Reformer tubes have a severe impact on plant replacement costs, since they constitute up to 30% of the total reformer price. [1, 3] Furthermore, reformer tubes can be a major cause of plant unavailability, especially in case of unexpected failures.

Due to higher hydrogen demand in the last decades, size of recent primary reformer units has increased significantly. A typical primary reformer for a 400 MTPD (441 STPD) ammonia plant designed in the 1960s had less than 200 reformer tubes, a modern reformer unit for a 3,000 MTPD (3,307 STPD) ammonia plant has around 400 reformer tubes. [1] Nowadays, the average heat flux of a reformer tube is often above 116.2 kW/m^2 (36,760 BTU/ft^{2}h), approximately twice the heat flux that was industry practice 20 years ago. [2] Process and material advances, like our latest development Centralloy® G4852 Micro R, have made possible such tube production capacity increases.

The primary failure mechanism of reformer tubes is still creep damage. Nevertheless, other life limiting mechanisms can play a key role in reformer tube failures as reformer units became larger and operate at more severe conditions. [3] This paper will describe some of the most relevant failure mechanisms observed in primary reformer tubes and how such damages can be addressed, minimizing plant potential unexpected failures and increasing process reliability. Furthermore, this paper describes remaining life assessment methodologies used by Schmidt+Clemens, emphasizing creep damage assessment approaches, following different methods available.

Creep Damage

Creep is the slow deformation of a material under a stress that leads to a permanent change in shape. Creep typically takes place at deformation rates less than 1.0%/min. [4]

When steam reformer tubes operate at high temperatures, thermal stresses across tube wall and internal pressure lead to creep damage. Progressive alteration (ageing) of the alloy matrix creates dislocation paths that gather at grain boundaries causing the distinctive voids in the structure that are typical for creep damage incidence (Figure 1). As damage progresses, creep voids align, leading to micro crack formation and finally, tube material failure. [3]

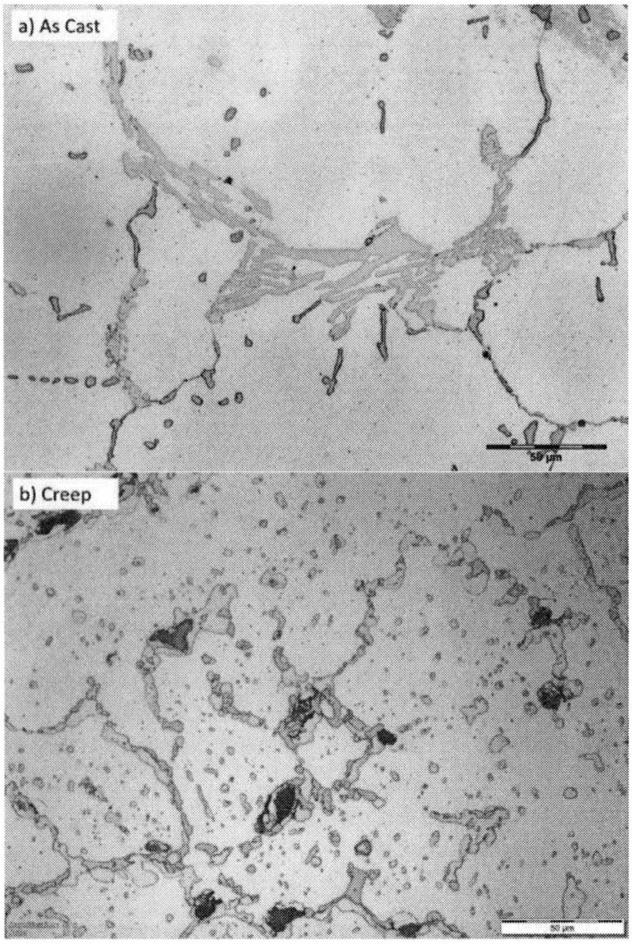

Figure 1. Microstructure of HP-Nb MA material a) As cast, b) Aged, creep damage

Stress concentration across tube wall thickness causes creep to start near the inner surface of the tube. Creep damage typically occurs over a large part of the circumference and the axial part of the tube. Leading to the distinctive tube diametric expansion and final tube rupture in a longitudinal direction. [3, 5, 6]

Centrifugally cast materials are the main choice for steam reformer tubes. Creep resistance in cast alloys is achieved by the presence of around 0.4 wt% of carbon in the alloy, forming a complex network of carbide structures in the austenitic matrix. Typical centricast materials used in steam reformer tubes are based on alloy HP-Nb (25Cr35NiNb), Schmidt+Clemens (S+C) Centralloy® G4852. [5]

Further improvements in creep resistant materials were the "micro alloys" (HP-Nb MA), like S+C's Centralloy® G4852 Micro, where additions of strong carbide forming elements (Ti, Zr and others), create an added strengthening effect due to the formation of additional carbides.

S+C developed the latest generation of high creep resistant materials, "Micro R materials" or Centralloy® G4852 Micro R. In this improved material, small controlled additions of strong carbide forming elements and balancing several key alloy element contents, produce the precipitation of very fine, nano sized precipitates (Nb, Ti) (C,N) that are uniformly dispersed in the matrix. [3]

Higher creep resistant materials, like Centralloy® G4852 Micro R, can significantly increase reformer tube life under creep regime, over previous materials, minimizing failure risks and increasing operational flexibility in reformer units.

Effect of Tube Chemical Composition on Creep Resistance.

As noted before, creep resistance relies on the formation of carbides in the alloy structure. Microstructure consists in austenite dendrites surrounded by eutectic carbides in the interdendritic region (chromium and niobium carbides). In addition to this network of primary carbides formed on grain boundaries during tube material solidification, small carbide precipitates (secondary carbides), generate in the austenitic matrix during reformer tube operation. [5]

Consequently, it is important to adapt the composition of micro-alloying element additions (Ti, Zr, etc.) and other alloying elements like Nb to obtain the best and most durable high temperature creep resistance. Moreover, since an optimal carbide precipitation pattern is required, it is desirable to optimize the composition and nature of these secondary carbide precipitates so they remain stable and unchanged during long periods of operation. [5]

Unbalanced alloy chemical compositions might lead to lower creep resistance. Main elements like Ni, Cr and Si also have a strong impact on high temperature creep resistance. The next example addresses the importance of a correct chemical composition in order to improve reformer tube reliability.

Figure 2 shows the microstructure of two ex-service HP-Nb MA tube samples. S+C supplied one sample (a), while a competitor supplied the other (b). Both tubes were overheated after 1.5 years in service and S+C performed a remaining life assessment of them.

Even though both microstructures appear to be similar, G4852 Micro structure has a finer and more evenly distributed precipitation of secondary carbides compared with the material from other producer. Additionally, Ti rich structures (TiCN) are visible in S+C sample. No such structures are visible on the other sample.

Chemical analyses of samples detected that the S+C sample was complied with the composition requirements of the tube designer. The competitor's sample was not within compositional requirements:

- Ni contents 1-2 wt% lower than specified.
- Si contents higher than 1.8 wt%.
- High Mn contents to compensate lower weldability due to high Si levels.
- No relevant Ti contents.

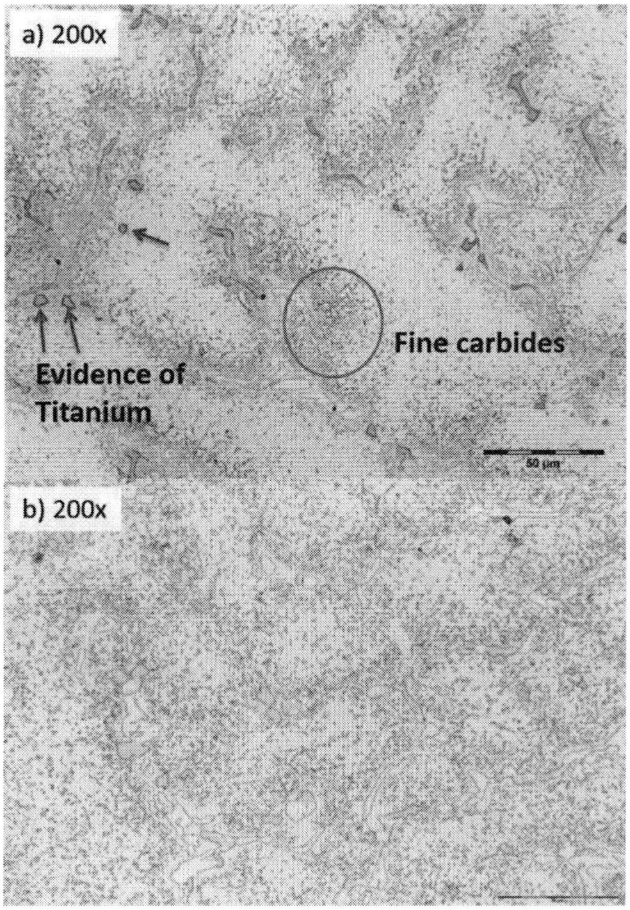

Figure 2. Microstructure of aged HP-Nb MA samples -a) G4852 Micro, b) Different supplier material

Creep testing of samples found that the S+C sample was still within the minimum creep requirements of alloy G4852 Micro. The other sample was not (see Figure 3).

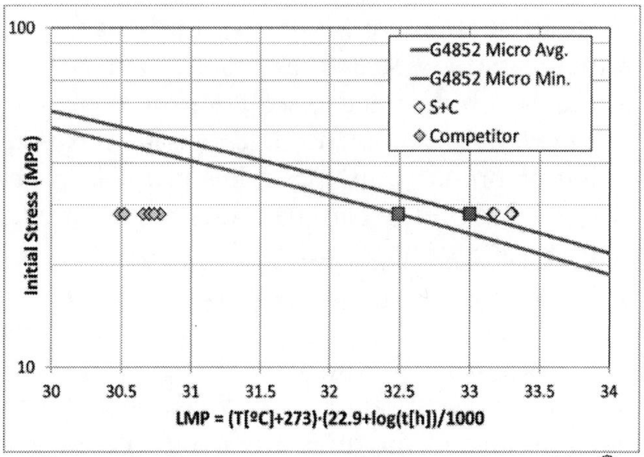

Figure 3. Creep test results over Centralloy® G4852 Micro parametrized creep strength curve

The sample from a different manufacturer failed in service after a short time in operation. S+C G4852 Micro tube, even considering the overheating effect, was still fit for service.

Overheating

Overheating is one of the most common failure modes found in steam reformer tubes, with several cases assessed by S+C each year. Tube metal temperature variations have a dramatic effect on reformer tube life, due to creep damage acceleration. A temperature increase of 10°C (18°F) over design temperature will reduce the reformer tube life approximately 40%. [3, 6]

Different operational factors can lead to tube overheating: Burner mechanical issues, over-firing during furnace start-up, catalyst deactivation/poisoning, lack of steam control, or inadequate burner regulation. Additionally, it is not easy to have a good thermal balance in large steam reformers that have both hot and cold areas in the reformer box.

Excessive reliance on furnace instrumentation is very dangerous if no specific tube metal temperature data is available. This is especially important during start-ups and plant upsets. Visual inspection and/or pyrometer temperature checking of reformer tubes is highly encouraged.

Alloy overheating induces a massive disruption of material structure (Figure 4). Early ageing of primary carbide structures (round and coarser primary carbides) and dissolution of secondary carbides. In more severe stages, partial re-dissolution of primary carbides takes place with abnormal/amorphous carbides structures formed. Extreme temperatures will lead to liquation of the austenitic matrix and fast part failure. [5, 6]

Furthermore, high temperature exposure leads to the rapid incubation of creep voids close to the outer surface of the tube, producing generalized cracking along grain boundaries and interdendritic zones, with damage progressing to the inner surface of the tube (Figure 4). Damage leads to a localized tube diametric expansion and final tube rupture on either longitudinal or transverse direction.

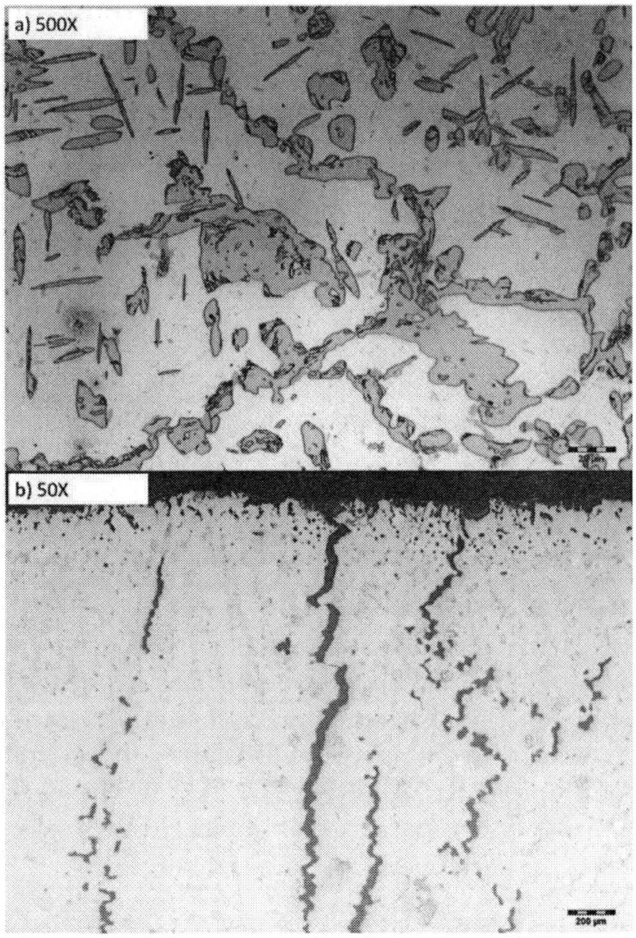

Figure 4. Microstructure of overheated HP-Nb MA samples -a) Deformed carbides, b) Creep related cracks

Metal Dusting

Metal Dusting is a kind of corrosion damage similar to oxidation attack of carburized metal. It is typically observed in industrial furnaces, boilers, and heat exchangers that operate in strongly reducing atmospheres. Compared with normal corrosion, it occurs randomly and progresses much faster. Ellipsoidal pits, found on the metal surface, are a characteristic indication of metal dusting (Figure 5). [6, 7]

Metal dusting generally takes place at temperatures from 480 to 815°C (900 to 1500°F), while the most critical temperature for metal dusting corrosion is around 600°C (1112°F). [7]

Steam reformer process gas, with its high carbon activity (a_c), can induce metal dusting corrosion attack on steam reformer tubes and outlet components. Modern steam reformer designs tend to avoid critical temperature areas where metal dusting attack might take place.

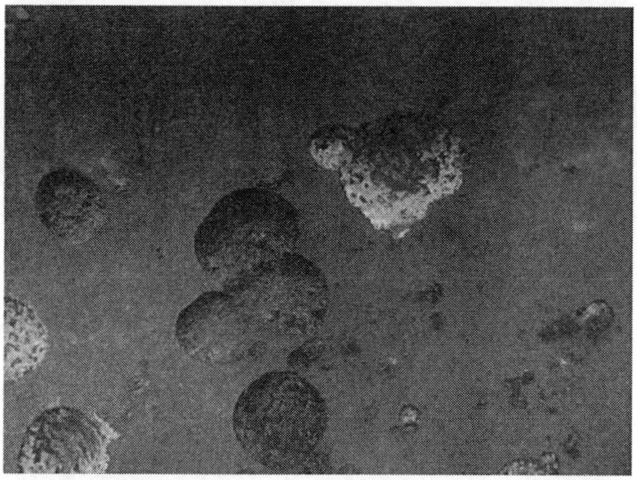

Figure 5. Metal Dusting "pitting" on a HP-Nb MA sample (maximum pit depth around 10 mm)

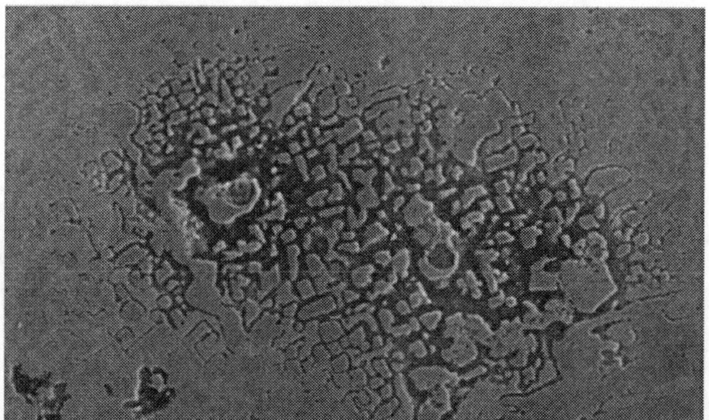

Figure 6. Metal dusting corrosion inside a HP-Nb MA sample (SEM image, transversal section)

Nevertheless, damage or deterioration of insulation equipment in the outlet portions of tubes can produce metal dusting corrosion attacks as seen in Figures 5 and 6 and incidence of metal dusting damage in reformer tube components still occurs occasionally.

Dissimilar-Metal Welds (DMW)

Dissimilar-Metal Weld (DMW) failures take place in weld joints between low alloy carbon steels and austenitic steels, mainly in boilers and heat exchangers. DMW failures occur in old steam reformer designs, containing weld joints between Cr-Mo steel flanges and austenitic tubes and S+C examined several cases in the past years.

There are three root causes of DMW failures: carbon migration from the heat-affected zone (HAZ) of the ferritic part into the weld, thermal expansion differences between ferritic and austenitic steels and differences in corrosion resistance to flue/process gases between these two metals. [7]

Some steam reformer designs have areas, typically around flanges, where the process gas temperature reaches the steam dew point, leading to the formation of carbonic acid (resulting from carbon dioxide dissolving in water). Differences in corrosion resistance between the weld material and a Cr-Mo alloy flange generates a galvanic cell that leads to the corrosion of the flange, creating an oxide wedge that penetrates the flange. As carbon migrates from the flange to the weld material, a continuous and thermally stable carbide layer forms between the weld and the flange. Carbides precipitated in the fusion line, help the progression of corrosion damage (Figures 7 and 8).

Martensite forms between ferrite band and weld metal due to the dilution of alloying elements from the steel and the weld metal. This layer acts as a stress raiser compromising the structural integrity of the weld joint (Figure 8). Differences in thermal expansion coefficient between flange, weld and tube materials, tend to further increase stresses on the weakened area.

Figure 7. Dissimilar-metal weld failure between austenitic tube section and ferritic flange

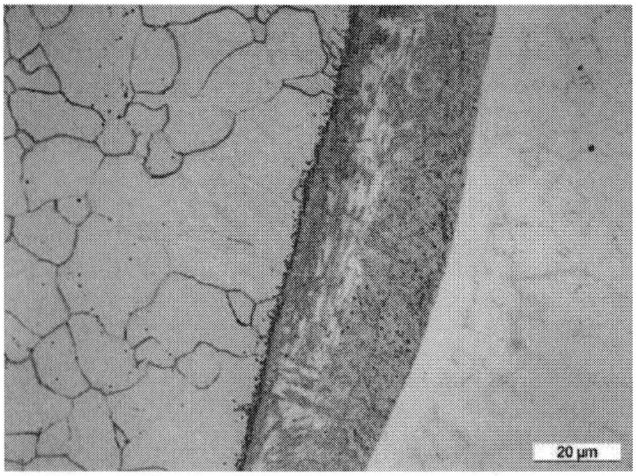

Figure 8. Carbide precipitation and martensite formation on the weld fusion line (material ASTM A182 Grade F11)

Considering these facts, a metallurgical notch takes place, inducing material cracking along carbide layer of the fusion line. In order to avoid such problems, it is recommended to replace Cr-Mo flanges with stainless steel ones.

Sulphur and Sodium Corrosion

Hot corrosion damage found in steam reformers is quite similar to oxidation damage. However, damage evolution is significantly faster than normal oxidation processes. Such cases still happen occasionally and S+C has performed several investigation reports related with massive reformer tube failures.

The most common cause of hot corrosion in steam reformer tubes is the contamination of dilution steam due to issues with the pH control base compound (caustic or amine). [8]

Inadequate pH control in the dilution steam drum can produce carry-over of liquid droplets or even slugs of liquid into the reformer tubes. Caustic carry-over can severely damage steam reformer tubes and designers recommend limiting sodium (125 to 500 ppb) in the combined feed and dilution steam, in order to prevent caustic corrosion. [8]

The presence of sulfur enhances corrosion attack by the formation of corrosive Na/S compounds (typically Na_2S). Molten compounds are formed which disrupt and destroy the formation of protective oxide layers (Figure 9). Additionally, sulfide scales formed are not protective, accelerating corrosion kinetics by depleting the chromium content of the tube. [7]

Figure 9. Sulfur corrosion attack on outer tube surface (material HP-Nb MA)

Impurity issues can also arise on the combustion side of the furnace. Fuel gas is susceptible to contamination from recirculating flows and cross-connection with streams added into the fuel system from other plant sections. Avoiding such contaminant gradients is the only way to prevent this type of corrosion damage.

Stress Corrosion Cracking (SCC)

Stress Corrosion Cracking (SCC) is a failure mode related to the simultaneous presence of tensile stresses, an aggressive environment, and a susceptible material. Two theories explain the mechanisms of SCC: electrochemical mechanism and stress-sorption mechanism. According to the first, a galvanic cell creates a corrosive environment that together with applied stresses open a notch on the material. Further corrosion and stress incidence lead to a generalized cracking. [7]

Stress-sorption theory indicates that damage incubation takes place by an initial notch that can be either mechanical or corrosion related. This initial notch progresses by applied stresses and by the incidence of corrosion damage, which prevents material self-healing. [7]

Crack propagation in stress corrosion cracking undergoes extensive crack branching and follows, in general, a perpendicular path to the applied stresses. Cracking can propagate either intergranularly or transgranularly depending on the specific material/environment combination (Figure 10). [7]

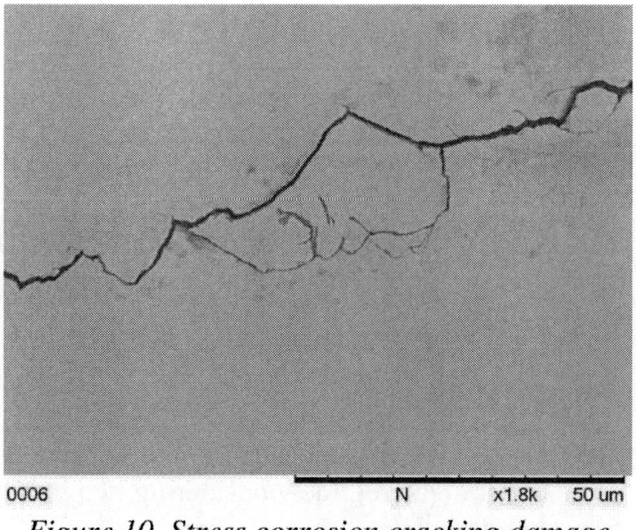

Figure 10. Stress corrosion cracking damage (SEM image, alloy 304H)

S+C inspected several cases of SCC damage in steam reformers, taking place in low alloy components purchased from external suppliers like top pipes and flanges, linked with deterioration of insulation and the presence of external contaminants in the process gas (like Zn, Na, Cl, Al, Mg, P and Ca). Insulation damage, leads to steam condensation in these areas, forming an acidic condensate in which gas contaminants converge. This situation creates a highly corrosive environment, initiating corrosion damage on these low alloy components. Internal pressure adds the required stresses needed for SCC damage propagation.

Removal of these external contaminants and an adequate maintenance of these critical items (top pipes and flanges) are the most effective ways to prevent SCC damage during operation.

Remaining Life Assessment (RLA)

Typically, design codes and specifications used for the manufacturing of reformer tubes, do not cover life assessment of components that have been in operation for a certain period. Nowadays, API 579-1/ASME FFS-1 Fitness for Service Code includes a remaining life assessment (RLA) section for components operating in the creep range. [9]

Schmidt+Clemens has provided remaining life assessment (RLA) services for more than 40 years for high alloy materials used in steam reformer applications. S+C follows a strict methodology for evaluating material deterioration, with different methods used to characterize material status: dimensional control, chemical composition, macrostructure inspection, microstructure inspection (optical), microstructure inspection (SEM/EDS), tensile tests at room or high temperature, accelerated creep tests, and carbon/nitrogen content profiles.

Most important point in performing accurate lifetime estimations is an adequate sample selection, able to represent the overall status of the furnace.

Without a representative sample, it is impossible to provide an accurate and realistic evaluation of the status of the entire steam reformer unit.

Careful analysis and evaluation of all tests performed, based on our long experience in material inspection, helps provide a remaining life estimate. In many cases, our material characterizations detected hidden problems within the reformer units inspected.

Accelerated creep tests are probably the most decisive information considered for reformer tube life assessments. Unfortunately, these tests have limitations as an excessive test acceleration might over estimate actual creep damage. Test sample preparation, with the removal of outer areas, might underestimate actual creep damage, if this damage is located on those external sample areas.

S+C uses two different methodologies to evaluate remaining sample life, based on accelerated creep tests: Larson-Miller method and Omega (Ω) method.

Larson-Miller Method

The Larson-Miller Method (LMP) is one of only two methods, together with the Omega method, available to model creep behavior according to API 579-1/ASME FFS-1 remaining life assessment code. [9]

Manufacturers like S+C provide stress to rupture values for their alloys, parametrized in models like Larson-Miller. These plots represent the relation between applied stresses, temperature and time to rupture by using a parameter named Larson-Miller Parameter (LMP):

$$LMP = f(\sigma) = T \cdot (\log t_r + C)$$

where:
T= Temperature in Kelvin
t_r= Time to rupture in hours
C= Material constant

Parametric stress-rupture strength representations, consider both an average value and a minimum one or lower scatter band at a 95% confidence level. Lifetime estimations normally use the minimum value, in order to be on the "safe side", as recommended by API 579-1. [9]

Based on this parametrized model, it is possible to calculate the theoretical time to rupture of a non-serviced (as-cast) creep test sample, at a given temperature and stress level. Ex-service sample tested at the same temperature, stress level and taken to rupture, provides the information required to perform a lifetime estimation. Simpler stress to rupture tests (STR) provide this data and there is no need to perform more costly and technically complicated creep tests (with continuous creep elongation measurement).

Lifetime estimation follows the "life-fraction rule" or "linear cumulative damage rule", considering that the amount of total life consumed in operation by creep damage follows [6]:

$$\%\,total\,life = \frac{\begin{array}{c} rupture\ time\ of \\ ex-service\ sample \end{array}}{\begin{array}{c} rupture\ time \\ \text{"new" sample} \end{array}} x100$$

Remaining life assessment following this method is intuitive, easy to perform and its accuracy can be excellent with an adequate selection of testing conditions. Typically, target time to rupture values must exceed 500 hr in order to provide an accurate estimation. As a "rule of thumb", longer target time to rupture test conditions provide a much better representation of actual sample creep damage.

Omega Method

The Omega (Ω) Method assesses life of components under creep regime, considering two material characteristics: alloy creep rate and a coefficient for the shape of creep-curve.

This method simplifies the three-stage creep model by the assumption that after a short "primary creep" stage, strain rate (creep rate) is continuously increasing and material strain rate can be defined by a series of material measured parameters and a dimensionless factor named Ω, which describes the evolution of creep strain with time (factor dependent of stress, temperature and material). [10]

Based on this model, creep rupture takes place when strain rate becomes infinite, while it is possible to evaluate remaining life by measuring creep rate strain at service stress and temperature on a creep tested ex-service sample. In the meantime, omega value (Ω) can be determined from material test data available. The following equations describe the Omega method [10]:

$$\dot{\varepsilon}_{os} = \frac{\dot{\varepsilon}}{\dot{\varepsilon} t_s \Omega_p + 1} \qquad LFC = \frac{\dot{\varepsilon} t_s \Omega_p}{\dot{\varepsilon} t_s \Omega_p + 1}$$

where:

$\dot{\varepsilon}_{os}$= Original strain rate of material in service

$\dot{\varepsilon}$ = Strain rate of ex-service material

t_s = Time in service

Ω_p= Material factor, test condition dependent

LCF= Life Consumed Fraction (0 means no life consumed, while 1 indicates no remaining life left)

The main limitation of the Omega method is the lack of an extensive creep rate database in comparison with well-characterized LMP type models. Additionally, results are dependent on creep strain values, strongly affected by testing equipment mismatches. In addition, creep tests are typically more expensive and technically complicated than simpler stress to rupture tests (STR).

Use of Creep Test Information for RLA

S+C uses the LMP and Omega methods for RLA in steam reformer samples. The next examples show the use of both methods.

Case 1. Two ex-service samples (A and B) on alloy HP-Nb. Creep test conditions: temperature 800°C (1472°F) and stress of 50.5 MPa (7324.4 psi).

Under these conditions, the "new" Centralloy® G4852 sample has an average time to rupture of 1945 hr. and a minimum value of 1000 hr.

Table 1 shows test results as time to rupture (t_r) and calculated strain rate ($\dot{\varepsilon}$), LCF represents Omega Life Consumed fraction, while LMP Min. and LMP Avg. are consumed life time percentage based on minimum and average LMP creep curves respectively. Both methods provide comparable results that were in accordance with other inspections indicating that both samples have mostly consumed its life in operation.

Sample	tr (hr)	$\dot{\varepsilon}$ (%/hr)	LCF	LMP Min.	LMP Avg.
A-1	110	0.1459	0.93	89.0%	94.3%
A-2	95	0.1506	0.93	90.5%	95.1%
B-1	92	0.1985	0.95	90.8%	95.3%
B-2	83.5	0.1824	0.94	91.7%	95.7%

Table 1. Case 1. RLA of ex-service samples, using LMP and Omega methods

Case 2. Four ex-service samples (A to D) on alloy G4852 Micro (HP-Nb MA). Creep test conditions: temperature 900°C (1652°F) and stress of 47 MPa (6820 psi).

Under these conditions, "new" Centralloy® G4852 Micro sample has an average time to rupture of 2610 hr. and a minimum value of 940 hr. Table 2 shows test results of these creep tests.

In this case, additional tests performed could not detect any relevant deterioration, with the exception of some minor creep damage found in sample B. Some of the creep tests planned were terminated after 1540 hr testing, as requested by the client. Like in previous case, data obtained show comparable results between both models, with a better alignment of the average result of sample B in the LMP with the Omega, though Omega method provides more information that LMP in the non-ruptured tests.

Sample	tr (hr)	$\dot{\varepsilon}$ (%/hr)	LCF	LMP Min.	LMP Avg.
A-1	1180	9.0E-4	<0.1	-	54.8%
A-2	>1540	6.4E-4	<0.1	-	<41%
B-1	1305	1.3E-3	0.31	-	50.0%
B-2	931	1.9E-3	0.53	1%	64.3%
C-1	>1540	4.5E-4	<0.1	-	<41%
C-2	>1540	8.5E-4	<0.1	-	<41%
D-1	>1540	7.7E-4	<0.1	-	<41%
D-2	>1540	4.3E-4	<0.1	-	<41%

Table 2. Case 2. RLA of ex-service samples, using LMP and Omega methods

Both cases indicate that Omega results are more comparable with LMP average values than LMP minimum, which is the design basis for reformer tubes. Therefore, in some cases, Omega method might overestimate existing damage. In addition, creep strain rates of "as cast" materials are not so well characterized as parametrized time to rupture models like Larson-Miller, limiting Omega method potential and accuracy.

Conclusions

Creep damage is the primary cause of reformer tube failure in steam methane reformers (SMR). This paper describes other damage mechanisms that could lead to component failures, emphasizing the importance of a proper control of process parameters, reformer equipment and stream flow contamination.

Alloy material creep resistance depends on a balanced and controlled chemical composition. If these requirements are not met, reformer tubes would not be able to withstand the severe conditions of steam reformer operation.

Both Omega and LMP methods can provide a reliable remaining life assessment on reformer materials under creep. Omega method application has a strong potential, but still requires a better material characterization in order to improve its effectiveness.

References

1. *Reduce Stress and Increase Throughput of Reformer Tubes.* Richter, P., Zecevic, N. s.l.: Nitrogen + Syngas 2019.
2. *Reformer Follows Function.* Nielsen, N.K., Andersen, N.U., Landeira, M.A. s.l.: Hydrocarbon Engineering, Feb. 2013.
3. *Centricast Materials for High Temperature Service.* Jakobi, D., Huber, J. s.l.: Nitrogen + Syngas 2010.
4. *Superior Spun Cast Material for Steam Reformer furnaces: Alloy Centralloy® 4852 Micro R.* Jakobi, D., Karduk, P., Von Richthofen, A. s.l.: 55th Annual Safety in Ammonia Plants and Related Facilities Symposium 2010.
5. *State-of-the-Art of Materials and Inspection Strategies for Reformer Tubes and Outlet Components.* Gommans, R., Jakobi, D., Jiménez, J.L. s.l.: 46th Annual Safety in Ammonia Plants and Related Facilities Symposium 2001.
6. *Atlas of Creep and Stress-Rupture Curves.* Boyer H. E. (Ed.): ASM International, 1988.
7. *ASM Handbook Volume 11-Failure and Prevention.* Powell, G.W., Mahmoud, S.E. (Ed.): ASM International, 1998.
8. *An Overview of Self-Inflicted Contaminants.* De Haan, S. s.l.: 18th Ethylene Producers Conference, AIChE Spring National Meeting, 2006.
9. *Fitness-For-Service,* Code API 579-1 /ASME FFS-1, 2016.
10. *The Omega Method and Effective Method for Life and Damage Prediction in Creep Tests and Service.* Prager, M., in Strength of Materials, Oikawa et al. (eds.): Japan Institute of Metals, 1994.

Predictive Corrosion and Mechanical Failure of Static Equipment

In the past years, the Chemelot Site experienced several process safety incidents which resulted in a decrease in the manufacturing plants' reliability, availability, and an overall increase in process safety issues. The primary root causes of these incidents were corrosion and mechanical deterioration failures of static equipment and piping. This led to two main questions: "How to improve field inspections and (continuous) monitoring in order to achieve predictive end of life and maintenance?", and "Where and how to collect all the necessary information for managing Assets Integrity and Performance?" To answer these questions Sitech developed its own pragmatic approach of risk-based inspection methodology. Applying this methodology delivers continuous quality improvement and a lower quantity of information needed for predicting end of life and on time maintenance of static equipment and piping. The quality and quantity of data are frequently reviewed by an interdisciplinary team of subject matter experts in combination with a suitable IT-infrastructure in line with the Sitech risk-based inspection methodology best practice.

Ahmed Mhaoui
Sitech Manufacturing Services C.V., The Netherlands

Introduction

Located at the Chemelot Site in Geleen in the Netherlands, with a long history of hands-on experience and professional knowledge, Sitech Manufacturing Services is the leading Innovative Technology Partner for the manufacturing plants on the Chemelot site. Our goal is to help our customers to ensure that everything works in good harmony and to achieve cost-effectiveness while optimizing operations. We strive for a combination of expertise, reliability, integrity, and constant innovation.

Sitech manages services for Turnaround, Maintenance, Plant Modifications, and includes a Technology group. This Technology group consists of a pool of highly qualified experts with process, equipment, plant strategy, and hands-on asset knowledge.

In the past years, the Chemelot site experienced several Process Safety Incidents, which resulted in a decrease in the manufacturing plants' reliability, availability, and an overall increase in process safety issues. There were serious concerns about the plants' integrity, reliability, and availability. The primary root cause of these process safety incidents were corrosion and mechanical deterioration failures of static equipment and piping. Despite the extensive safety studies, residual life studies, and regular (legal) inspections, these deterioration failures continued to occur.

Other challenges for the manufacturing plants are aging equipment and piping, changing business operations, plant modifications, and high staff turnover (loss of experience).

Our customers wanted support and help to improve the overall site operation. Having a reactive approach and explaining why incidents occurred did not help predict "potential" incidents before they occurred. A shift in focus and mindset was necessary to proactively identify risks created by equipment failure.

The development of our own pragmatic approach for Risk-Based Inspection (RBI) started in 2015. In-depth research was done on what is known around the world about RBI. Several best practices were investigated, including their differences, the successes and pitfalls combined with our own (hands-on) experience on the Chemelot site. Also, the best available innovative tools and technologies were included. The collected knowledge was tested to make it pragmatic and sustainable for use on the Chemelot site. The methodology developed is still being applied and improved.

One of our customers on the Chemelot site is OCI Nitrogen. They have also experienced the necessary challenges in guaranteeing reliable, available, and safe operations. Because of those challenges, OCI Nitrogen conducted a Reliability study in collaboration with Sitech.

For Ammonia plants reliability and availability is key. Each shutdown and/or startup has a serious impact on operations. Therefore, it is very important to monitor and guarantee reliable operations and overall integrity.

In addition, Ammonia is the raw material for several other manufacturing plants at the Chemelot site. When incidents happen within the Ammonia plants, the reliability, as well as integrity, will be compromised within the other "connected" manufacturing plants.

At the request of OCI Nitrogen, a "Mechanical Integrity Assessment" was initiated in 2017. The assessment focused on scenarios related to mechanical integrity and turned out to be very time consuming.

The assessment was indeed a way forward to success, but it was still not able to predict end of life or on time maintenance during equipment (items) life cycle. There were uncertainties and assumptions about the damage state of equipment (items) due to corrosion and mechanical deterioration failures.

In January 2019, a RBI project was started by OCI Nitrogen on one of the ammonia plants to increase its reliability and integrity. Sitech was selected for this project because of our knowledge and experience.

Methodology

Risk-Based Inspection (RBI)

RBI has been known for almost 30 years. The methodology for optimizing the inspection planning of static equipment and piping has made several developments over the years. Since 2005, a lot of changes and new insights have been made to enable implementation of an effective RBI program. The methodology was always difficult to understand and use in practice. History has shown us many times how complex RBI can be. Due to these difficulties over the past years, several best practices have been developed worldwide.

All of these best practices contain "jigsaw" pieces about how to implement a suitable RBI program, but these separated pieces never provided an overarching methodology. Still, RBI was difficult to implement due to aging plants, changes in business operations, and the quality of workmanship in the field.

Several analysis methods were developed in the past, namely:

- full quantitative analysis,
- full qualitative analysis,
- semi-quantitative analysis.

Sitech's pragmatic approach is based on semi-qualitative analysis. The risk assessment puts the focus on:

- In-depth knowledge of damage mechanisms;
- In-depth knowledge of the damage state of equipment (items) and piping;

- In-depth knowledge and experience of an interdisciplinary team consisting of subject matter experts from Sitech, manufacturing plants, and other experts, when necessary.

Sitech's RBI model, shown in Figure 1, illustrates the workflow, with the yellow-shaded steps marking the differences compared to API 580 RBI Planning Process (Figure 2).

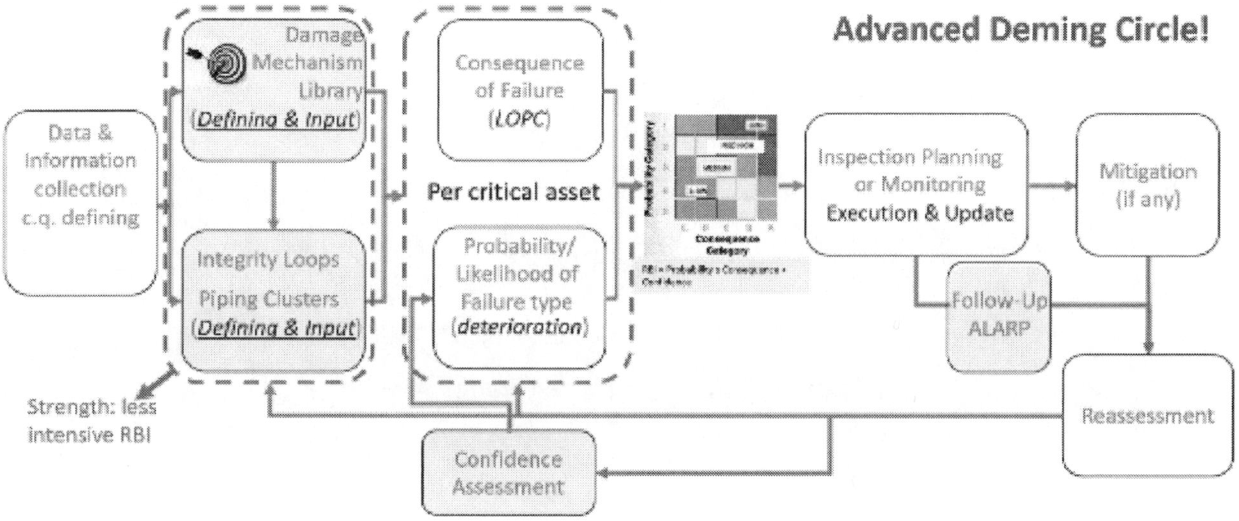

Figure 1. Workflow Sitech RBI Approach

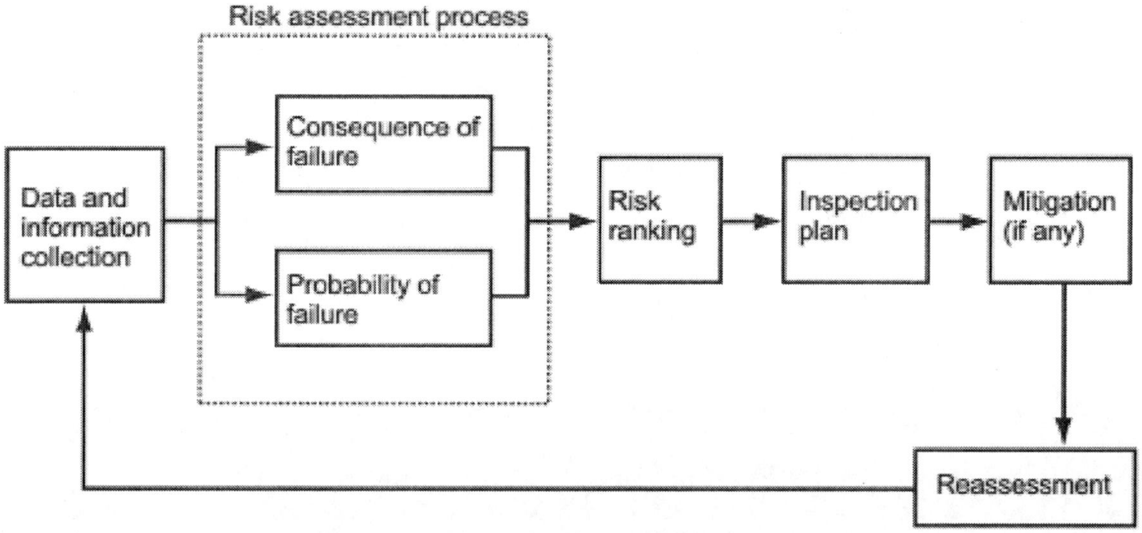

Figure 2. API 580 RBI Planning Process

Change management

When implementing Sitech's Pragmatic RBI approach, OCI Nitrogen and Sitech were aware that a change of mindset and behavior was also needed to complete the success.

For the change approach, the Lawson & Price Model, as shown in Figure 3, was used.

To monitor the change of mindset and behavior a combination of the Implementation Quotient (IMQ) Curve and the Change Curve, as shown in Figure 4, was used.

To make sure the maturity growth was in place OCI Nitrogen hired an independent third party to support and monitor the change management.

Figure 3. Lawson & Price Change Model

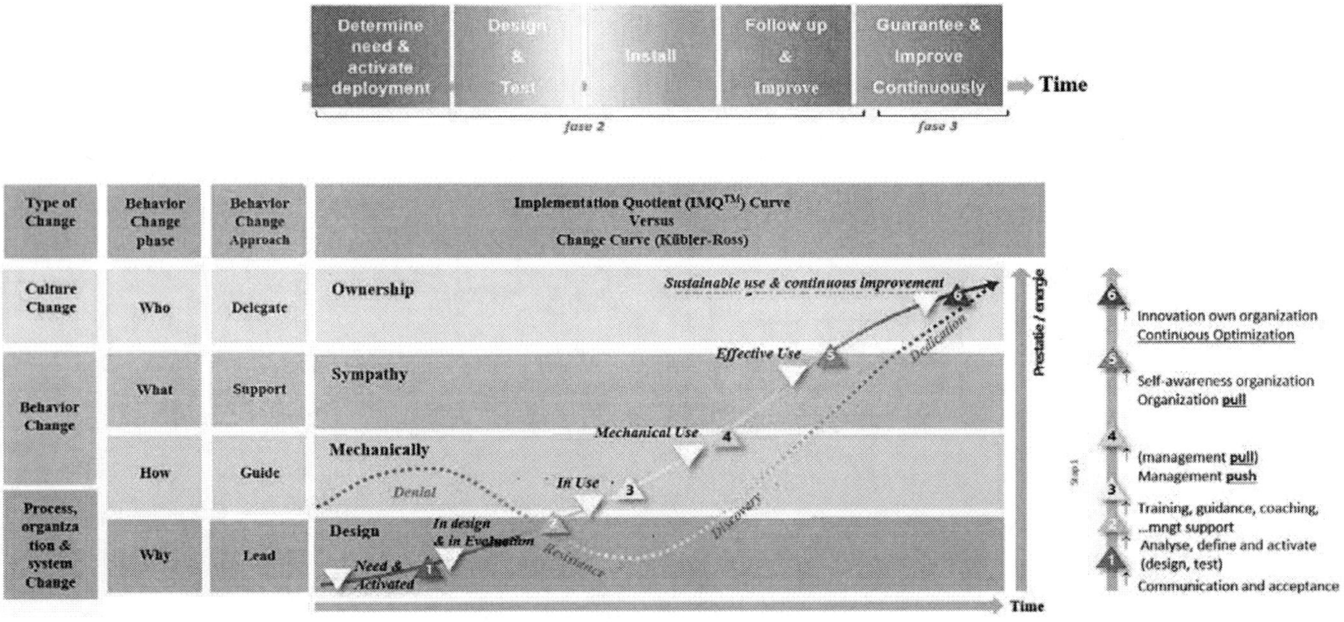

Figure 4. Combined IMQ & Change Curve

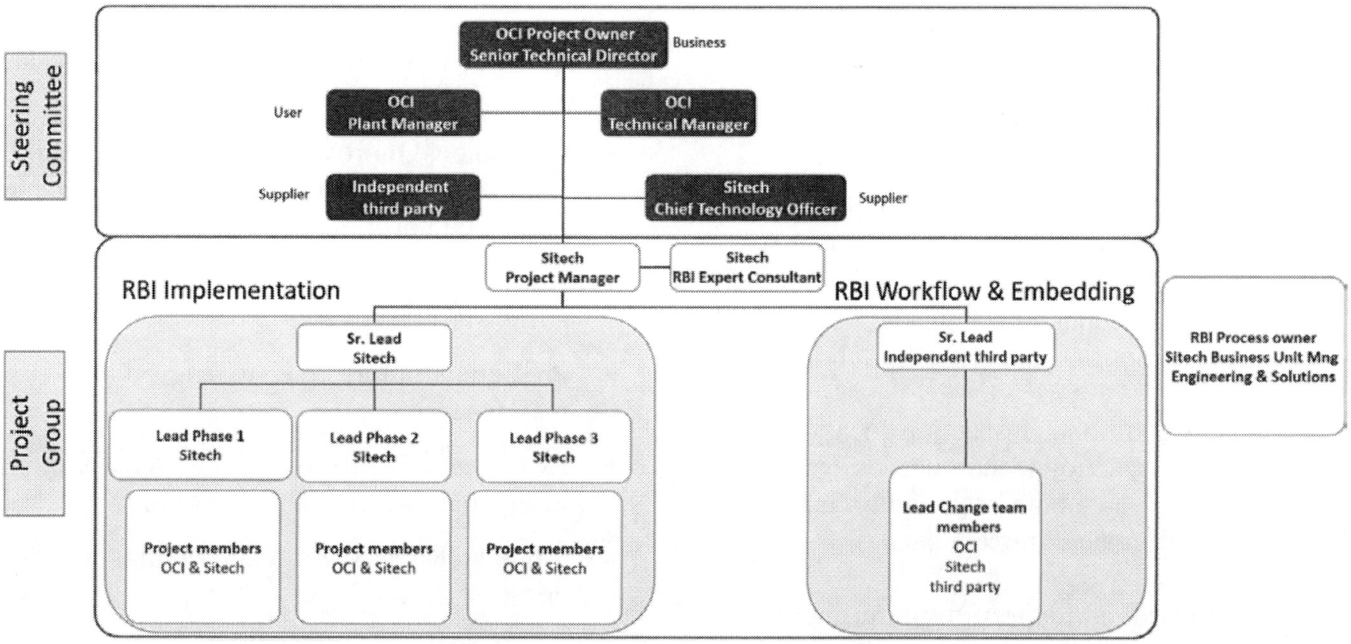

Figure 5. RBI Joint Project Org Chart

Implementation

The project was managed, advised, implemented, and audited by OCI Nitrogen, Sitech, and a third-party. This process is illustrated in Figure 5.

Due to the complexity of the project, it was divided into two simultaneous stages, namely:

1. Implementing Sitech's pragmatic approach of RBI;
2. Transition from a pragmatic approach RBI to running business and RBI for Life (DNA) by using a suitable work process, which was needed to embed the Sitech best practice in daily operations.

Stage 1 was divided into three phases, which are briefly described in the following sections.

Phase 1. Ammonia Damage Mechanism Library

Ammonia plants have a history of corrosion and mechanical deterioration failures, but the information was only available on paper documents and in the mind of the subject matter experts. During the project the history of corrosion and mechanical deterioration failures were combined with the Sitech Damage Mechanism Library knowledge. This resulted in an Ammonia Damage Mechanism Library with known and unknown deterioration failures for the ammonia plants to be used for evaluation and inspection of their equipment and piping.

Phase 2. Integrity Loops and equipment items

A Mechanical Integrity Assessment was conducted on one of the OCI Nitrogen plants from 2017 until early 2019. The information and results were used as "Front End Loading" data, together with the knowledge gathered from the Phase 1 "Ammonia Damage Mechanism Library", for the Phase 2 implementation.

Integrity Loops are a combination of corrosion loops (systematization of equipment items and piping with same process fluid, chemical resistance of used materials and similar consequence when failure should occur) with mechanical circuits (systematization of equipment items with similar construction, mechanical parameters, failure effect, etc.). By setting up Integrity

Loops it is possible to have a better understanding of potential damage mechanisms with the same likelihood, susceptibility, and a combined consequence.

With an interdisciplinary team of subject matter experts, each selected integrity loop, with systematized equipment items and piping, were reviewed. This includes checking the details, the facts, the assumptions, and the uncertainties.

As a result, additional inspection locations, revised use of inspection techniques, online "early warnings" process monitoring, and online material deterioration were implemented.

Phase 3. Improved inspection planning and monitoring

For inspection planning a list of equipment and piping was used which were scheduled for one of the ammonia plants to be maintained and inspected in the upcoming turnaround of November 2020.

Due to reviewing of the integrity loops and their systematized equipment items and piping, almost all of the inspection plans for the selected equipment had to be revised before use. Different inspection locations were added or removed. For example, the field inspection locations were changed into monitoring locations (like Integrity Operating Window, Design Window, or Safety Window). Several field shutdown inspection locations were changed into in-service field monitoring. For inaccessible locations, the use of Non-Intrusive Robotics was planned. Some field shutdown or in-service field monitoring were changed to Real-time material deterioration monitoring.

RBI Database

All the data collected, generated, evaluated, and reported have been gathered into a RBI Database. Sitech has different possibilities to gather RBI data in different tools for their customers. An IT-infrastructure was already in place. With some minor changes to the IT-infrastructure, the RBI best practice was able to be installed quickly. The benefits of this database include:

- Continuous improvement and single point of truth of generated information used for predicting end of life and maintenance;
- Knowledge assurance regardless of experienced staff turnover;
- Predicting failure, end of life and on time maintenance;
- Real-time critical process parameters monitoring and material deterioration;
- Suitable Risk Assessment in line with the best practice.

RBI Communications and reviews

During the implementation of RBI in the context of the change approach, these main structural RBI meetings were introduced:

- Plant Integrity Meeting consisting of an interdisciplinary team of subject matter experts. Meeting topics included:
  - Operational changes;
  - IOW alarms;
  - Incidents (LOPC's);
  - Short term and long term reviews;
  - Inspection & monitoring results and evaluations for update planning
- Management Integrity Meeting consisting of decision-making managers. Main subjects included:
  - Important announcements, proposals, and requests from Plant Integrity Meeting team;
  - Evaluation summary of inspections and monitoring;
  - Investments of additional inspection, monitoring, repairs, and replacements;
  - Management of Change items

During these meetings, the RBI database was used for evaluation and updates were made.

Results and Lessons Learned

During the project, the ammonia plant organization learned the benefits of having an interdisciplinary team of subject matter experts. The quality of end of life predictions and the required predictive maintenance schedule was improved using the team concept. The team was able to expand their knowledge in some areas through collaboration and interaction. The team also identified areas where their knowledge was limited and more evaluation was needed.

The efforts of the team resulted in mapping some unknown damage mechanisms which were not initially focused on, and identified known damage mechanisms with new inspection methods and monitoring locations. The quality and level of detail of reviewing the integrity loops, with their systematized equipment items and piping, identified the so-called "weak spots" and some previously unknown locations for potential failure of the equipment and piping. Overall, the probability of detection was checked for each damage mechanism and that information was applied to the damage state of equipment and piping. In this way, we learned that the expressions "assumption is the mother of all mishaps" and "it's all in the details" are true and determine the quality of the review process. Good communication is just as important as understanding the methodology.

Through the effort of understanding the methodology and implementing a change of mindset and behavior, it was possible for a mature organization to work together and successfully complete this project. Follow-up projects were initiated to implement the methodology for other equipment in the remaining ammonia plants and other facilities.

Examples by using best practice

Predictive unknown damage mechanism

OCI Nitrogen is investigating the use of biogas combined with the natural gas for fuel gas. The quality of natural gas changes. In combination with biogas, the chemical composition of the fuel gas deviates from standard values, which is found by applying chemical analysis techniques. The amount of acid forming components can be higher than with natural gas alone. One of the acid forming chemical compound is hydrogen cyanide, which provides an example of a damage mechanism not normally experienced in ammonia plants.

The interdisciplinary team of subject matter experts was asked if hydrogen cyanide could cause future failures, and if so what changes should be considered during project design engineering to prevent future failures.

In condensation and under deposit areas, high concentration of cyanide salts can develop. In these areas, there is a high likelihood of cyanide acid corrosion and cyanide stress corrosion cracking. The cyanide reacts with the iron (Fe^{2+}/Fe^{3+}) and forms a cyanide-iron complex. These affected areas can be recognized by cyanide bluish discolorations. In these areas, loss of wall thickness occurs or branched fine cracks initiate.

Both damage mechanisms are likely to occur especially when the selected material is carbon steel or low alloy steel. In this case, both the supply pipeline as well the fuel gas pipeline were chosen to be engineered in carbon or low alloy steel.

For cyanide stress corrosion cracking the likelihood is greater when the condensation or deposits occur around stress-oriented locations. Figures 6 and 7 show an example of a cyanide damage state, from a non-ammonia process application.

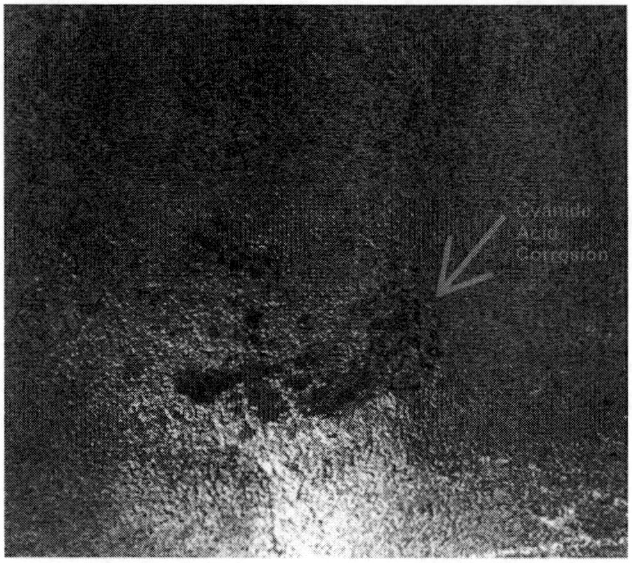

Figure 6. Cyanide Acid Corrosion

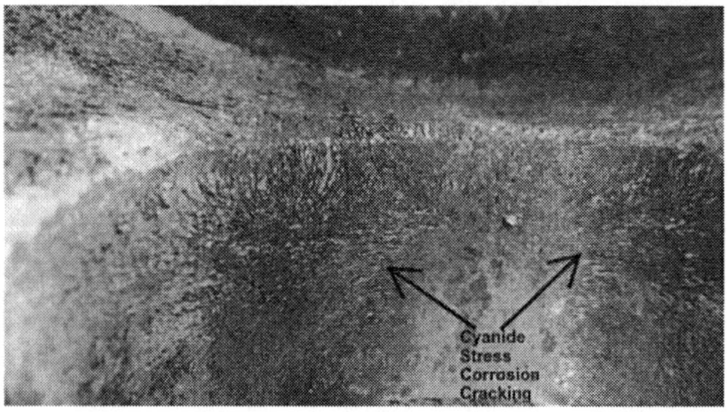

Figure 7. Cyanide SCC

Cyanide acid corrosion does not initiate quickly, propagates slowly, and causes local loss of wall thickness. The probability of failure in a short time of operation is low and the probability of failure after a long time of operation is very high.

Cyanide stress corrosion cracking initiates rapidly (within hours), propagates very slowly (takes years), and causes transgranular branched fine crack formation through the wall thickness. The probability of failure in a short time of operation is very low and the probability of failure after a long time of operation is high.

The consequence of failure in a short time is very low and the consequence of failure in a long time is a leak before break. If allowed to continue, there is a possible break before leak due to transgranular branched fine cracks that grow towards each other.

The probability of detection during field inspections is medium using Selenium isotope radiography and high using the combination of Time of Flight Diffraction (TOFD) or Phased Array Ultrasonic Testing (PAUT) techniques. The probability of detection using these techniques increases when correct preparations are performed in advance of the inspection, to get better results during field inspections.

In Figures 8 and 9 other examples of non-ammonia related examples of the cyanide damage state discovered during field inspections. The cracks are easy to detect due to correct preparations in advance.

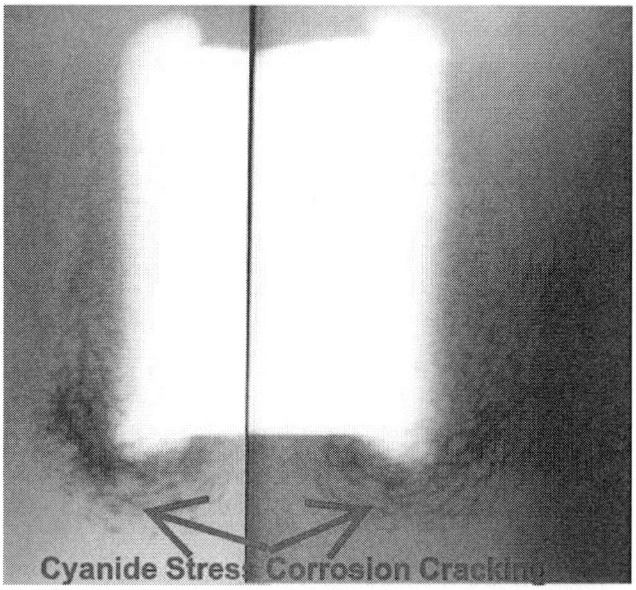

Figure 8. Field Radiographic Cyanide SCC

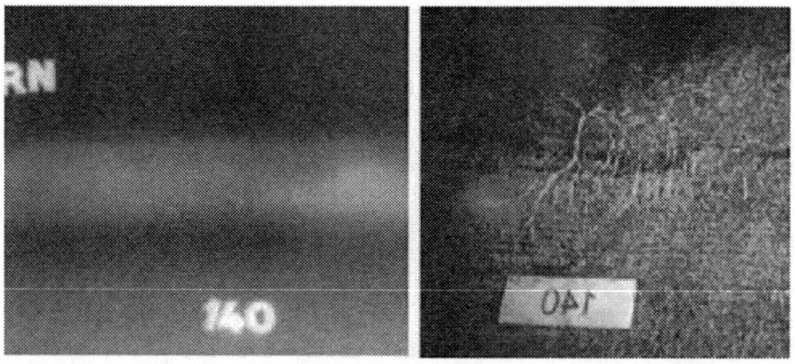

Figure 9. Field Radiographic versus Magnetic Particle Inspection (MPI)

As result of this, the following recommendations were made:

- Use non-susceptible materials, such as stainless steels, to mitigate the risk, eliminating the need to monitor early warnings or track degradation.
- When possible, remove hydrogen cyanide before it gets into the supply pipeline and beyond to the fuel gas pipelines.
- Limit the number of chemical compounds of hydrogen cyanides and monitor the amounts in relation to the damage mechanisms (Integrity Operating Window).

Preventive corrosion deterioration

During the project, several carbon and low alloy steel synthesis gas pipelines were reviewed. Previous investigations found some of these pipelines had carbon steel welds that did not have the correct heat treatment to prevent susceptibility to high temperature hydrogen attack (HTHA). The carbon steel piping was operated intermittently within the recommended criteria defined by the "Nelson Curve" found in API RP 941. The operating conditions (intermittently ~300°C (572°F) and ~35 barg (508 psig) with 75% H_2) are indicated in Figure 10.

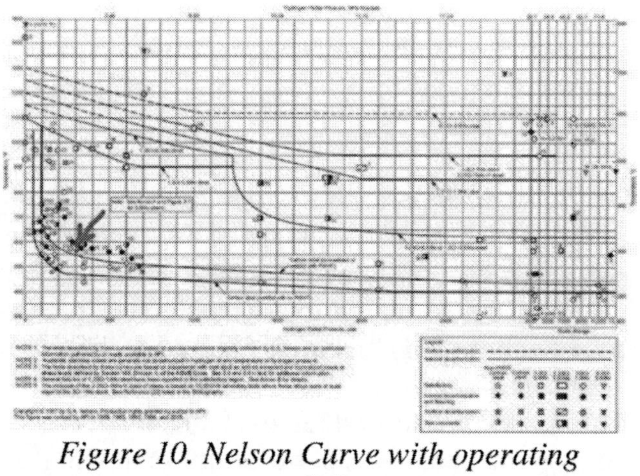

Figure 10. Nelson Curve with operating conditions

The interdisciplinary team of subject matter experts reviewed this information and found that not only was HTHA likely to occur in these pipelines, but there was also a risk of hydrogen embrittlement and thermal fatigue. Therefore, several additional locations were marked to for investigation, to understand more about the current damage state.

The strategic replacement of every affected pipeline to non-susceptible material was advised. Before the field investigations and the replacements were completed, a leak occurred near one of the welds (Figure 11).

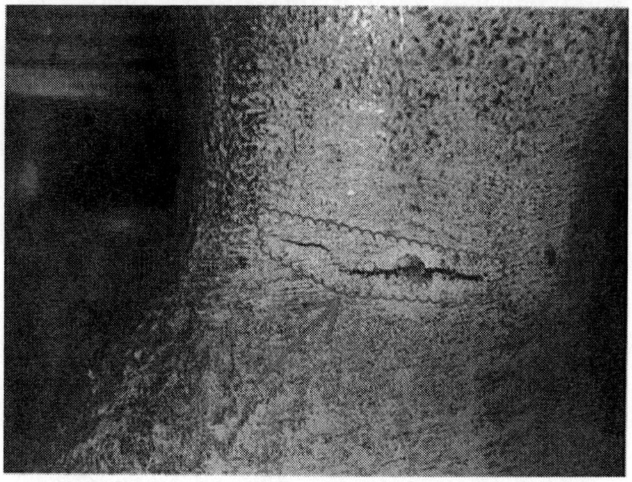

Figure 11. Leakage spot due to cracking

The cracked weld was successfully repaired, but, unfortunately, it was not possible to perform a fractographical investigation to understand the cause of the visual crack. Before welding, the area was soaked (hydrogen heat treatment, 30 minutes at 250°C (482°F) and after welding a Hydrogen Post Weld Heat Treatment (30 minutes at over 650°C (1202°F)) was done to reduce susceptibility to HTHA.

Because it was possible to repair the weld, the failure is not likely due to HTHA. If it was HTHA methane (CH_4) would be present in the molten weld, which could not be soaked out, and it would be visually seen.

Because of this event, it was not possible to wait until the turnaround to investigate the other welds. Therefore, multiple in-service NDT investigations were started, and are still in progress. Early strategical replacements to resistant materials is planned, in order to mitigate these damage mechanisms, and consequently the risks.

Predictive corrosion deterioration

For years, corrosion under insulation (CUI) has been a challenge to manage, control, and predict. Several questions arise when preparing for CUI inspection and maintenance, including:

- When and where to replace insulation?
- How to prevent corrosion from occurring or causing failure?
- How and when to perform correct inspection and monitoring?

In ammonia plants, the equipment is protected by different types of insulation (e.g. cold, hot, personnel protection, etc.), and the same challenges are in place as mentioned before. Ammonia plant personnel only think of the cost to remove the insulation (whether partially or completely) for maintenance and inspections. CUI costs a lot of money and delaying or not removing insulation is not an option. Additionally, it is not possible to make judgements based on a few local areas. Therefore, large amounts of insulation must be removed to evaluate CUI effectively.

As a technology partner, Sitech provide different kinds of inspection and monitoring best practice techniques. For one of our clients on the Chemelot site, a predictive CUI monitoring system was provided, in cooperation with a third-party.

With this real-time monitoring, it is now possible to detect early moisture and early corrosion (formation of iron-oxides) under insulation (Figure 12). The loss of wall thickness can also be monitored. This means on time maintenance, less insulation removal for visual inspections or NDT, and less NDT on insulation.

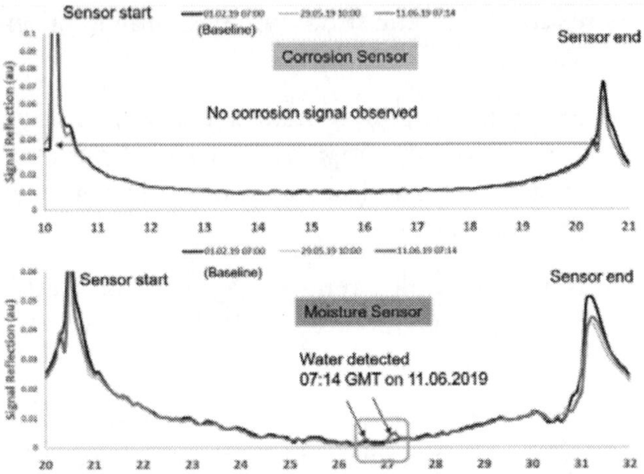

Figure 12. Moisture and Corrosion detection

OCI Nitrogen Plants were advised to use this technique when it was necessary to remove insulation on critical areas for inspection, for maintenance, or in case of repair or replacement. Otherwise, field investigations can be performed while the equipment is in-service or during shutdowns.

Conclusion

The RBI methodology best practice may seem like it is all known or it has already been done! However, that is not true. Despite the extensive safety studies, residual life studies, and regular (legal) inspections, equipment and piping deterioration failures continue to occur. Other challenges for the manufacturing plants were aging equipment and piping, changing business operations, plant modifications, and high staff turnover (loss of experience).

The best practice brings structure to the assessments, ensures that knowledge is secured, decreases time consuming studies, and increases the depth of the matter to act much more predictively rather than react after failures occur.

The examples show that by focusing more on damage mechanisms, periodically reviewing the damage mechanisms, and regularly reviewing the Integrity Loops with their systematized equipment items and piping, fewer unexpected failures will occur, both in the short and long term.

The strength of the interdisciplinary team is that knowledge not only comes together, but there is also interaction to separate assumptions from the facts. In this way, more attention is paid to converting assumptions into facts about the matter or the available history in order to be more predictive.

The in-depth review of damage mechanisms increases the likelihood of detection, and allows other techniques to be deployed during shutdown, as well as in-service inspections and real-time monitoring.

By understanding the methodology, getting RBI into people's DNA as well as a suitable RBI database, we close the continuous improvement Deming circle time after time.

References

1. Mhaoui, A. *Sitech Risk Based Inspection Methodology Best Practice.*
2. Schmitz, P. *Mechanical Integrity of Process Installations. An Assessment based on Bow-Ties.*
3. OCI Nitrogen. *Reliability study.*

Low-NO_x Retrofit Project to Meet New EU Emissions Directives and Increase Efficiency

To meet new EU directives on emissions and to increase throughput/capacity, a European fertilizer plant needed to replace 105 dual fuel (oil and gas) burners. The original burners had been operating in the facility's down-fired ammonia reformer since 1973. The plant required the new burners to fire a single fuel, such as natural gas or a fuel mix of natural gas and off gas, at NO_x levels less than 120 mg/Nm^3 (60 ppm), CO levels less than 10mg/Nm^3 (8 ppm), and noise levels less than 85 dB(A). EU directives on emissions require NO_x levels not to exceed 100 mg/Nm^3 (49 ppm). Zeeco met all criteria and achieved NO_x levels less than 100 mg/Nm^3(49 ppm) by retrofitting internal components of the facility's existing burners and reusing existing external components to minimize downtime and save on both installation and up-front capital costs. Internal components replaced included front plates, burner tiles, and gas gun assemblies. External components repurposed included wind boxes, pilots, and scanners. Zeeco tested two burners simultaneously under more rigorous conditions than those existing in the field to verify lack of flame impingement or flame-to-flame interaction. This paper will demonstrate an economical, efficient burner retrofit solution to meet EU emissions directives and company efficiency goals.

Nigel Palfreeman, Ashley Lynk
Zeeco Europe, Ltd.

Introduction

With global emissions regulations becoming ever more stringent, process heater operators constantly seek new ways to lower NO_x emissions while maintaining many of the benefits that traditional burner technology afford. In addition, operators continue to prioritize lowering overall operating costs.

Within the chemical industry, steam reforming is the most widely used technique for the large-scale production of ammonia. The largest process heater incorporated within a typical ammonia plant is the primary reformer. Located at the heart of the plant, its maintenance and operation represent a significant facility cost. Many ammonia plants located in Europe or across the globe have been in service for decades but installing new burner technology into aged heaters while looking to meet modern emissions standards can present an interesting set of challenges. In the past, Zeeco has used a range of Ultra-Low NO_x burner technologies, including Zeeco's GLSF Free-Jet and GLSF Min-Emissions burners, for new primary reformer installations and retrofit projects. Both burner types utilize Internal Flue Gas Recirculation (IFGR) as their primary method of NO_x reduction. More recently, Zeeco has developed the GB Single-Jet Ultra Low NO_x burner, which also uses IFGR to lower NO_x in flue gases from process heaters but does so in a simpler and more compact package.

Summary of Application

Zeeco recently completed the successful installation of a burner retrofit project within the European market. This involved the retrofitting of GB

Single-Jet burner technology into a 45-year-old primary reformer containing one hundred and five (105) down-fired burners. Zeeco was tasked with replacing the outdated burner technology which was producing between 250-300 mg/Nm^3 (120-150 ppm) NO_x at the furnace stack. A key performance issue with the burners was the high level of excess Oxygen (O_2) needed to ensure the complete combustion of fuel in the heater. This was leading to low overall fuel efficiency within the primary reformer's radiant section, resulting in increased operating costs.

To satisfy a short site turnaround window, the GB Single-Jet burner technology was installed into the existing burner plenum (see Figure 1). Zeeco achieved this by using only a new burner front plate, complete with fuel gas riser, gas tip, and cone assembly. The existing front plate and riser systems were directly replaced by Zeeco's equipment, but the existing burner plenums remained in place. New burner tiles were also installed without having to modify the refractory lining within the primary reformer. Reusing the existing wind-box and pilot assembly meant fewer modifications to the burner and furnace, and substantial cost savings for the customer. Overall, this approach also saved valuable time during the turnaround while achieving the main project goal; reducing NO_x emissions to less than 120 mg/Nm^3 (60 ppm).

Figure 1 shows the general arrangement detail for the solution implemented in both side view and a view looking up at the burner from the heater floor. The GB Single-Jet solution utilizes proven technology in use for more than 70 years (i.e. simple gas tip located in a stabilization cone) to produce very low NO_x emissions. Use of these simple components that have been trusted since the original construction date of the primary reformer furnace allow retrofit installations without compromises on burner mounting, performance, flame shape, NO_x emissions, Carbon Dioxide (CO_2) emissions, and temperature distribution along the tubes. In other words, a simple design is used to reduce the overall scope and to achieve complex performance requirements while saving substantial costs and valuable time.

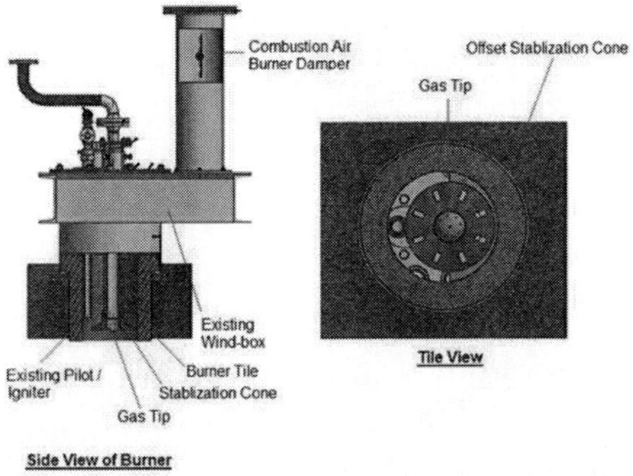

Figure 1. Zeeco's GB Single-Jet burner retrofit solution for down-fired primary reformer

Figure 2 shows the top view of the furnace burner layout. For this application, there are seven (7) rows of fifteen (15) burners for a total of one hundred and five (105) down-fired burners located in the primary reformer. The burner-to-burner spacing between the seven (7) rows is 2.2 m (7.2 ft) and the burner-to-burner spacing within each row is 1.05 m (3.4 ft). The roof area where the down-fired burners are located is approximately 229.6 m^2 (2470 ft^2).

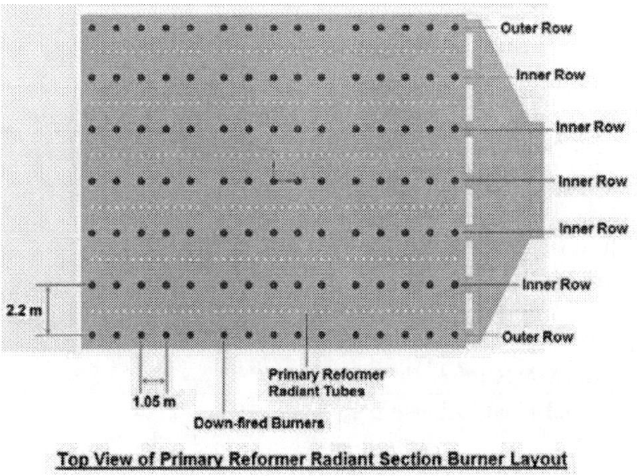

Figure 2. Primary reformer

Information on the primary reformer design, burner data, and fuel gas composition is shown in Tables 1,2, and 3.

Description	Primary Reformer
Type of Reformer Furnace	Ammonia Reformer
Type of Draft	Forced
Direction of Firing	Down Fired
Direction of Flame Shape	Downward
Primary Reformer Width, m (ft)	14 (46)
Primary Reformer Length, m (ft)	16.4 (53.8)
Primary Reformer Roof Area, m^2 (ft^2)	229.6 (2471)
Total Number of Burner Rows	7
Distance Tube to Tube Between Tube Rows, m (ft)	2.2 (7.2)
Total Maximum Heat Release for Reformer, MW (MMBtu/hr)	158.19 (539.77)
Total Normal Heat Release for Reformer, MW (MMBtu/hr)	131.84 (449.86)
Total Minimum Heat Release for Reformer, MW (MMBtu/hr)	27.42 (93.56)
Heat Release per Roof Area at Maximum Duty, MW/m^2 (MMBtu/hr*ft2)	0.689 (0.218)
Heat Release per Roof Area at Normal Duty, MW/m^2 (MMBtu/hr*ft^2)	0.574 (0.182)
Heat Release per Roof Area at Minimum Duty, MW/m^2 (MMBtu/hr*ft^2)	0.119 (0.038)

Table 1. Primary reformer data

Description	Inner Burners	Outer Burners
Type of Burner	Zeeco GB Single-Jet	Zeeco GB Single-Jet
Total Number of Burners	75	30
Number of Burners per Row	15	15
Number of Rows of Burners	5	2
Maximum Heat Release per Burner, MW (MMBtu/hr)	1.674 (5.712)	1.088 (3.712)
Normal Heat Release per Burner, MW (MMBtu/hr)	1.395 (4.760)	0.907 (3.095)
Minimum Heat Release per Burner, MW (MMBtu/hr)	0.290 (0.990)	0.189 (0.645)
Turndown	6:1	6:1
Available Combustion Air Pressure Drop, mbar (psi)	25 (0.36)	25 (0.36)
Design Excess Combustion Air, %	10%	10%
Combustion Air Temperature, °C (°F)	300 (572)	300 (572)
Furnace Temperature, °C (°F)	1200 (2192)	1200 (2192)
Normal Relative Humidity, %	79%	79%
Ambient Air Temperature, °C (°F)	16 (61)	16 (61)
Plant Site Elevation Above Sea Level, m (ft)	253 (830)	253 (830)
Total Heat Release at Maximum Duty, MW	125.55	32.64
Percentage of Total Heat Release, %	79.4%	20.6%
Total Heat Release at Normal Duty, MW	104.63	27.21
Total Heat Release at Minimum Duty, MW	21.75	5.67

Table 2. Burner data

Description, (Volume %)	Natural Gas	Mixed Gas	Off Gas
Methane, CH_4	94.57%	65.10%	37.9%
Ethane, C_2H_6	1.7%	0.82%	
Propane, C_3H_8	0.49%	0.24%	
Butane, C_4H_{10}	0.29%	0.14%	
Pentane, C_5H_{12}	0.74%	0.36%	
Hexane, C_6H_{14}	0.02%	0.01%	
Carbon Dioxide, CO_2	0.55%	0.26%	
Nitrogen, N_2	0.93%	6.27%	11.2%
Hydrogen, H_2		19.29%	37.1%
Argon, AR		7.18%	13.8%

Table 3. Fuel gas data

NO_x Emissions Achieved at the Jobsite

This solution achieved NO_x emissions levels of 80-90 mg/Nm^3 (40-45 ppm) at 2.0% excess O_2 in the client's furnace. Additionally, the burners were proven capable of operating at even lower excess air levels in Zeeco's test furnace. At 1.1% excess O_2, the burners produced nearly 0 mg/Nm^3 carbon monoxide (CO), eliminating another client concern. Even at these very low excess O_2 levels, the GB Single-Jet remained reliably stable with well-defined and consistent flame shapes.

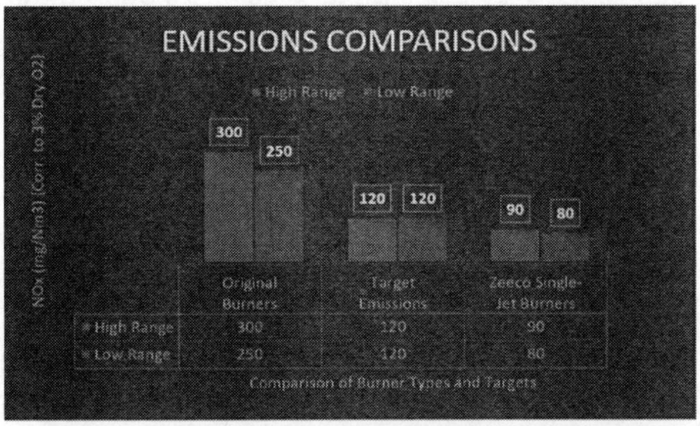

Figure 3. Emissions comparisons between Zeeco GB Single-Jet and original burners

How Does a GB Single-Jet Achieve Ultra-Low-NO_x Emissions?

Zeeco's GB Single-Jet burner, shown in Figure 4, was developed using the skeleton of a conventional raw gas burner which has a single riser, single tip, and stabilization cone. This design offsets the gas tip from the center of the tile throat to the tile edge to achieve low NO_x emissions while retaining the simplicity and easy-to-maintain components of a raw gas burner.

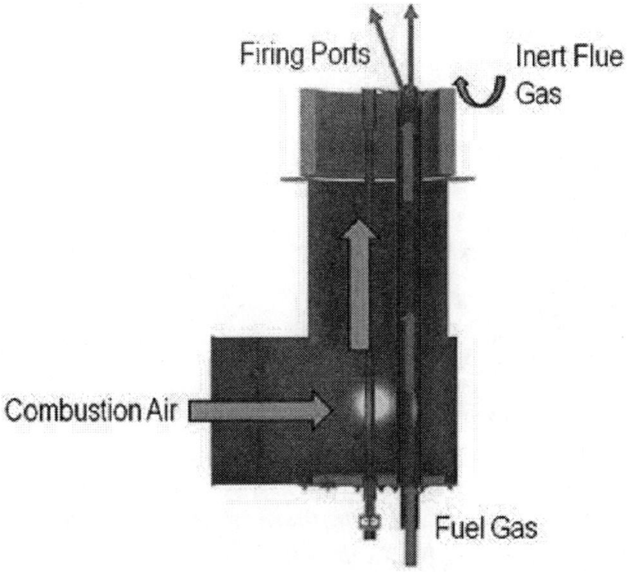

Figure 4. Zeeco's GB Single-Jet ultra-low-NO_x burner using IFGR NO_x reduction

Offsetting the gas tip allows separation of the fuel gas from the combustion air stream. A small portion of the fuel gas is injected directly into the combustion air stream for flame stabilization while the majority is injected upwards. At the center of the tip, gas is directed straight forward creating a staging effect with respect to the other gas fired at an angle. This staging effect of the gas slows the combustion reaction time leading to a reduced peak flame temperature and lower thermal NO_x production.

Positioning the gas tip at the edge of the tile allows the burner to exploit flue gas recirculation patterns within the furnace (see Figure 6). The momentum of the fuel gas that is injected forward entrains flue gas into the unburned fuel gas and into the initial flame. Reconditioning the fuel gas with inert products of combustion lowers the peak flame temperature resulting in lower production of thermal NO_x.

To illustrate the processes that are occurring in the flame envelope, we break-up the flame into three distinct zones. As shown in Figure 5, the IFGR zone is where the inert products of combustion air are pulled into the fuel gas and initial flame. As the fuel gas exits the gas tip and mixes with the inert products of combustion, a new, reconditioned fuel gas composition forms, resulting in combustion at a reduced peak flame temperature. The combustion air zone provides the combustion air necessary to initiate combustion and for the ignition of the reconditioned gas to stabilize on the cone. The combustion in this zone is considered fuel rich since there is more unburned fuel mixed with inert products of combustion than there is combustion air necessary for complete combustion of the entire mixture. IFGR is the primary NO_x reduction function for approximately the first half of the flame length as measured from the cone.

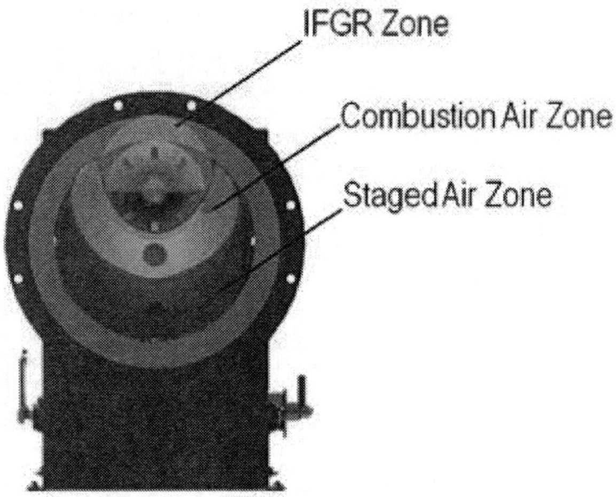

Figure 5: Zeeco's GB Single-Jet Ultra-Low-NO_x burner – showing flame zones

The staged air zone provides the remaining amount of combustion air to mix with the remaining unburnt, reconditioned fuel mixture. This is the primary NO_x reduction function for approximately the second half of the flame length as measured from the cone. The major advantage of using staged air in this region is that if there is flame-to-flame interaction in this region, the emissions are not as impacted as with other methods of NO_x reduction since the mixing of the internal flue gas recirculation is already complete at this stage of combustion. Therefore, any interaction in this zone will cause the unburnt mixture of reconditioned fuel to run into the reconditioned fuel of another burner, both of which move away from the burner cone until the mixture meets enough staged air for complete combustion. This feature allows burners to be retrofitted in many applications that would be very difficult for a standard staged fuel gas or Ultra-Low-NO_x IFGR only burner, both of which rely on mixing the inert products of combustion with the staged fuel gas throughout most of the burner flame length to meet the desired NO_x emissions.

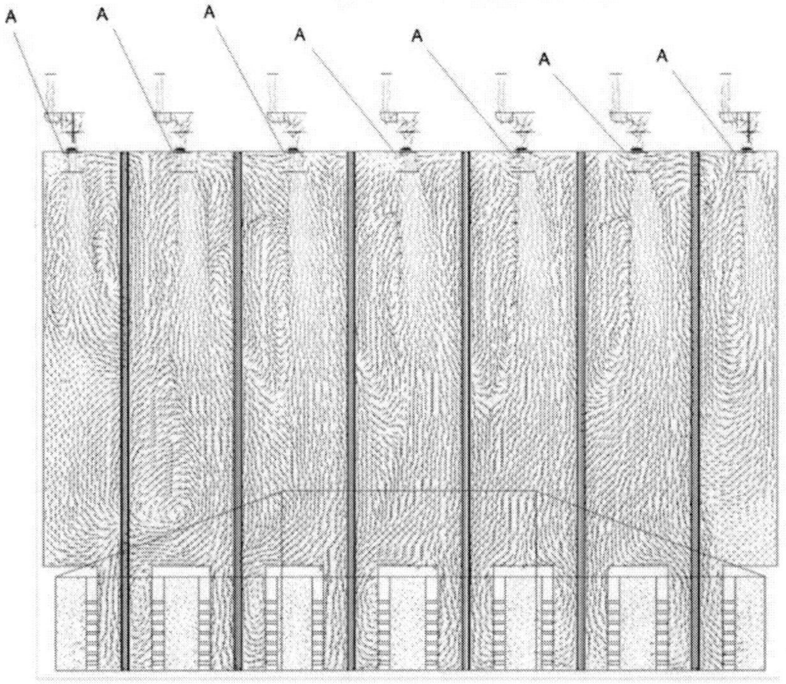

Figure 6. Velocity vectors show that the products of combustion are pulled into the gas tip location of the burner to mix with the fuel gas exiting the tip. This reconditioned fuel composition burns at a lower peak flame temperature and produces less thermal NO_x emissions.

GB Single-Jet Impact on Primary Reformer Tube Temperatures

With any Ultra-Low NO_x burner retrofit, a computational fluid dynamics (CFD) study is recommended to determine flame characteristics and uniformity of heat transfer. Historically, the client had issues with flame impingement and uneven tube outlet temperatures, so Zeeco modeled temperatures of the furnace gases as they could be expected to perform after the retrofit. The gas temperature contour shown in Figures 7 and 8 confirms mostly uniform gas temperature across the primary reformer, indicating uniform tube outlet temperatures. This is a critical parameter for efficient operation of a reformer, as uneven temperature can lead to premature catalyst degradation. Figure 7 shows slightly higher temperatures in the top 40% of the tube length closest to the burners located in the furnace roof section.

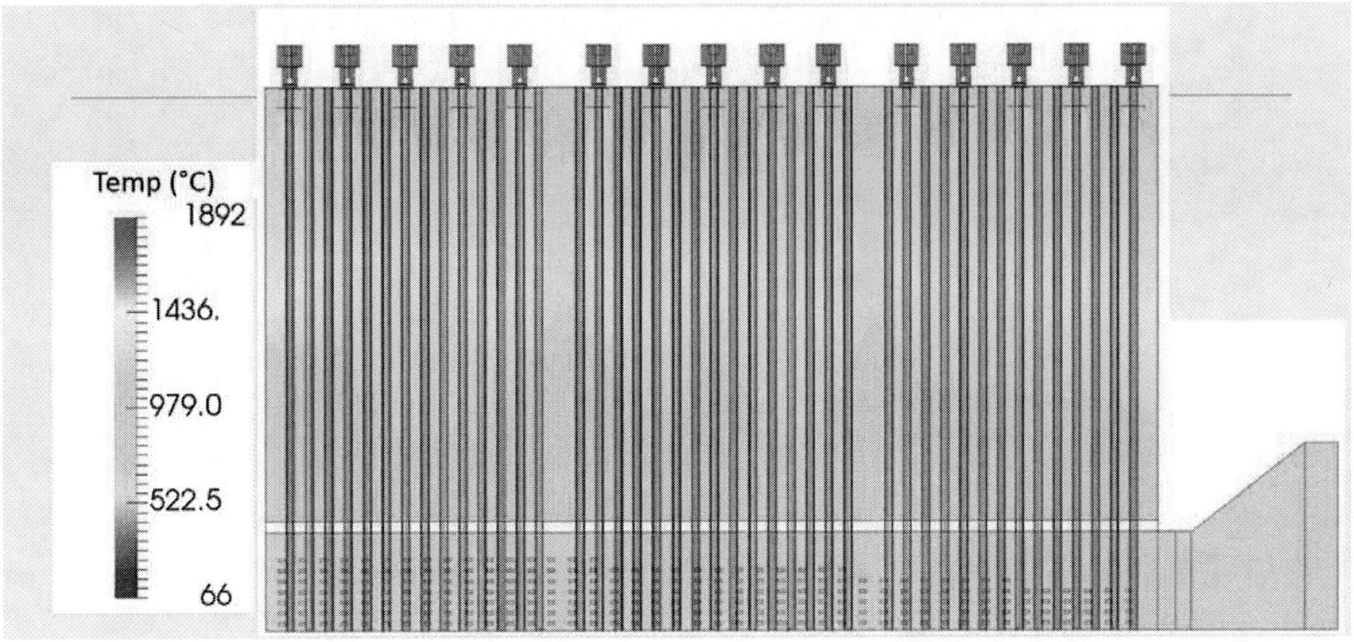

Figure 7. Temperature contour within primary reformer post GB Single-Jet burner retrofit.

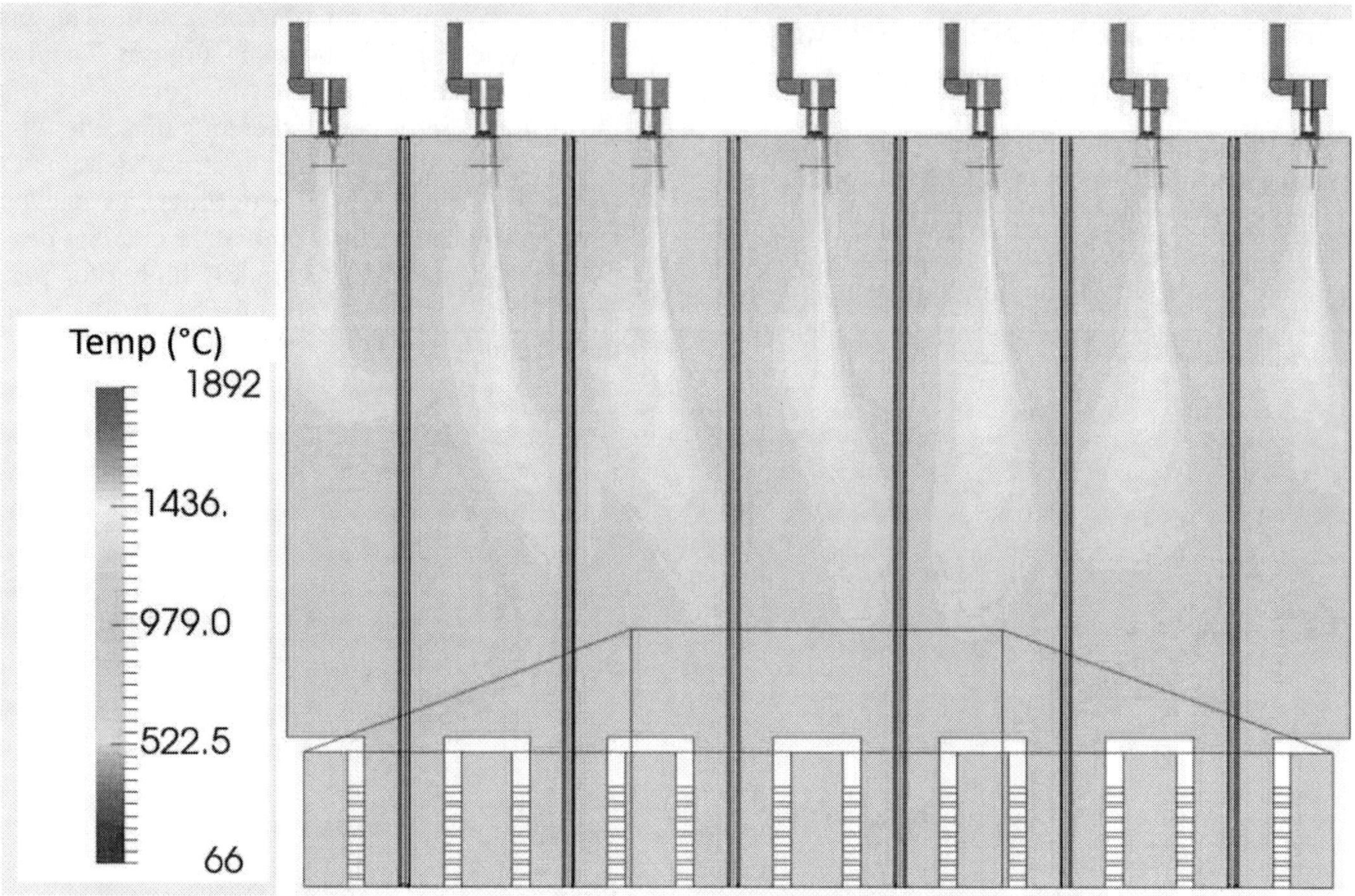

Figure 8. Temperature contour within primary reformer containing GB Single-Jet burner retrofit.

GB Single-Jet Impact Flame-To-Flame Interaction and Furnace Tube Impingement

The client's existing burners were tightly spaced, so it was necessary for Zeeco to verify there would be no adverse impact from flame-to-flame interaction – a common issue with other Ultra-Low NO_x burner flames when retrofitted into older heaters. The CO contours shown in Figure 9 indicate that the flame shapes are compact and very well defined. There is no indication of flame-to-flame interaction adversely affecting the flame in such a manner as to cause flame impingement upon the furnace tubes (see Figure 9). In addition, as detailed earlier and shown in Figure 5, for the portion of the flame furthest from the gas tip, the flame's NO_x emission reduction mechanism is staged air, so NO_x emissions are less impacted when flame-to-flame interaction occurs. Figure 10 shows the flame-to-flame interaction is in the furthest half of the flame from the tip, indicating no adverse effect on thermal NO_x emissions. Hence, Zeeco was able to install the burners with added confidence that NO_x emissions targets would be met and there would be no visible flame impingement on the tubes. The lack of any issues due to flame-to-flame interaction was then confirmed when the system was started up in the field as there was no visible flame impingement and the NO_x emissions achieved were less than required.

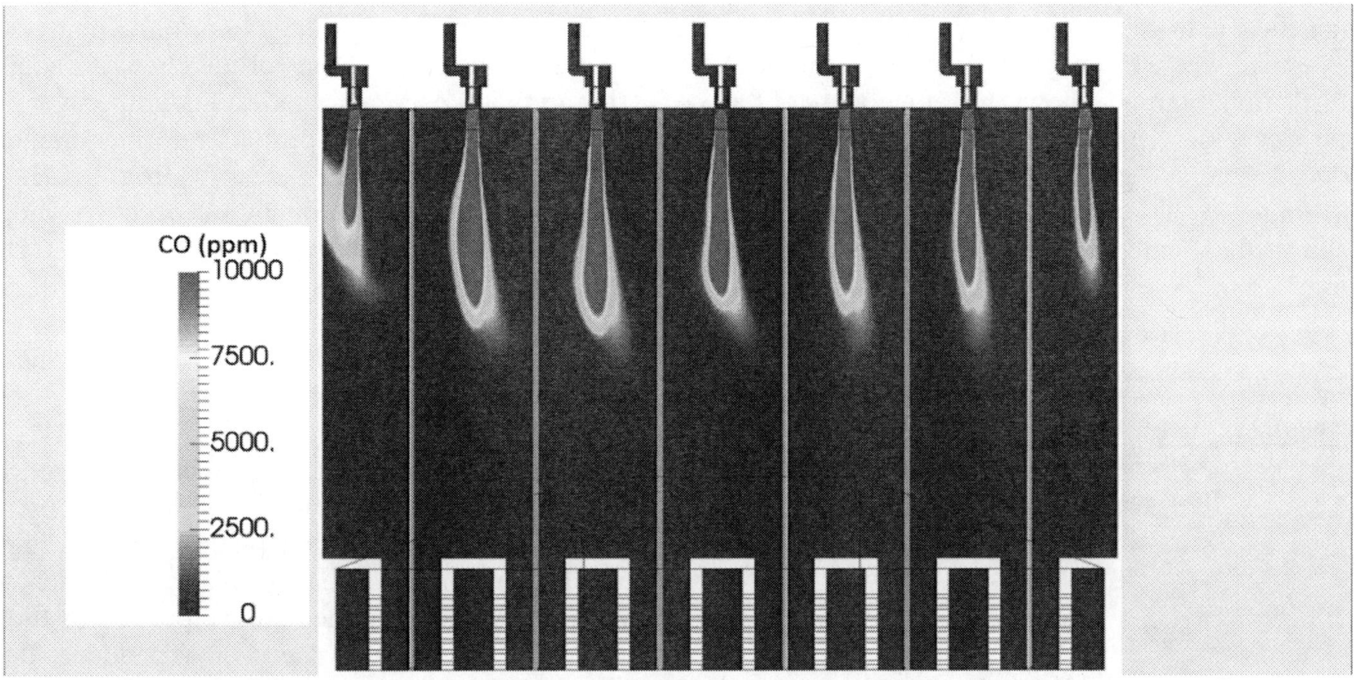

Figure 9. CO contour within primary reformer containing GB Single-Jet burner retrofit.

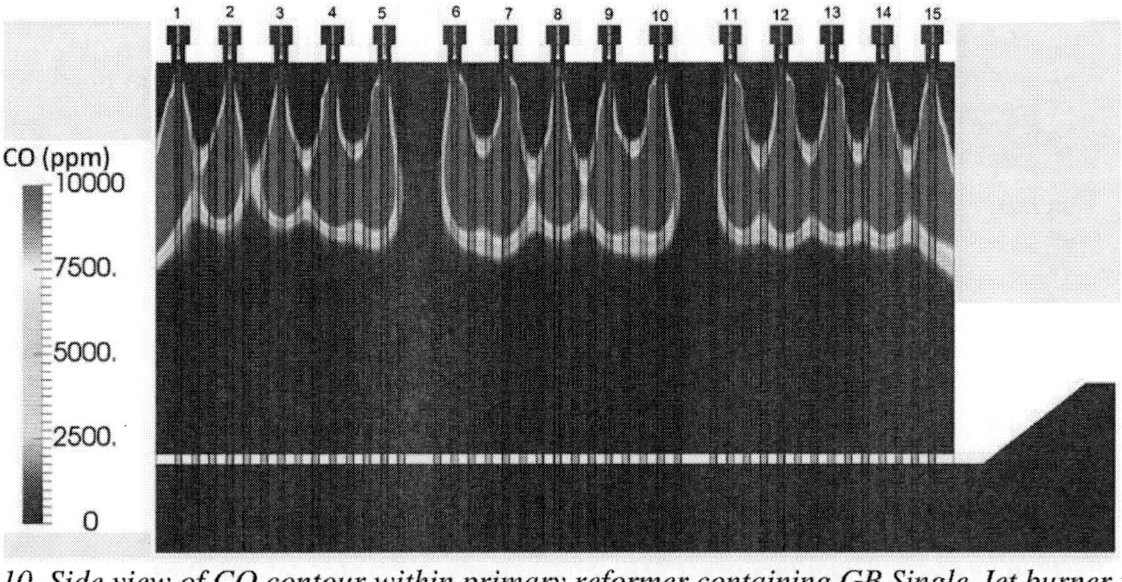

Figure 10. Side view of CO contour within primary reformer containing GB Single-Jet burner retrofit. From the above, the flames are very uniform and approximately 40% of the length of the reformer tube length.

GB Single-Jet Impact Upon Uniform Flame Lengths

One of the most important characteristics of the burners in a primary reformer application is their ability to maintain uniform flame lengths and shapes, ensuring consistent gas temperature distribution along the length of the furnace's radiant tubes. Figure 7 shows slightly higher temperatures in the 40% of the tube length closest to the burners in the furnace roof section. If compared to Figure 10, this slightly elevated temperature is matched with the results of CO formation. For example, where Figure 10 shows 2000 ppmv CO formation, Figure 7 shows the temperature of the furnace gases (products of combustion) are reduced from a slightly elevated level back to the average in a corresponding pattern. Table 4 shows the flame length as determined by Figure 10 and the points at which the CO formation is approximately 2000 ppmv.

Burner Number	Flame Length	Deviation	Deviation
	% of Furnace Tube Length	% of Furnace Tube Length	% of Avg Flame Length
1	43.1%	2.6%	6.5%
2	37.9%	-2.5%	-6.3%
3	37.9%	-2.5%	-6.3%
4	39.7%	-0.8%	-2.0%
5	41.4%	0.9%	2.3%
6	41.4%	0.9%	2.3%
7	41.4%	0.9%	2.3%
8	39.7%	-0.8%	-2.0%
9	41.4%	0.9%	2.3%
10	41.4%	0.9%	2.3%
11	39.7%	-0.8%	-2.0%
12	39.7%	-0.8%	-2.0%
13	39.7%	-0.8%	-2.0%
14	39.7%	-0.8%	-2.0%
15	43.1%	2.6%	6.5%
Average	40.5%		

Table 4. Flame lengths as determined from the CFD side view of CO contour (see Figure 10) within primary reformer containing GB Single-Jet burners. The average flame length is 40.5% of the length of the process tubes and the deviation as a function of % flame length is + 6.5% to – 6.3% which is better than the target of ±10%.

GB Single-Jet Technology – Efficiency, Longevity, NO_X and CO_2 Emissions Reduction

Zeeco's client requested the new burner installation primarily due to emissions reduction legislation within the European Union. The reduction of emissions from approximately 250-300 mg/Nm^3 (120-150 ppm) to 80-90 mg/Nm^3 (40-45 ppm) far exceeded the expectations of the client. In addition to that, Zeeco managed to lower the overall fuel consumption of the primary reformer radiant section by reducing the excess O_2 from 6-7% to 2.0%. The significant reduction in excess oxygen levels in the radiant section also reduces CO_2 emissions. The GB Single-Jet technology does not need to operate at higher excess air levels to ensure low CO emissions or to maintain a high level of flame quality or stability.

Thermal efficiency within a fired heater is a function of the stack temperature and excess air concentration used in the combustion process. Figure 11 shows the area of highest efficiency overall, representing the region close to stoichiometric combustion. In Zeeco's experience, the zone of highest thermal efficiency is between 1 and 2% excess O_2. Overall, there is a trade-off between low excess O_2, high CO levels and wanting to avoid a high level of Unburnt Hydrocarbons (UHC's) in the effort to maximize thermal efficiency.

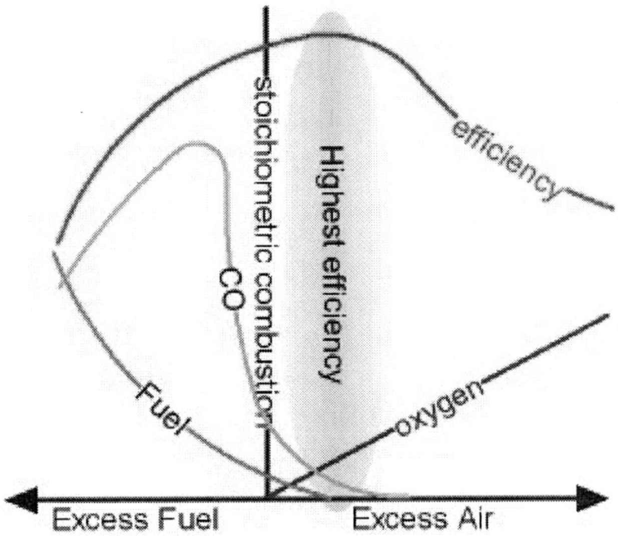

Figure 11. Combustion efficiency vs excess O_2

Conclusion

Zeeco supplied one hundred and five (105) burner retrofit burner kits for installation on a primary reformer in Europe. Rigorous testing was done both by computational and physical testing to confirm key burner performance criteria, and actual NO_x production is meeting EU legislation targets and guaranteed values. Burners are operating with lower excess air than the previously installed burners, increasing thermal efficiency, saving fuel gas, and reducing NO_x and CO_2 emissions.

Key Conclusions from the Project:

- Achieved less than 100 mg/Nm^3 (49 ppm) NO_x emission which was less than the 120 mg/Nm^3 required.
- Achieved lower excess oxygen levels (2% compared to previous 6-7%).
- Achieved lower fuel consumption for the primary reformer radiant section.
- Achieved lower CO_2 emissions since lower fuel consumption was achieved.
- No flame impingement on furnace tubes.
- Flame length of approximately the upper 40% of furnace length.
- Deviation in flame length was +6.5% to -6.3%, better than the ±10% desired. Achieved a very uniform temperature distribution along furnace tubes.
- Reused existing wind-box and pilot equipment.
- Significantly less burner and furnace modification.
- Less time required for turnaround since required modification was significantly reduced.
- Lower overall cost since significantly less furnace and burner modification were required.

Successful Execution of a 2200 MTPD Synloop in Freeport, Texas

In early 2015 KBR executed an EPC contract with Yara International and BASF Group's new joint production company to provide ammonia synthesis licensed technology and engineering, procurement, and construction (EPC) services to build a 2200 MTPD world-scale ammonia synthesis loop at BASF's site in Freeport, Texas. The plant uses hydrogen from nearby chemical facilities and high purity nitrogen from an air separation unit as its feeds for high purity liquid ammonia.

Given its location and feedstock, there are several unique features to the process flowsheet – both the synthesis gas compressor and refrigeration compressor are motor driven, waste heat recovery generates superheated steam for export to the BASF facility, and the plant liquid ammonia storage is located some fourteen kilometers from the production unit, with several existing ammonia users drawing from the pipeline to the storage facility.

Ved Gupta
Yara

Robert McAdams, Omar Ismail
BASF

George Colman
KBR

Amy Mayer, Scott Osborne, Taylor Archer
Clariant

Introduction

The joint venture (JV) of Yara International (68%) and BASF Group (32%) constructed a new world-scale ammonia production plant at the latter's Freeport Texas production site in the United States. While YARA managed the project activities, BASF operates the plant and the integrated export terminal.

Plans to form the joint venture were initiated in 2013, the announcement was made in May 2014, and it was officially formed in February 2015. The plant, then known as the Yara Peony project, broke ground in July 2015 and began production in May 2018.

The project generated approximately 550 jobs at the peak of construction and 35 full-time jobs when it came online.

The new plant has an ammonia production capacity of 750,000 t per year (826,000 ST per year) with nameplate capacity of 2200 MTPD (2425 STPD). The plant uses pure hydrogen and nitrogen, supplied by separate pipeline, as feedstock. The offtake agreement provides ammonia to Yara and BASF at a rate proportional to their respective equity share in the JV.

The technology licensor and engineering, procurement and construction (EPC) contractor for the project was KBR, based in Houston. Texas.

The technology using pipeline hydrogen and nitrogen as feedstock, without the conventional front end, significantly reduces the plant's capital expenditures, maintenance, and local CO_2 emissions compared to traditional natural gas-based ammonia plants. As part of the design the possibility that a conventional front end for the plant could be retrofitted has been considered.

Yara BASF ammonia plant construction

In addition to the construction of the facilities for the new plant, KBR also built a new ammonia tank at the existing terminal, while BASF upgraded its pipeline infrastructure.

Owing to the demands of the new ammonia plant, Praxair, the feedstock supplier, also decided to increase its hydrogen and nitrogen production capacity and extend its existing Gulf Coast pipeline systems by approximately 74 km (46 mile) from Texas City to the Yara BASF JV project site in Freeport.

BASF's Freeport site, which came online in 1958, was its first manufacturing site located outside of Europe. The site covers an area of approximately 575 acres, 275 acres of which were vacant for development in December 2014. It currently houses 25 plants producing 24 different products, mainly basic chemicals, intermediates, and fiber intermediates. The site employs more than 850 workers and 840 contractors.

Feed Stocks, Storage and Shipping

The JV signed a 20 year agreement with Praxair, the largest industrial gas company in North America, to supply 170 MMSCFD (190 kNm^3/h) of hydrogen and 55 MMSCFD (61 kNm^3/hr) of nitrogen for the new plant.

The design feedstock composition, using import hydrogen and nitrogen, is almost inert free. The normal feed compositions are given in Table 1.

Components	Mole Fraction
Hydrogen Feed	99.99%
N_2	0.01%
$CO+CO_2+O_2$	<3 ppmv
Hydrocarbons	<10 ppmv
Nitrogen Feed	99.97%
Argon	0.03%
O2	< 10 ppmv

Table 1. Normal Feed Compositions

Design of the compressors and other equipment have also considered a makeup gas composition based upon a conventional steam methane reformer (SMR) based frontend, in the event one is installed in the future.

Besides the ammonia production facilities, the project also included additional utilities, such as cooling water systems, start-up heater, flare, and a new 20,000 t (22,000 ST) ammonia storage tank next to an existing 30,000 t (33,000 ST) tank

Project Execution

Development of the project was managed by YARA and KBR. Comprehensive and effective management was achieved by adopting international good practices in contracting and project management. A key feature of the project execution strategy was to ensure that a safe, efficient, and reliable plant was designed and constructed on the strength of proven technology using experienced contractors and suppliers.

To ensure safety, quality, and consistency of the overall plant design, BASF also supported YARA in activities such as review of the offsite utilities design, detailed engineering design, HAZOP, and 3D model review. KBR provided EPC services, as well as commissioning and start-up advisory services.

Throughout all the phases of the project, YARA, BASF, and KBR worked together with the common goal of designing and constructing a safe, reliable, and energy efficient world class ammonia synthesis loop. To ensure a safe, timely, and efficient startup of the ammonia plant, KBR also provided training to BASF's operating team using a dynamic Operating Training Simulator (OTS). During the commissioning of the ammonia plant, KBR provided remote plant monitoring (RPM) services using its proprietary "Ammonia InSite" technology.

Ammonia Plant Process Design Features

The BASF-YARA ammonia project had several key objectives - high efficiency and reliability, integration into the existing brownfield site, and potential future expansion to include a conventional SMR based make-up gas production facility.

A unique feature of the plant facility is that it has no onsite storage capacity. Ammonia produced at the plant is exported via pipeline to customers and the terminal storage facility 14.5 km (9 mile) from the plant. The warm, liquid ammonia is transported via pipeline and chilled prior to storage at the terminal. The pipeline belongs to BASF and is designed for temperatures of 5°C to 20°C (40°F to 68°F), hence ammonia to the pipeline must be heated. Once at the storage facility ammonia must be chilled to -33°C (-27°F). Facilities also exist at the storage area that allow ammonia heating and pipeline injection. Both terminals are operated by BASF, and shown in Figure 1.

Figure 1 Ammonia Storage Terminals

Since this plant is a standalone ammonia synloop, there is a limited steam system within the facility. Superheated steam is produced at 46 barg (670 psig) (export pressure requirement) with heat from the ammonia converter effluent. This steam is used to operate a Boiler Feedwater (BFW) pump, with the balance exported to the BASF facility.

The need to superheat the steam led to an innovative and proprietary design, combining both the boiler and superheater into one unit, thus eliminating a nozzle.

The limited steam production from the synloop led to the use of electric drives on both the synthesis gas compressor and refrigeration compressor.

For the synthesis gas machine, a three case MHI compressor driven by a 34.5 kV motor with variable speed drive was selected. Given the low suction pressure, 20 barg (290 psig), and future design requirements for a conventional natural gas-based make up gas, the rated power of the machine was 28690 kW (38340 hp). For the refrigeration train, a two case MHI compressor with a constant speed 13.8 kV motor rated at 5445 kW (7300 hp) was used.

KBR's synloop technology uses an efficient and compact ammonia synthesis configuration, as shown in Figure 2.

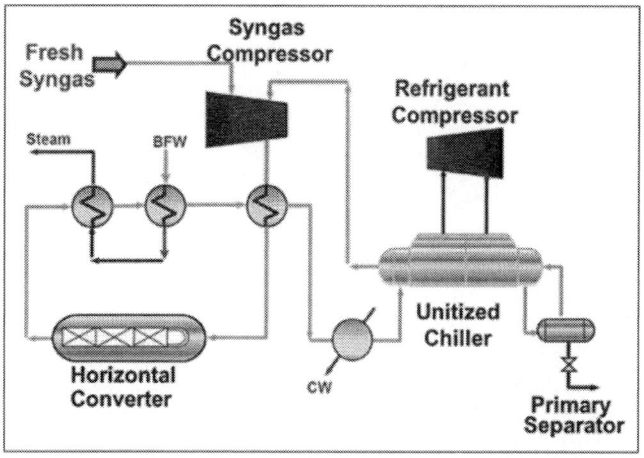

Figure 2 Ammonia Synthesis Loop

This synloop features the KBR horizontal converter (Figure 3) and unitized chiller (Figure 4). KBR's proprietary design of the cold wall horizontal converter eliminates the need for a heavy-duty crane for basket and catalyst installation. Clariant's AmoMax®10 RS ammonia synthesis catalyst was chosen for this project due to its high activity and low water generation during reduction

Figure 3. Typical KBR Converter

Figure 4. Typical Unitized Chiller

The unitized chiller has been used in KBR plants for synthesis loop refrigeration since the early 1980's. There are over fifty references in operation.

With a total of only seven pieces of equipment in the high-pressure synthesis loop, the result is lower cost, smaller plot area, and lower pressure drop in the loop.

Figure 5 shows an overall view of the production unit during construction, showing the cooling tower, compressor building, control room, and synloop.

Figure 5. Ammonia Production unit

Since the compressor deck is sized for future additions, such as a process air compressor and feed gas compressor, it is quite spacious for now. A view of the syngas and refrigeration compressor machines with their motor drives is provided in Figure 6.

Figure 6. Synloop Compressor deck

Project Milestones

The major project milestones are listed in Table 2.

MajorEvents	Date
Project Development Begins	2013
Contract Signed	December 2014
Construction started	July 2015
Precommissioning phase	31 Jan 2018
Commissioning started	15 May 2018
Ammonia Catalyst reduction and first ammonia	17-18 May 2018
Ten day production test completed	10 Jun 2018
Performance Test Complete	**11 Nov 2018**

Table 2. Key Milestones of theProject

Plant Performance

Though operation was achieved in mid-2018, the plant performance guarantee test was delayed due to product pump capacity problems that prevented operation at greater than 85% capacity, unless both pumps were running. There were also issues with the remote storage tank boil off gas recovery system. The full performance test was conducted in November 2018 over a five-day period. The plant demonstrated that all performance guarantees, including capacity, product specification, and feed consumption were met.

Table 3 shows the comparison of the ammonia plant performance against the design. The measured values were average measured data over the test run period.

	Guarantee	Measured
Ammonia production, MTPD (STPD)	**2200 (2425)**	**2227 (2455)**
Ammonia product quality: NH_3,% (after water inj.) H_2O,% Oil, ppmw Inert Gases, max	 99.5 min. 0.2 to 0.5 5.0 max 10 ppmw	 99.6 0.4 <0.1 <5
Specific energy Import GJ/MT (MMBtu/ST)	3.79 (3.26)	3.08 (2.65)
Specific energy Export GJ/MT (MMBtu/ST)	3.31 (2.85)	3.26 (2.80)
H2 Consumption kg/MT (lb/ST)	180 (360)	179.5 (359)
N2 Consumption kg/MT (lb/ST)	835 (1670)	823 (1646)

Table 3. Comparison between the expected and the average measured values over the Test Run

Challenges and Lessons Learned

Successful commissioning of an ammonia plant requires careful planning and utilization of lessons learned from other successful startups. Below are some of the main challenges and lessons learned during plant commissioning and start-up.

Site Environmental Restrictions

Given the existing plant permitting, BASF was keen to limit any additional pollutants from the synthesis loop. This included NOx (from the start-up heater and flare, ammonia in any waste water and ammonia in the working environment (<25 ppm). Noise was also limited to 85 dB(A).

Delay due to refrigeration compressor suction strainer collapsing

During initial commissioning of the refrigeration compressor, high pressure drop on the suction strainers was noted, causing an immediate shutdown. In one strainer, part of a pipe "choker" was found but others were also damaged as the construction itself was found to be weak and screens were damaged. At the initial machine run-in, the installed 80 mesh fine screen created a higher than expected pressure drop across each element which in some cases caused the element to deform, become dislodged from the element retaining system, and in one case the element travelled down the suction pipe almost entering the compressor case. As a result, these screen elements had to be replaced. However, since the unit was already charged with ammonia, it was difficult to clear the fumes for access. The ammonia system had to be evacuated, cleared, and refilled to modify the suction strainers. Lack of isolation valves to isolate the suction strainers delayed this repair as the entire refrigeration system had to be cleared.

Overheating of the syngas motor drive cabinet

The ambient air temperatures in Freeport can exceed 38°C (100°F). During summer operation, there were issues related both to insufficient cooling water flow and restricted cooling air recirculation within the electrical motor drive control cabinets that led to a short outage of the unit. Improper line cleaning resulted in early fouling of the drive cabinet cooler and limited plant production due to the drive cabinet temperature limit when operating at full load.

Catalyst reduction water limitations

There were strict and unique ammonia product quality restrictions during catalyst reduction, due to customers receiving ammonia directly from a bidirectional pipeline. Due to these constraints, accepting product from the ammonia plant with an ammonia purity less than 98.6%w/w was not possible. The following measures were implemented to ensure that on-specification product continued to be supplied to pipeline consumers while the convertor catalyst reduction was performed:

- The convertor catalyst charge consisted of 100% pre-reduced and stabilized Clariant AmoMax® 10 RS catalyst, since this version produces ~90% less water during the reduction as compared to the oxidized version.
- Temporary ammonia truck loading/unloading systems were constructed at both the plant and terminal, designed to handle 0-99% aqua ammonia.
- To minimize the quantity of off specification ammonia transported by truck during the catalyst reduction, the reduction of the first two beds was performed at ~60 barg (870 psig), instead of the normal 85 barg (1230 psig). Plants that either have limited aqua ammonia handling capabilities or are simply trying to reduce the rate of production of off-spec ammonia during a startup should consider this method as an option. As demonstrated, the simple modification of lowering the reduction pressure proved effective in reducing the rate of aqua ammonia production during startup.
- The high activity of AmoMax 10 RS at low pressure was key in the success of the reduction.
- Shortly after the reduction of the second bed was complete, the purity of the ammonia product was measured at 98.9% pure, indicating that the pressure could be increased to complete the reduction more efficiently. The ammonia product was then injected directly into the pipeline.
- The remainder of the reduction was completed in just over two days, with a brief interruption due to loss of hydrogen feed.
- Regular water content readings were taken throughout the reduction. All but two were well below 1000 ppm, and the outliers were deemed erroneous due to a malfunction of a wet test meter indicator. In total, the reduction took slightly less than four days, which was longer than typically expected due to the need to minimize product water levels. This very challenging catalyst reduction was successfully completed by the project team

through careful and detailed cooperation between Clariant, KBR and Yara/BASF.

Ammonia Product pumps

During the performance test it became necessary to operate both product pumps in parallel above 85% load because of issues with the pump suction head losses. The pump manufacturer later resolved this problem by modifying the pumps' casings.

Summary

The successful start-up of the new ammonia synthesis loop is a major accomplishment for BASF, YARA and KBR. The fact that it is one of the few large ammonia plants that does not contribute CO_2 to greenhouse gas emissions is a significant achievement

Despite some of the initial problems encountered, the ammonia plant runs very smoothly and successfully demonstrated that its performance was better than the guarantees in terms of feedstock and energy consumption, ammonia production, and product quality. Subsequent plant rate tests have demonstrated the ability to achieve production rates in excess of 128.5% of name plate capacity, or 2830 MTPD (3120 STPD).

Acknowledgements

Authors would like to acknowledge the untiring efforts put in by YARA, BASF, KBR and Clariant in turning this synloop project into a successful operating plant.

Secondary Reformer Revamp in Ammonia Plants with Improved Safety, Efficiency, and Operational Reliability

The paper summarizes KBR's key experiences on design, modeling, inspecting, repairing, and revamping KBR's secondary reformer for C. F. Braun type of hot surface dual layer refractory designs and introducing the highly reliable KBR water jacketed cold wall single layer refractory design. The water jacketed design, with its mechanical robustness, and ability to withstand severe operational flow and thermal conditions contributes to sustained high process performance without increasing the need for maintenance and entailing shutdowns because of refractory failure.

By combining a KBR process study, optimization, and advanced CFD simulation technology to include KBR's advanced combustion chamber design used in grassroots facilities since the 1980's, the revamped secondary reformer will provide superior operational safety and reliability.

John Li, Kaushal Bhatt, Robert Lin
KBR Technology

Introduction

The secondary reformer is one of the key pieces of equipment in the ammonia plant. This paper summarizes KBR's know-how on design, modeling, inspecting, repairing, and revamping the secondary reformer. KBR offers the C. F. Braun (Braun) type of dual layer refractory designed secondary reformer as well as the highly reliable and proven KBR water jacketed, single layer refractory design. . KBR has basically four different designs which includes double layer refractory, single layer refractory with burner design, single layer refractory with insert piece and single layer refractory Auto Thermal Reactor (ATR) type. All three single layer designs are water jacketed. The jacketed design, with its mechanical robustness, and ability to withstand severe operational flow and thermal conditions contributes to sustained high process performance without increasing the need for maintenance and entailing shutdowns because of refractory failure

The paper highlights the salient feature of the jacketed and non-jacketed designs and KBR's long-standing history with providing reliable secondary reformer design. The dual layer design is known for its double layer refractory and presence of thermal indicative paint on its outer surface. The water jacketed design, can withstand severe operational flow and thermal conditions that contributes to sustained high process performance without increasing the need for maintenance or entailing shutdowns because of refractory failure. KBR's advanced combustion chamber design has been used in grassroots facilities since the 1980's and its innovative design eliminating the need for a burner will provide the superior operational safety and reliability.

Process Description

In an ammonia plant, the secondary reformer is responsible for providing the most important element in the ammonia product, nitrogen. The typical Ammonia Process is shown in Figure 1.

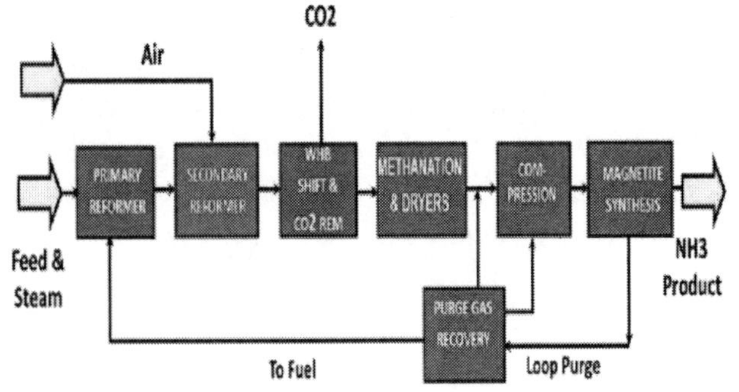

Figure 1. Typical Ammonia Production Process

In a typical ammonia plant, the process air is compressed with a process air compressor and preheated to 620°C (1150°F), and then routed to the secondary reformer.

In a conventional secondary reformer, the quantity of air is controlled to produce a three-to-one molar ratio of hydrogen to nitrogen in the synthesis gas. . In the KBR Purifier process, about fifty percent more air is used than in a conventional process. This results in a normal hydrogen/nitrogen molar ratio at the exit of methanation and feed to the cold box of 2.0. The extra air provides additional reaction heat in the secondary reformer and allows the temperature of the gas exiting the primary reformer be about 100°C (180°F) lower than conventional plants.

During secondary reforming, the oxygen in the air combusts some of the process gas from the primary reformer, leading to a high temperature (approximately 1300°C to 1360°C, or 2370°F to 2480°F) in a special mixing and combustion chamber above the catalyst bed. The hot gas from this combustion passes down through a bed of nickel reforming catalyst where it reacts to produce more hydrogen in a similar manner to the primary reformer, but without outside heat input.

Due to the overall endothermic nature of the reforming reaction, the gas temperature leaving the secondary reformer is reduced to about 870°C - 1000°C (1600°F – 1830°F). Typical operating parameters of the secondary reformer are shown in Figure 2.

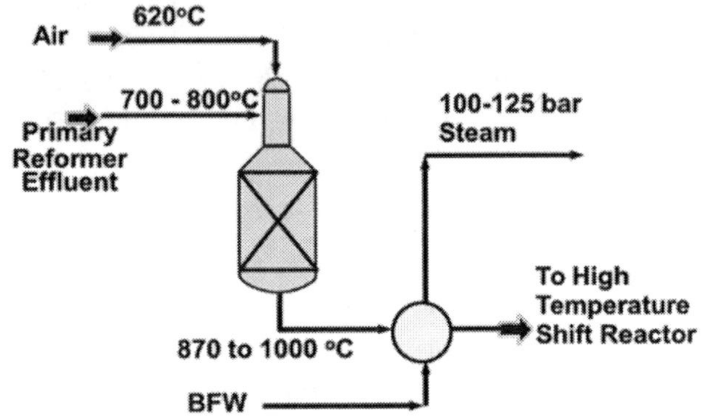

Figure 2. Purifier Secondary Reformer Operations

KBR's Combustion Chamber Design for Secondary Reforming

Because of the high temperature inside the secondary reformer, the secondary reformer must be internally refractory lined.

For the CFB design, the air enters from the side through nozzle sparger and the flue gas enter from the top through the transfer line.

KBR has been using a combustion chamber since 1983 in all the global ammonia plants. For the water jacketed design the process air is introduced at high velocity axially at the top in a refractory cylinder and the reformed gas from the primary reformer enters radially perpendicular to the air flow. The high velocity air induces instantaneous combustion, product mixing, and dissipation through the cylinder. The process gas will have uniform temperature and composition as it enters the catalyst bed of the secondary reformer.

The very old style secondary reformer design had both air and product gas entering from the side.

The original KBR Secondary Reformer with combustion Chamber design is shown in Figure 3.

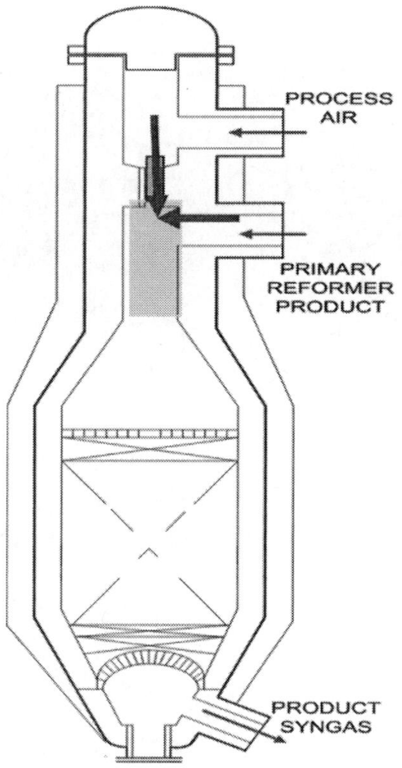

Figure 3. Original KBR Secondary Reformer with Combustion Chamber

KBR's Secondary Reformer – Innovated Combustion Chamber

KBR's new ATR (Auto Thermal Reformer) is a proven and reliable design in the ammonia industry. The newer ATR based secondary does not have a burner nor an air insert piece. The combustion zone at the large cone of the secondary reformer is where the temperature is in the adiabatic zone.

The latest KBR secondary reformer uses the same advanced combustion chamber technology but eliminates any metallic parts. The basic design is shown in Figure 4.

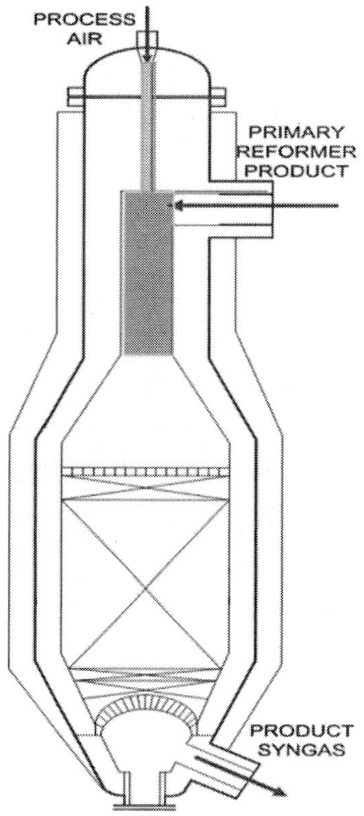

Figure 4. Secondary Reformer with Innovated Combustion Chamber

There is little wear and tear of the internals in the combustion area over prolonged operation, and the equipment is basically maintenance free. In both designs, the primary reformer effluent gas and process air streams are combusted and mixed in this chamber. Only a well-mixed and uniformly distributed gas passes over the secondary reformer catalyst bed.

KBRs metal free combustion chamber has been in use since 1994. This design has the following advantages:

- All refractory construction;
- Basic functionality same as original; and
- Can withstand a wider range of flame temperatures, without concern of wear/tear.

Mechanical Design Features

Hot wall Design

The legacy C. F. Braun design is a standard hot wall design offered in earlier Purifier plants prior to the merger with M. W. Kellogg. This design does not have a water jacket system. Typically secondary reformers of this type will use a dual layer refractory system, which means there are two layers of refractory. The first layer is called a hot face (the exposed layer to the hot gas providing structural integrity and erosion protection) and, the second layer is called a backup face (a layer of lower density which provides thermal insulation).

A newer and improved dual layer refractory design should have both a short and long anchor system present inside the vessel to hold the two refractory layers.

The short anchors are provided in the back up layer and are responsible to ensure sturdiness of the thermal insulation layer. The long anchors are provided to ensure structural integrity of the hot face. The thickness of the back up layer is generally double (2X) the hot face layer, or slightly less, to provide the required thermal protection. When designing this type of secondary reformed one must ensure the hot face and backup refractory interface temperatures are correctly calculated, considering the porosity of the refractory and hydrogen content of the process gas.

The insulating back up refractory layer protects the vessel wall from excessive temperature. It is usually less dense than hot face refractory to provide better insulating properties. The hot face, on the other hand, is a denser refractory which is more compact and provides strength to the back up layer. It is basically present for structural integrity of the total refractory installation.

The key feature of this hot wall dual layer refractory design is that shell is generally 1.25Cr-0.5Mo material (or higher) to ensure the vessel design meets API 941 requirements regarding hydrogen partial pressure and temperature. In addition, the vessel external surface is always coated with thermal indicative paint, which notifies the operator of a hot spot by changing color, from blue to white, as shown in Figure 5. The color change occurs when the surface temperature reaches a predetermined value and is permanent.

Figure 5. Secondary Reformer with Thermal Indicative Paint, Originally Colored Blue

In design practice, a slightly higher metal design temperature is preferred to ensure that the vessel works within desired operational fluctuating range. The thermal indicating paint on the external vessel is the first line of equipment protection.

The dome design is an arch type with a lower angle comparing with other ATR type secondary reformers. The arch structure is a free supporting dome and, when assembled and installed correctly, it will last a long time without structural integrity issues.

Metal Temperature Indicators (MTI) can also be installed on the body of the vessel to ensure that the temperature is constantly monitored. MTIs provide additional monitoring if any hotspot is developing in those refractory damaged/defect areas as the metal temperature will rise. The plant operator can also periodically measure the external vessel surface temperature using a temperature gun in case the temperature indicating paint fails and/or thermocouples at the MTI location provide false alarms.

Overall, this is a very safe design with many KBR legacy ammonia plants still in operation around the world. We have also improved refractory anchor system designs. In older plants only long anchors were provided and short anchors were absent. In newer plants we have included long and short anchors for hot face and back up face respectively. The improved anchor design will increase the refractory integrity for long-term reliable plant operations.

Cold Wall Design

The basic process gas flow path is similar to hot wall vessels, the process air enters through the top of the secondary reformer and the Steam Methane Reformer (SMR) gas coming through the reformer transfer line enters from the side of the secondary reformer into the top mixing and combustion zone. The salient feature of this design is the presence of water jacket around the pressure shell which ensures that the vessel is always protected, even if there is a refractory failure during operation.

KBR has used the water jacketed secondary reformer since the 1950's, even though KBR had improved the combustion chamber design without the air insert. An example is shown in Figure 6.

Another basic difference of the cold wall secondary reformer is the use of a single refractory layer, generally with thickness close to 12 inches. The refractory material selected is a denser refractory which provides strength as well as ensuring thermal insulation. The presence of the water jacket provides a secondary form of protection to keep the vessel's pressure shell cool. This proven arrangement allows the pressure shell to be constructed using carbon steel instead of low chrome materials used for hot wall designs.

Figure 6. KBR Cold-Wall Secondary Reformer Erected in the Field

There are also MTI's installed over the shell at various heights to provide skin temperature measurement. As the jacket water is also near boiling point, the temperatures will read in range of 100°C-110°C (212°F-230°F). In case there is refractory failure and formation of a hot spot, the measured consumption of water in the water jacket increases due to higher steam generation, which will alarm the operators. Even with increased steam generation the vessel shell can still operate safely for a time, as long as water flow can be maintained.

KBR's cold wall water jacketed design has three distinct advantages:

1) Material selection: Due to presence of water jacket the metallurgy can be carbon steel. The pressure shell always remains cool regardless of the refractory condition. In case of a refractory failure, the shell is still protected with emergency jacket-water flow rate activated and subsequent increase in steam production. The cost of the secondary reformer is comparatively cheaper to other design due to carbon steel construction of the shell. Since the refractory is single layer monolithic casting it is easier to install and only a single dry out is needed to cure the installed refractory at the site.

2) Mechanical Design: The mechanical design of the secondary reformer is much more robust and reliable compared to any other designs. The fabrication of carbon steel pressure vessel is relatively simple but will involve postweld heat treatment (PWHT) of the whole vessel due to thickness exceeding ASME code requirements. The service PWHT is not required per process. The vessel is designed per ASME Section VIII Division 1.

3) Refractory Design: The refractory design for the cold wall secondary reformer is based on a single layer monolithic design. With long and short anchors providing assertive strength to the ceramic structures, the refractory layer is very reliable with a proven track record for many years at various KBR licensed plants around the world.

   The key feature of the design being the short and long anchor presence in the monolithic refractory, which provides compressive mechanical strength to the refractory ensuring refractory hold up in operation without issue.

Additionally, the Incoloy anchors are found to be reliable and have very few issues (broken, thermal tear off) during operation of the equipment.

The secondary reformer bottom arch structure, alumina balls (1 inch, and 2 inch), top hexagonal target tiles, and circular bricks also have been used over the years from various proven suppliers with successful result, see Figure 7 below for catalyst Bed and Dome Structure.

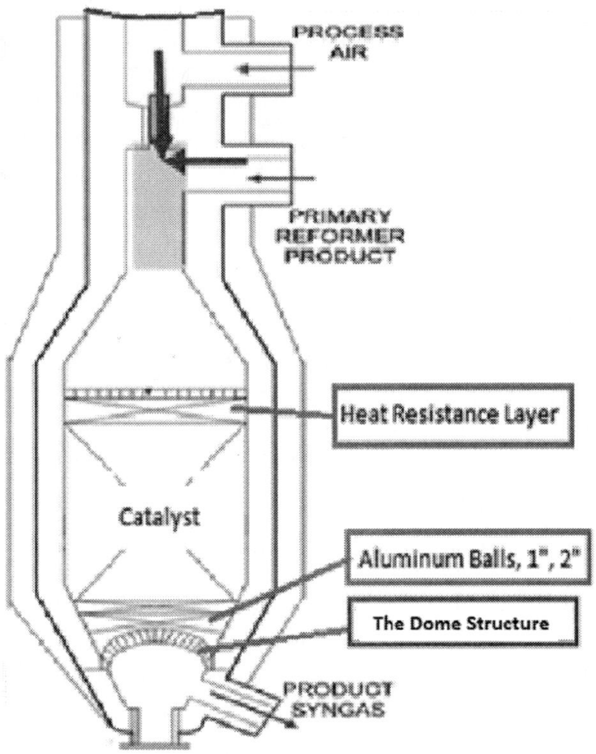

Figure 7. Cold Wall 2nd Reformer Catalyst Bed

In KBR's designs the catalyst is supported by an arch shaped, freely self-supported dome of refractory brick. The dome design is very reliable with few incidences of failure. The dome structure has been designed to be sturdy enough to hold all the catalyst dead weight along with the load due to pressure drop. The typical KBR dome assembly is shown in Figure 8.

Figure 8. KBR Cold-Wall Secondary Reformer Erected in the Field-Dome Assembly

With more than 150 successfully installed KBR water-jacketed secondary reformers globally, this cold wall single layer refractory design is proven very reliable and safe with minimum maintenance requirement.

Revamp Design Confirmation Checks With CFD Simulation

The secondary reformer computational fluid dynamics (CFD) models are developed in-house by KBR using ANSYS Fluent modelling software. Dimensions of the Secondary Reformer were extracted from mechanical vessel drawings. Key input details and dimensions from these drawings were combined into a scaled elevation drawing to aid in visualization of the existing vessel for model development and interpretation.

The effects of the catalyst bed above the dome structure on gas flow profiles were incorporated by modelling it as a porous media with a resistance of the calculated catalyst pressure drop and bed depth.

The elevation of the top of the bed (top of the hexagonal target tiles) is set per the original vessel design but the target tiles were not modelled separately from the catalyst since their pressure drop impact is small relative to the catalyst bed. The CFD model was based on all surfaces being adiabatic (i.e., no heat transfer out of the domain). Flows and physical properties for the model were based on simulations developed in-house by KBR with input provided by Client.

Combustion and the resultant flame pattern and temperatures are of key interest in the model. Three combustion reactions were included:

1. $2H_2 + O_2 \rightarrow 2H_2O$
2. $2CH_4 + 3O_2 \rightarrow 2CO + 4H_2O$
3. $2CO + O_2 \rightarrow 2CO_2$

Reaction rate parameters for these combustions were taken from the built-in ANSYS Fluent database and are the same values used in previous KBR secondary reformer CFD models with similar feed gas compositions. The effects of catalytic reforming in the bed were not included in the model since the focus is on gas flows, mixing, and interactions within the combustion chamber and conical transition to the catalyst bed.

Modelling begins by creating the grid and running cases with flowrates between minimum flows and maximum flows to bracket the range under consideration. The modeling results including patterns of temperature, velocity, and gas species inside the secondary reformer.

The simulation example, shown in in Figures 9A and 9B, is of a vintage hot wall (Braun) design taken from a recent revamp project. It is based on the reformer dimensions including external 180° turn in transfer line. The primary purpose is CFD simulation is to ensure the flow and heat distributions are even in order to protect the internal refractory surface from flow erosion and excessive hot spot formation. In an exemplary study case, a series of burner/spargers are used for air injection. The nozzle size on each sparger had been increased from 13.5 mm to 14.5 mm to accommodate the 20% production increase. This is extremely important for hot-wall safety of the reformer vessel and reliable long-term operations of the ammonia plant. The plot showing is the external loop has impact on the velocity distribution inside the combustion chamber where twelve process air/steam injection spargers are located. The 180° turn impact on velocity distribution can be seen in Figures 9A and 9B below.

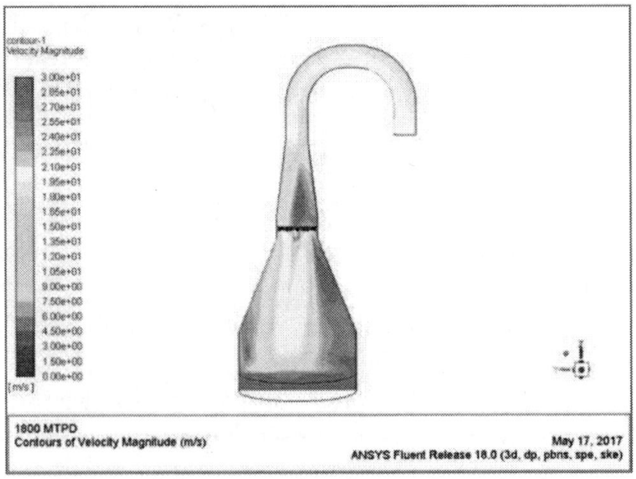

Figure 9A. Vertical Velocity Distribution

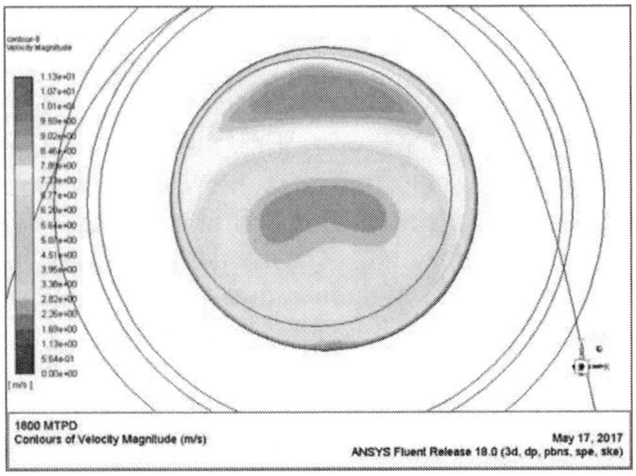

Figure 9B. Horizontal Velocity Distribution

Although velocity distribution above the combustion area is distorted above the air/steam spargers (Figure 9A and 9B), the effect of combustion of process gas with air is to have a suction effect which helps to even out the profile somewhat. The small pressure drop across the twelve air/steam sparger pipes also aids in mitigating process gas velocity distortion. This improvement can be visualized in Figures 10A and 10B. In Figure 10A, just below the air/steam spargers, it can be seen that there is no impingement of flame on the combustion chamber area of Secondary Reformer. Figure 10B shows the velocity variation above the catalyst bed, which is quite subdued.

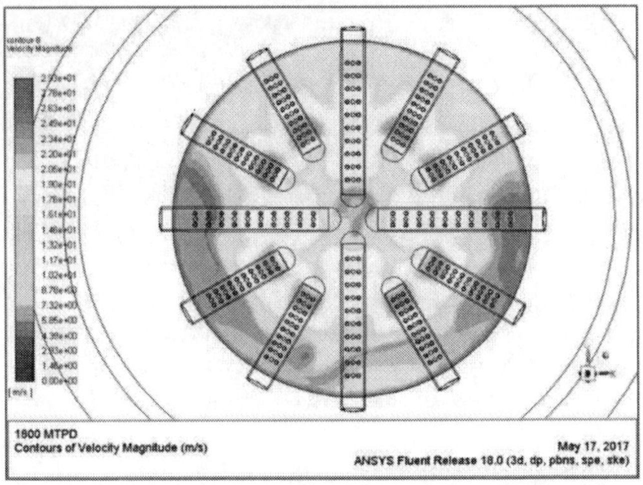

Figure 10A. Velocity distribution just below the Spargers, No Impingement

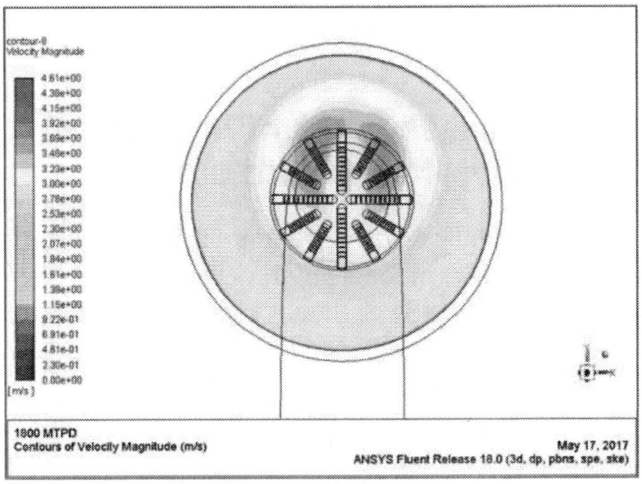

Figure 10B. Velocity distribution above the Catalyst Bed – Even Distribution

The temperature profile along the Reformer transition area is shown in Figure 11A. The results show there is no impingement of high temperature on the refractory. There is a temperature variation of 800°C to 1400°C (1470°F to 2550°F) from one side of the vessel to other.

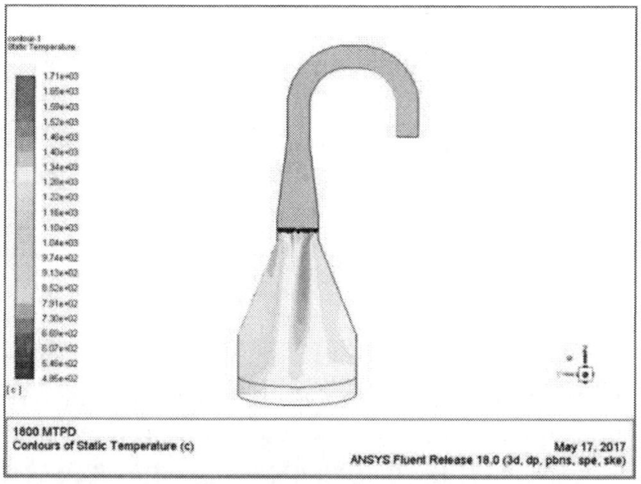

Figure 11A. Temperature profile in vertical plane – Incoming Transfer line Effect

A similar variation is seen in the cross sectional plot, Figure 11B, of temperature above the catalyst bed. This is mainly due to the effect of 180° turn in the transfer line on the velocity variation of incoming process gas.

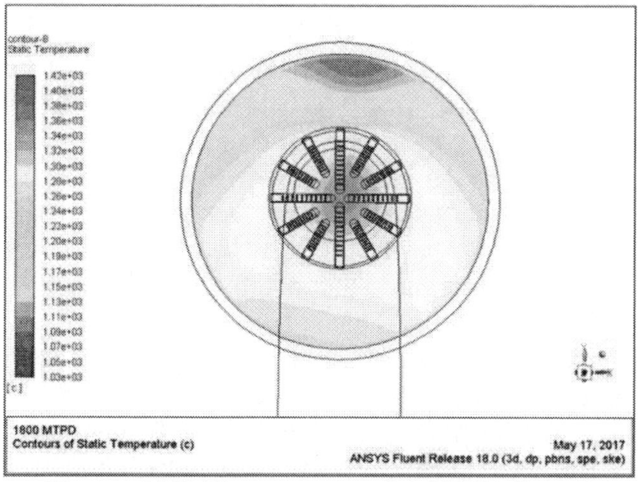

Figure 11B. Temperature profile in Cross-Sectional, above Catalyst Bed

After many iterations of CFD simulations with incorporated design changes, the analytical results of the redesigned secondary reformer show no direct impingement of hot flame/gas on the refractory walls around the burner spargers or on the refractory surface of the vessel. The purpose is to eliminate any possible refractory damage which may lead to channeling of hot combustion air, penetrating the refractory to the vessel internal metal surface. The variation of velocity as well as temperature across the vessel cross-section is expected.

The complete combustion is evidenced by looking at the concentration profile of oxygen just above catalyst bed in Figure 12. This CFD analysis is similar to one of a similarly designed secondary reformer, which is operating satisfactorily without any operating issues.

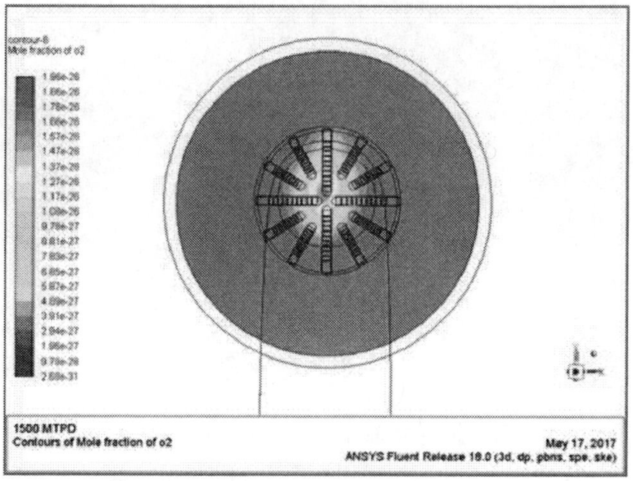

Figure 12. Concentration profile of Oxygen, just above the Catalyst Bed

The CFD simulation case discussed above is based on a major client revamp project. The project delivered a 25% increase in capacity by improving production and reliability.

Case Study - a Legacy C. F. Braun Plant

In one of the recent revamp projects, KBR had come cross to a replacement requirement of the secondary reformer. Given the history of the existing secondary reformer it was diagnosed with hot spot problems and had been repaired several times, with a major repair being made in 2008. In this instance there was an excessive bulge area on the vessel wall due to the hot spot, and a section of shell material had to be cut off and replaced with new metal (Figure 13).

The plant was established in 1999 or 2000 and had seen various refractory issues in the secondary reformer in the first few years of service. Refractory was replaced on various points; albeit at different locations.

Figure 13. Secondary Reformer with Bulge Area

During the major revamp in 2019, a final decision was made to replace the existing secondary with new one of similar design.

While closely monitoring all the major contributing factors, the key to successful operation of any refractory lined vessel is the installation of refractory during original erection. Various parameters, like refractory mixing, amount of water, temperature of mixing and pouring method etc., affect the installation of refractory. Dry out of the installed refractory is also a major parameter for the final quality of the refractory layer.

Refractory installation was made at site, starting from erection of vessel to the point when piping was completed and vessel was ready for service.

To improve the quality control of the revamp refractory installation increased supervision from KBR was provided at various hold points at site ensuring that refractory met or exceeded the design specifications.

During project execution stage, KBR had ensured that all engineering procedures were revisited, and improvements were made. An example was increasing the number of anchors in the vessel, which was almost doubled. Original vessel had presence of only long anchors, while new vessel was designed with latest KBR's general practice to include short anchors.

Fabrication is another key point in how the equipment performs in operation. KBR selects only high-quality fabricators with an approved subvendor list for critical and proprietary items. Selecting a high-quality vendor would increase equipment reliability for the client. The selected vendors have expertise on building KBR critical and proprietary equipment. They usually possess high-class engineering and fabrication facilities to develop fabrication drawings and assembly of the vessels in shop while following welding, PWHT and nondestructive evaluation (NDE) requirements of codes and standards. KBR has also increased quality vigilance at the shop by dispatching engineers and quality inspectors from KBR to make specific visits at designated hold points during Pre-Inspection meetings.

At last, after basic equipment design, engineering and fabrication, it is very important for KBR to supervise and guide the installation of vessel and refractory at site. This is to ensure vessel is correctly installed per specifications at client site including refractory.

KBR has successfully worked with clients and helped develop project specific site inspection and test plan (ITP) and inclusion of hold points where KBR makes site visits as necessary. These holding points helped client completing site work with less quality issues and make the whole installation process is much smoother. The presence of KBR engineering field experts is also a vital point ensuring quality work during installation of refractory and vessels. KBR engineering field experts have sound knowledge of fabrication, refractory, welding and have the required know how to avoid field issues as they have visited numerous sites around the world and have been exposed to typical problems encountered at site for these critical equipment

By doing so, KBR has helped the client significantly reduce the problems and issues on the equipment installation, in situ refractory line fabrication, etc. As shown in Figure 14, on site refractory installation saved revamp turnaround time and associated costs with improved long-term equipment reliability.

Figure 14. Refractory Pouring at Site

The revamp plant was commissioned in August 2019 and the secondary reformer had no hot spots or other issues previously found in the past.

Lessons Learned and Recommendations

For any hot wall design secondary reformer in service more than 25 years we recommend clients take temperature measurements in close proximity to the vessel. If the outside metal shell is exposed to elevated temperature (beyond the metal design temperature) because of possible refractory failures the equipment may be on a path to catastrophic failure. Previous calculations by KBR on carbon steel show that it could happen in as little as 15~30 minutes for complete refractory failure over a 300 mm x 300 mm (11.8 inch by 11.8 inch) area. Because of this, the unit should have been shut down immediately when high wall temperature is discovered.

Although re-use of an existing vessel may sound more economical, the operator may be faced with the same or worse experience since the equipment has been under the excessive thermal and mechanical load for more than 20 years and under multiple repairs during the mechanical life cycle. Also, there are significant problems in placing a water jacket on a vessel designed with a dome support that was intended to be used without a water jacket. As the non-water jacked vessel is much hotter at the dome support allowing for circumferential expansion of the dome. Adding a water jacket to the exiting vessel, restricts this and violates the current dome design.

As a result, we recommend either:

1. A KBR water jacked (current generation) secondary reformer. This may be the most expensive option and may not be workable due to piping constraints and revamp schedules which must be evaluated.

2. Replacing the unit in kind (1.25Cr-0.5Mo) but paying much attention to refractory design, installation (mixing and anchor welding), and curing as well as dry out quality controls. However, the client has to understand that these materials are not easily repaired due to the rigorous welding and PWHT requirements.

In addition, the transfer line header should be replaced together with transfer line in-between the primary reformer and the secondary reformer. This is because the original refractory pour of the transfer line might not have been performed as accurately as was done for the new secondary reformer. This can ensure integrity and increase service life for vessels, especially if a non-water jacketed vessel is used for revamp optimization.

KBR recommends, instead of high temperature painting, the temperature monitoring device that it be wound with continuous fiber optic thermocouples on 150 mm (5.9 inch) centers. This will provide earlier and more accurate alarms and trends without requiring personnel to approach a vessel that is in danger of a local refractory failure and vessel over heating event. KBR has used this technique on various projects, where it has proved invaluable.

Conclusion

KBR water jacketed secondary reformers are robust, reliable and provide safe solution for reforming. The designs are proven with experience of many years and around hundreds of ammonia plants in the world.

This is largely because of the advantages of KBR's advanced ammonia technology where more than 200 ammonia plants have been licensed.

While many secondary reformers are at or near the end stage of lifetime, or in need of replacement because of the revamp requirement, it is recommended to replace them with KBR's innovative all-refractory combustion chamber design for its mechanical robustness. The KBR engineered secondary reformer, implemented with our in-house knowledge on fabrication and refractory installation, has the ability to withstand severe operational flow and thermal conditions contributes to sustained high process performance without increasing the need for maintenance and entailing shut-downs because of refractory failure, so the safety and reliability of the ammonia plant operation can be largely improved.

Loss Prevention Learnings from Beirut and Similar Ammonium Nitrate Explosions

The recent catastrophic explosion in Beirut highlights how the safe handling and storage of solid ammonium nitrate (AN) is of paramount importance. This article discusses the explosion hazard of solid AN from a fundamental and historical perspective and focuses on the main mechanisms that contributed to past events, including fire, contamination, product degradation, and shock. Critical loss prevention recommendations are summarized, referencing the recent major update to FM Global Loss Prevention Data Sheet 7-89 from January 2021. Lastly, a method to predict the explosive yield for explosions of solid AN in storage is presented.

Lorenz R. Boeck and Patrick W. Mahan
FM Global

Introduction

Ammonium Nitrate (AN) is a common industrial chemical used predominately as fertilizer and as a component in the manufacturing of blasting agents. If handled properly, it is a stable material with manageable risk. History shows, however, that poor storage and handling practices of solid AN severely increase the risk of major incidents, particularly explosions, such as the recent Beirut disaster.

The 2020 Beirut AN blast and many other catastrophic AN explosions could have been prevented by following well-established safety guidelines. Implementing safe practices along the entire AN supply chain is a critical task for safety professionals worldwide. Safety recommendations, such as those contained in FM Global Loss Prevention Data Sheet 7-89 – *Ammonium Nitrate and Mixed Fertilizers Containing Ammonium Nitrate* [1], have been available for many years. Significant revisions to this data sheet were made in January 2021, including updated loss prevention recommendations.

This paper examines AN explosions from both fundamental and historical perspectives and discusses how future explosions can be prevented. We first review the critical mechanisms that control AN's propensity to initiate and sustain detonations and discuss how these mechanisms affect the severity of past events, including Beirut. We then summarize major recommendations for preventing future AN explosions that have been included in FM Global Data Sheet 7-89. Lastly, we present an empirical method for predicting the yield of AN explosions for consequence analysis, based on a review of past events.

The Ammonium Nitrate Explosion Hazard

AN can detonate through rapid thermal decomposition [2], forming a detonation wave composed of a leading shock wave and a closely coupled reaction zone. This section of the paper discusses the detonability of AN, known effects that can sensitize AN, and the mechanisms that can initiate detonations.

Detonation characteristics of AN have been studied over several decades [3–9] and depend on many factors including bulk density, porosity, temperature, physical state (solid or liquid), and the presence of blending substances, conditioners, or contaminants. Generally, the higher-density AN used as fertilizer (FGAN – fertilizer-grade ammonium nitrate, density typically greater than 0.85 g/cm^3 (53 lb/ft^3)) is less sensitive to detonation than the lower-density AN (TGAN – technical-grade ammonium nitrate, density typically less than 0.85 g/cm^3 (53 lb/ft^3)) used for the manufacturing of blasting agents.

Under normal conditions, solid AN and especially FGAN is relatively insensitive to detonation, which is reflected by its large critical diameter. This parameter corresponds to the minimum diameter of a cylindrical charge that can sustain steady-state detonation. King and Bauer [10, 11] determined a critical diameter of about 1 m (40 in) for FGAN prills. Recent experiments [6] with TGAN showed critical diameters between 55 mm and 81 mm (2.2–3.2 in) under weak confinement, and between 8 mm and 12 mm (0.3–0.5 in) under strong confinement. For comparison, critical diameters of common secondary explosives, such as RDX or TNT, are on the order of 1–10 mm (0.04–0.4 in) [12].

Several mechanisms are known, however, that can sensitize AN, which has two significant practical consequences. First, it becomes 'easier' to initiate detonation in sensitized AN, i.e., a lower-energy source can achieve initiation. Second, once detonation is initiated within a sensitized mass of AN, the detonation is more likely to produce a higher explosive yield.

<u>Mechanisms that can sensitize AN include:</u>

Sensitization by heat, e.g., due to fire exposure

Intense heat from a fire can sensitize AN and is a well-known mechanism that has contributed to most past AN explosions [13]. As temperature increases, and especially at temperatures above AN's melting point of 170°C (338°F), it becomes more sensitive to detonation. King and Bauer [10, 11] reported a decrease in the critical diameter with increasing temperature, from 95–120 mm (3.7–4.7 in) at 220°C (428°F) to less than 46 mm (1.8 in) at 260°C (500°F). Molten AN can flow and accumulate in cavities that provide confinement, supporting detonation initiation and propagation [14].

Sensitization by contamination

When mixed with AN, several groups of substances are known to enhance its detonability. These include but are not limited to combustible solids and dusts; ignitable liquids; charcoal; organic compounds; acids; corrosive materials; chlorides; phosphorus; sulfur; and metals such as chromium or copper and its alloys [15–18].

Sensitization by thermal cycling

Repeated temperature cycles, especially across 32°C (90°F), cause changes in the crystalline structure of AN, irreversibly decreasing its bulk density and increasing its sensitivity [19, 20].

Sensitization by moisture absorption

AN is hygroscopic, and the absorption of moisture can lead to caking, self-compression, and self-confinement, enhancing its detonability [18].

<u>Mechanisms that can initiate detonation in AN include:</u>

Initiation by shock waves

Strong shock waves from sources such as explosives can initiate AN detonations. The shock-sensitivity of AN increases with increasing temperature, which Van Dolah et al. [21] demonstrated using card-gap tests. The authors suggested minimum shock-initiation pressures for solid FGAN on the order of 0.5–2.0 GPa (73,000–290,000 psi). For molten AN, King and Bauer

[10, 11] obtained shock-initiation pressures as low as 0.65 GPa (94,000 psi) at 260°C (500°F).

Initiation by projectiles / falling objects

Projectiles impacting on AN produce shock waves that in turn can initiate detonation. Van Dolah et al. [21] exposed AN to bullets fired from rifles at speeds of 800–1,080 m/s (2,620–3,540 ft/s). Detonations were successfully initiated at a bullet velocity of 1,080 m/s (3,540 ft/s) for solid AN at a temperature of 140°C (284°F); initiation of molten AN at 260°C (500°F) was successful at a bullet velocity of 823 m/s (2,700 ft/s).

Heemskerk and Schuurman [22] investigated the ability of falling objects to initiate detonation in molten AN. No detonations were recorded for steel objects dropping from heights up to 12 m (39 ft) into molten AN at temperatures up to 270°C (518°F), causing peak pressures as high as 24.8 MPa (3,600 psi). Higher impact pressures than investigated in this study may, however, accomplish initiation [23].

Initiation by runaway thermal decomposition

AN decomposes at temperatures higher than about 150°C (300°F) through a mechanism of reactions recently reviewed by Babrauskas and Leggett [2]. This mechanism shows a temperature-limiting behavior first described by Feick and Hainer [24], due to the competition of exothermal and endothermal reactions: The reaction temperature is self-limited to 292°C (558°F) at a pressure of 100 kPa (14.7 psi), and increases as a function of pressure, e.g., to 315°C (599°F) at 200 kPa (29.4 psi). This self-limiting mechanism may prevent unconfined AN from undergoing the transition to detonation. Confinement, however, prevents the escape of gaseous decomposition products, resulting in a rise of pressure and temperature, accelerating decomposition rates, and potentially resulting in an explosion. Literature studies have attempted to predict whether the self-confinement provided by tall piles [21] or deep pools of molten AN [25] can be sufficient to overcome this limiting mechanism and allow for continued local pressure rise and transition to detonation. These and other studies have not determined whether this mechanism is credible under realistic conditions and has played a role in any past events.

Initiation by incompatible materials

As discussed earlier, incompatible materials can increase the sensitivity of AN. Incompatible materials can also initiate spontaneous detonation of AN. A known example is sodium dichloroisocyanurate, a water disinfectant/pool chemical. Once mixed with AN in the presence of water, nitrogen trichloride forms, which is highly unstable, and spontaneous detonations have been observed in laboratory tests [26]. This mechanism has been proposed as a possible cause of the 2001 Toulouse explosion.

Ultimately, loss prevention efforts should focus on averting the sensitization and detonation initiation mechanisms discussed in this section. Unfortunately, these mechanisms have contributed to a long series of past AN explosion accidents.

History and Causes of Ammonium Nitrate Explosions

Explosions of solid AN range among the most severe industrial accidents in history. Table 1 identifies 27 AN explosion events that occurred between 1916 and 2020, not including explosions of ANFO (ammonium-nitrate-fuel-oil). Fifteen events occurred in transport and twelve in storage.

Three major explosion causes can be identified:

1. Twenty-three explosions—about 85% of all events—were preceded by fire. As discussed earlier, heating from fire sensitizes AN and increases its propensity to detonate.

2. Three explosions were caused by the detonations of nearby explosives. Specifically, blasting used to disaggregate caked material has led to two significant accidents (Oppau, Tessenderlo).

3. One explosion was attributed to contamination with an incompatible material (Toulouse). Several other potential causes have been proposed for this event, and a final consensus was never reached.

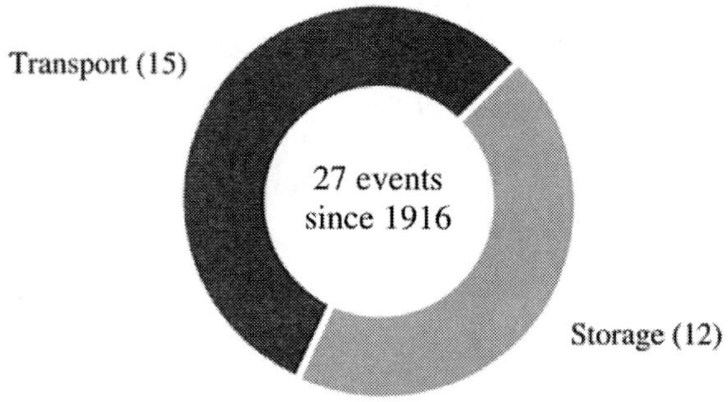

Figure 1. Categories of past explosions of solid AN.

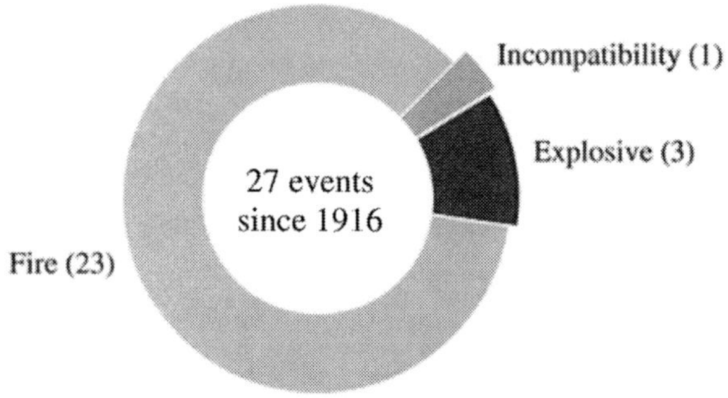

Figure 2. Causes of past explosions of solid AN.

Location	Country	Date	Category	Cause
Faversham, Kent	UK	4/2/1916	Storage	Fire
Kriewald	DE/PL	7/26/1921	Train	Explosive
Oppau	DE	9/21/1921	Storage	Explosive
Nixon, NJ	US	3/1/1924	Storage	Fire
Muscle Shoals, AL	US	4/4/1925	Train	Fire
Miramas	FR	8/5/1940	Train	Fire
Tessenderlo	BE	4/29/1942	Storage	Explosive
Texas City, TX	US	4/16/1947	Ship	Fire
Brest	FR	7/28/1947	Ship	Fire
Red Sea		1/23/1953	Ship	Fire
Traskwood, AR	US	12/17/1960	Train	Fire
Norton, VA	US	12/27/1961	Storage	Fire
Mount Vernon, MO	US	11/9/1966	Storage	Fire
Taroom	AU	8/30/1972	Truck	Fire
Pryor Creek, OK	US	1/17/1973	Storage	Fire
Toulouse	FR	9/21/2001	Storage	Material incompatibility
St. Romain en Jarez	FR	10/2/2013	Storage	Fire
Neyshabur	IR	2/18/2004	Train	Fire
Barracas	ES	3/9/2004	Truck	Fire
Mihăilești, Buzău	RO	5/24/2004	Truck	Fire
Ryongchŏn	KP	4/22/2004	Train	Fire
Pernik	BG	3/6/2007	Truck	Fire
West, TX	US	4/17/2013	Storage	Fire
Charleville	AU	9/5/2014	Truck	Fire
Ti Tree	AU	11/18/2014	Truck	Fire
Tianjin Port	CN	8/12/2015	Storage	Fire
Beirut	LB	8/4/2020	Storage	Fire

Table 1. Explosions of solid AN between 1916 and 2020.

The 2020 Beirut Explosion

The explosion at the Port of Beirut on August 4, 2020, is the most recent AN disaster and one of the largest chemical explosions in history. It claimed over 220 lives, caused more than 6,500 injuries, and left at least 300,000 people homeless [27]. Initial estimates of the total economic loss range up to 20 billion U.S. dollars [28]. It is a tragic example of how the combination of unsafe conditions can lead to a catastrophic AN explosion.

About 2,750 t (3,030 ST) of AN in 'big bags' arrived at the port in December 2013 on board the cargo vessel MV Rhosus. It was confiscated and stored in Hangar 12 at the port since October 2014. The bags were stored carelessly, with an estimated 70% of bags ripped open and AN spilling on the ground. Besides AN, several combustible or incompatible materials were reportedly stored in Hangar 12, including fireworks, wooden pallets, car tires, kerosene, oil, and hydrochloric acid. [29, 30]

On the day of the explosion, a fire at Hangar 12 was reported at around 5:54 PM. It has been suggested that the fire was caused by hot work [31]. At around 6:08 PM, the fire intensified, and a first smaller explosion occurred, which has been attributed to fireworks [32]. The main explosion of AN followed about 30 s later.

The AN explosion left a crater of 140 m (460 ft) in width. Within a radius of about 0.6 km (0.4 mi), mainly within the port area, buildings were completely destroyed. Structures up to about 1.3 km (0.8 mi) from the epicenter, including major Beirut high-rise buildings, sustained heavy damage. Damage to roofs, façades, and windows, was reported as far away as 9 km (6 mi) [33].

Several studies have published estimates of the explosive yield since the event. Table 2 summarizes these estimates, ranging between 100 and 1,120 t (110–1,230 ST) of TNT, with a median best estimate among all studies of 637 t (702 ST) of TNT. Our internal analyses of the explosion crater, air blast, and building damage, suggest a yield of 300–700 t (330–770 ST) of TNT, which agrees well with the published best estimate.

Ref	Method of yield estimation	Estimated yield (best estimate, in t TNT)	Estimated yield (range, in t TNT)
[34]	Seismic and far-field infrasound analysis	800	500–1,100
[35]	Video analysis of blast	500	100–1,120
[36]	Video analysis of blast	637	407–936
[37]	Video analysis of fireball	1,000	950–1,050
[38]	Video analysis of fireball	310	235–405
[23]	Crater, damage, and video analysis of blast	650	300–700/1,000
[39]	Video analysis of blast	500	150–700
	Median	**637**	

Table 2. Published yield estimates for the Beirut explosion.

Loss Prevention Recommendations

While the hazards involving AN can be severe, the risk can be managed with the appropriate measures. As part of our analysis of recent events, FM Global Loss Prevention Data Sheet 7-89 – *Ammonium Nitrate and Mixed Fertilizers Containing Ammonium Nitrate* has been updated. Data Sheet 7-89 provides guidance to ensure the safe handling and storage of AN. While there are many specific recommendations, the majority of the guidance is based on preventing exposure to fire, avoiding contamination, ensuring product stability, and eliminating exposure to shock.

Fire

Exposure to fire significantly increases the sensitivity of AN. As stated earlier, most AN detonations were preceded by an uncontrolled fire. To ensure safe handling, these exposure fires must be prevented by:

- Storing all AN in buildings of noncombustible construction. This includes limiting the use of flammable bins or partitions. While wood or plastic construction materials are desirable from a corrosion perspective, it introduces a fire hazard that should be avoided.
- Segregating AN storage from other combustible materials and not storing them in the same area or building. This is a particular concern for port facilities, logistics companies, and retail locations that handle AN and many other materials.

Fundamentally, AN should never be stored or handled in areas with a credible fire exposure. If it is, the potential fire exposure should be eliminated. The addition of fire protection (automatic sprinklers, etc.) is not an acceptable substitute for utilizing an inappropriate building design or comingling materials that should not be stored together.

Contamination / product stability

AN is incompatible with many materials, including oils, organic compounds, chlorates, copper salts, etc. Its material properties can also change over time. To ensure product stability and prevent contamination, the following measures should be taken:

- Only store AN in buildings or areas dedicated for this purpose. Do not store other incompatible materials in the same building or areas.
- Immediately remove damaged containers or contaminated materials to a safe area. Rework or eliminate promptly.
- Avoid contamination by water, including handling with wet equipment, roof leaks, poor drainage around buildings, condensation, etc.
- Avoid thermal cycling of AN, especially across 32°C (90°F) and prohibit storage of AN for longer than one year. This includes good stock rotation procedures (i.e., first in / first out) and limiting inventory carry-over from season to season.
- Maintain good housekeeping practices to avoid inadvertent contamination and material aging.

These measures are relatively easy to implement at manufacturing facilities, which produce these materials in large volumes and have dedicated infrastructure for their handling. However, significant challenges may arise at port facilities, logistics companies, and retail facilities. It is imperative that these down-stream facilities handle these materials with the appropriate discipline.

Shock

Exposure to other explosions or shock can result in the initiation of a detonation. AN should always be stored and handled in a manner that prevents this exposure.

- Do not use blasting agents (TNT, etc.) to disaggregate caked or solidified AN (this practice is now mostly discontinued).
- Do not store AN with or near other materials that present a detonation or shock hazard. This includes fireworks, munitions, blasting caps, AN Emulsions, etc.

These issues are rarely of concern at FGAN manufacturing sites. Sites making TGAN and also onsite explosives or explosive precursors (ANFO, emulsions, etc.) should conduct facility siting studies to ensure one material is not inadvertently exposed to another.

Recommendation summary

Every known significant event was the result of one or more of the above safeguards not being followed. The Beirut event is perhaps the worst of all, as the site had major deficiencies in every category:

1. Fire Exposure: The AN storage was in the same building as fireworks, pallets, and rubber tires. The event was preceded by a significant fire.
2. Contamination and Product Stability: The AN was stored in Beirut for 6 years without any environmental controls. Thermal cycling and humidity absorption would have degraded the material over time, and the FGAN would start to exhibit characteristics more like TGAN. This likely contributed to the high explosive yield.
3. Shock Exposure: The initial explosion is widely believed to have originated in the stored fireworks. This was a powerful initiation source for the AN that was already pre-sensitized by aging and fire exposure.

Estimating the Yield of AN Explosions in Storage

Predicting the consequences of AN explosions requires a method for estimating the explosive yield, which is then used as an input parameter for blast models. This section develops an empirical method for estimating the yields of solid AN explosions in storage.

The explosive yield, expressed as an equivalent mass of TNT that would cause comparable blast effects, M_{TNT}, can be estimated as the product of the available mass of AN, M_{AN}, its blast efficiency compared to TNT, EF, and the detonating fraction, DF, which describes the fraction of the material present participating in the detonation:

$$M_{TNT} = M_{AN} \cdot EF \cdot DF \qquad (1)$$

Previously, FM Global Data Sheet 7-89, revision 2013 [40], assumed a blast efficiency of $EF = 0.33$ and a detonating fraction of $DF = 0.1$. AN piles of 45–450 t (50–500 ST) were evaluated using Equation (1), and for larger piles the yield was assumed to remain constant. Predictions from this method are shown in Figure 3 (black dash-dotted line).These predictions were, however, found not to be conservative compared to major past events in storage (see Figure 3). To improve the method for future evaluations, the assumptions for blast efficiency and detonating fraction are adjusted as follows:

The blast efficiency is taken as $EF = 0.42$, which captures the theoretical maximum energy released by the detonation of AN [41]. This value has been adopted in several recent AN explosion studies [36, 42, 43] and is consistent with the Department of the Army Field Manual [44].

Among past events that were preceded by fires, see solid markers in Figure 3, detonating fractions as high as $DF = 1$ (solid red line) were observed. West and Tianjin Port are examples of explosions where yields are consistent with the

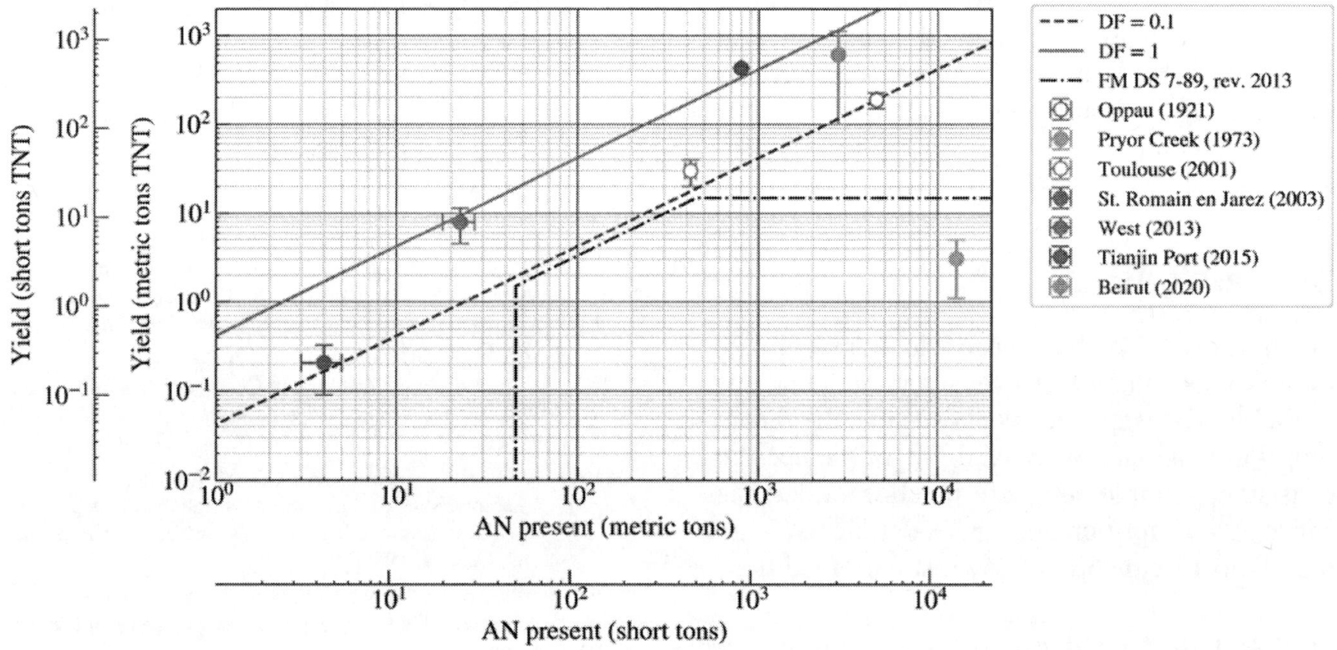

Figure 3. Yield estimates for past explosions as a function of the amount of AN present. Solid markers depict events with preceding fire; empty markers depict events without fire. Data from: Oppau [45]; Pryor Creek [46]; Toulouse [26, 47–49]; St. Romain en Jarez [50, 51]; West [14, 42, 43, 52, 53]; Tianjin Port [54–57]; Beirut [23, 34–39].

full participation of all AN present. Hence, a value of $DF = 1$ will be used to evaluate scenarios with a potential for fire engulfing a significant portion of the AN storage. This includes, for example, storage in buildings or bins of combustible construction or situations where significant amounts of other combustibles are present.

For past events without preceding fires, lower detonating fractions were observed, closer to $DF = 0.1$ (blue dashed line), consistent with the previous FM Global Data Sheet 7-89 assumption. Hence, a value of $DF = 0.1$ will be used for evaluating scenarios without significant fire exposure.

Where other conditions known to sensitize AN are present, such as gross contamination with incompatible materials or extended storage periods exceeding 2 years, $DF = 1$ would be assumed to capture the potential sensitization of the material.

Lastly, the revised method is not limited to a certain range of AN pile sizes. In contrast to the method in the previous FM Global Data Sheet 7-89, all pile sizes are evaluated using Equation (1). The revised method captures past events well, see Figure 3, and provides significantly improved predictions of AN explosion consequences.

Conclusions

The hazards of AN have been widely recognized for decades, and extensive safety guidance is available. Catastrophic events such as the recent Beirut explosion, however, highlight the need to consistently implement safe practices along the entire AN supply chain, where deficiencies are found most frequently at down-stream facilities.

Well-established safety practices should be followed, with focus on eliminating the main mechanisms that have contributed to past explosions, including fire, contamination, product degradation, and shock.

FM Global Loss Prevention Data Sheet 7-89 – *Ammonium Nitrate and Mixed Fertilizers Containing Ammonium Nitrate* was updated in January 2021, with emphasis on providing specific guidance for the safe handling and storage of AN.

References

[1] "FM Global Loss Prevention Data Sheet 7-89, Ammonium Nitrate and Mixed Fertilizers Containing Ammonium Nitrate," FM Global, Johnston, RI, USA, 2021.

[2] V. Babrauskas and D. Leggett, "Thermal Decomposition of Ammonium Nitrate," *Fire and Materials,* vol. 44, p. 250–268, 2020.

[3] S. Paterson and J. M. Davidson, "Detonation in Ammonium Nitrate," *Nature,* vol. 195, p. 277–278, 1962.

[4] M. A. Cook, E. B. Mayfield and W. S. Patridge, "Reaction Rates of Ammonium Nitrate in Detonation," *Journal of Physical Chemistry,* vol. 59, no. 8, p. 675–680, 1955.

[5] A. A. Deribas and V. A. Simonov, "Detonation Properties of Ammonium Nitrate," *Combustion, Explosion and Shock Waves,* vol. 35, no. 2, p. 203–205, 1999.

[6] H.-N. Presles, P. Vidal and B. Khasainov, "Experimental Study of the Detonation of Technical Grade Ammonium Nitrate," *C. R. Mechanique,* vol. 337, p. 755–760, 2009.

[7] A. W. King, A. Bauer and R. D. Heater, "The Detonation Properties of Ammonium Nitrate Prills," Canadian Fertilizer Institute and The Department of Mining Engineering, Queen's University, Kingston, ON, Canada, 1978.

[8] A. W. King, A. Bauer and R. D. Heater, "The Explosion Hazards of Ammonium Nitrate and Ammonium Nitrate-Based Fertilizer Compositions: A Summary," Canadian Fertilizer Institute and The Department of Mining Engineering, Queen's University, Kingston, ON, Canada, 1982.

[9] C. H. Winning, "Detonation Characteristics of Prilled Ammonium Nitrate," *Fire Technology,* vol. 1, p. 23–31, 1965.

[10] A. W. King and A. Bauer, "Shock Initiation Characteristics of Ammonium Nitrate," The Department of Mining Engineering, Queen's University, Kingston, ON, Canada, 1980.

[11] A. W. King, "Shock Initiation Characteristics of Ammonium Nitrate," Ph.D. Thesis, Queen's University, Kingston, ON, Canada, 1979.

[12] Z. X. Zhang, Rock Fracture and Blasting: Theory and Applications, Butterworth-Heinemann, 2016.

[13] V. Babrauskas, "Explosions of Ammonium Nitrate Fertilizer in Storage or Transportation are Preventable Accidents," *Journal of Hazardous Materials,* vol. 304, p. 134–149, 2016.

[14] "West Fertilizer Incident Support Services," Final Report, ABS Consulting, 2015.

[15] "Safety and Security Guidelines for the Storage and Transportation of Fertilizer Grade Ammonium Nitrate at Fertilizer Retail and Distribution Facilities," The Fertilizer Institute, 2017.

[16] "NFPA 400, Hazardous Material Code," National Fire Protection Association, Qunicy, MA, U.S., 2019.

[17] "Guidance for Compatibility of Fertilizer Blending Materials," Fertilizers Europe, Brussels, Belgium, 2005.

[18] "Chemical Advisory: Safe Storage, Handling, and Management of Solid Ammonium Nitrate Prills," EPA, OSHA, and ATF, 2015.

[19] G. Hansen and W. Berthold, "Sensitivity of Ammonium Nitrates to Detonation Impact," *Chemiker-Zeitung,* vol. 96, no. 8, p. 449, 1972.

[20] J.-F. Grainger, "AN Phase Transitions and their Impact on Plant Operation," in *Ammonium Nitrate - Nitric Acid (ANNA),* Little Rock, AR, U.S., 2009.

[21] R. W. Van Dolah, C. M. Mason, F. J. P. Perzak, J. E. Hay and D. Forshev, "Explosion of Ammonium Nitrate Under Fire Exposure," U.S. Department of the Interior, Bureau of Mines, 1966.

[22] A. H. Heemskerk and P. Schuurman, "Investigations into the Initiation of a Detonation of Molten Ammonium Nitrate by Falling Objects," in *Proceedings of the 6th International Syposium on Loss Prevention and Safety Promotion in the Process Industries,* Oslo, Norway, 1989.

[23] H. J. Pasman, C. Fouchier, S. Park, N. Quddus and D. Laboureur, "Beirut Ammonium Nitrate Explosion: Are not we Really Learning Anything?," *Process Safety Progress,* 2020.

[24] G. Feick and R. M. Hainer, "A Temperature-Limiting Mechanism in the Thermal Decomposition of Ammonium Nitrate," *Nature,* vol. 173, p. 1188–1189, 1954.

[25] W. G. Sykes, R. H. Johnson and R. M. Hainer, "Ammonium Nitrate Explosion Hazards," *Chemical Engineering Progress,* vol. 59, no. 1, p. 66–71, 1963.

[26] H. N. Presles, B. A. Khasainov, P. Vidal, V. Montassier, B. A. Ermolaev and A. A. Sulimov, "Phenomenological Description of the Spontaneous Detonation Mechanism in Moistened Ammonium Nitrate - Sodium Dichloroisocyanurate Mixture," in *15th International Detonation Symposium,* 2014.

[27] "International Medical Corps Situation Update," 9 September 2020. [Online]. Available: https://reliefweb.int/sites/reliefweb.int/files/resources/IntlMedCorps-BeirutExplosion_SitRep05.pdf.

[28] strategy&, "Beirut Explosion Impact Assessment," 8 September 2020. [Online]. Available: https://www.strategyand.pwc.com/m1/en/beirut-explosion/beirutimpactsssessment-statusupdate1.pdf.

[29] "The Beirut Explosion," Forensic Architecture, 2020. [Online]. Available: https://forensic-architecture.org/investigation/beirut-port-explosion.

[30] "How a Massive Bomb Came Together in Beirut's Port," New York Times, 09 09 2020. [Online]. Available: https://www.nytimes.com/interactive/2020/09/09/world/middleeast/beirut-explosion.html.

[31] "Beirut Blast Timeline: What we Know and What we Don't Know," The Guardian, 6 August 2020. [Online]. Available: https://www.theguardian.com/world/2020/aug/06/beirut-blast-timeline-what-we-know-and-what-we-dont-explosion-lebanon.

[32] "Fireworks, Ammonium Nitrate Likely Fueled Beirut Explosion," AP News, 5 August 2020. [Online]. Available: https://apnews.com/article/israel-ap-top-news-international-news-middle-east-lebanon-cbeb3263d6fc30a63a0300f588e7207b.

[33] "Beirut Explosion: What we know so far," BBC, 11 August 2020. [Online]. Available: https://www.bbc.com/news/world-middle-east-53668493.

[34] "Beirut Explosion Causes Strong Shock Waves - Infrasonic, Hydroacoustic and Seismic Signals Registered and Investigated at BGR," BGR, 4 August 2020. [Online]. Available: https://www.seismologie.bgr.de/sdac/erdbeben/big_quakes/beirut_200804_eng.html.

[35] S. E. Rigby, T. J. Lodge, S. Alotaibi, A. D. Barr, S. D. Clarke, G. S. Langdon and A. Tyas, "Preliminary Yield Estimation of the 2020 Beirut Explosion Using Video Footage From Social Media," *Shock Waves,* 2020.

[36] C. Stennett, S. Gaulter and J. Akhavan, "An Estimate of the TNT-Equivalent Net Explosive Quantity (NEQ) of the Beirut Explosion Using Publicly-Available Tools and Data," *Propellants, Explosives, Pyrotechnics,* 2020.

[37] J. S. Diaz, "Explosion Analysis from Images: Trinity and Beirut," *arXiv preprint arXiv:2009.05674,* 2020.

[38] C. J. Aouad, W. Chemissany, P. Mazzali, Y. Temsah and A. Jahami, "Beirut Explosion: Energy Yield From the Fireball Time Evolution in the First 230 Milliseconds," 10 2020. [Online]. Available: https://arxiv.org/abs/2010.13537.

[39] J. M. Dewey, "The TNT and ANFO Equivalences of the Beirut Explosion," *Shock Waves,* 2021.

[40] "FM Global Loss Prevention Data Sheet 7-89, Ammonium Nitrate and Mixed Fertilizers Containing Ammonium Nitrate," FM Global, Johnston, RI, USA, 2013.

[41] M. Braithwaite, "Ammonium Nitrate - Fertiliser, Oxidiser and Tertiary Explosive. A Review of Ammonium Nitrate Safety Issues based on Incidents, Research and Experience in the Safety Field," United Kingdom Explosion Liaison Group (UKELG), Loughborough, 2008.

[42] S. Davis, T. DeBold and C. Marsegan, "Investigation Findings and Lessons Learned in the West Fertilizer Explosion," *Journal of Fire Sciences,* vol. 35, no. 5, p. 379–395, 2017.

[43] D. M. Laboureur, Z. Han, B. Z. Harding, A. Pineda, W. C. Pittman, C. Rosas, J. Jiang and M. S. Mannan, "Case Study and Lessons Learned from the Ammonium Nitrate Explosion at the West Fertilizer Facility," *Journal of Hazardous Materials,* vol. 308, p. 164–172, 2016.

[44] "Department of the Army Field Manual 5–250, Explosives and Demolitions," Department of the Army, Washington DC, 1992.

[45] "Explosion in a Nitrogenous Fertilizer Plant," French Ministry for Sustainable Development, Report No. 14373, 2008.

[46] R. Freeman, "The Cherokee Ammonia Plant Explosion," *Chemical Engineering Progress,* vol. 71, no. 11, p. 71–74, 1975.

[47] F. Berthelemy, "Accident on the 21st September 2001 at a Factory Belonging to the Grande Paroisse Company in Toulouse, Produced Jointly with the Explosives Inspectorate and with Help from the INERIS," Report of the General Inspectorate for the Environment, Affair No. IGE/01/034, 2001.

[48] Y. Moulleau and N. Dechy, "Initial Analsis of the Damage Observed in Toulouse after the Incident that Occurred on the 21st September on the AZF Site of the Grande Paroisse Company," in *International ESMG Symposium on Process Safety and Industrial Explosion Protection*, Nurnberg, Germany, 2002.

[49] N. Dechy, T. Bordeaux, N. Ayrault, M. A. Kordek and J. C. Le Coze, "First Lessons of the Toulouse Ammonium Nitrate Disaster, 21st September 2001, AZF Plant, France," *Journal of Hazardous Materials,* vol. 111, no. 1-3, p. 131–138, 2004.

[50] "Fire Inside a Barn and Explosion of Fertiliser," French Ministry for Sustainable Development, Report No. 25669, 2010.

[51] M. A. Kordek, C. Michot and G. Malair, "Accident (Explosion Succeeding to Fire) at St Romain en Jarez (France), 2 October 2003," 2003. [Online]. Available: https://www.slideshare.net/watermon/ammonium-nitrate-explosion.

[52] "West Fertilizer Company Fire and Explosion," CSB Investigation Report 2013-02-I-TX, 2013.

[53] V. Babrauskas, "The Ammonium Nitrate Explosion at West, Texas: A Disaster that Could Have Been Avoided," *Fire and Materials,* vol. 42, no. 2, p. 164–172, 2018.

[54] "Accident Investigation Report of the Fire and Explosion in Tianjin Port," State Administration of Work Safety, China, 2016.

[55] B. Zhao, "Facts and Lessons Related to the Explosion Accident in Tianjin Port, China," *Natural Hazards,* vol. 84, p. 707–713, 2016.

[56] G. Fu, J. Wang and M. Yan, "Anatomy of Tianjin Port Fire and Explosion: Process and Causes," *Process Safety Progress,* vol. 35, no. 3, p. 216–220, 2016.

[57] P. Huang and J. Zhang, "Facts Related to August 12, 2015 Explosion Accident in Tianjin, China," *Process Safety Progress,* vol. 34, no. 4, p. 313–314, 2015.

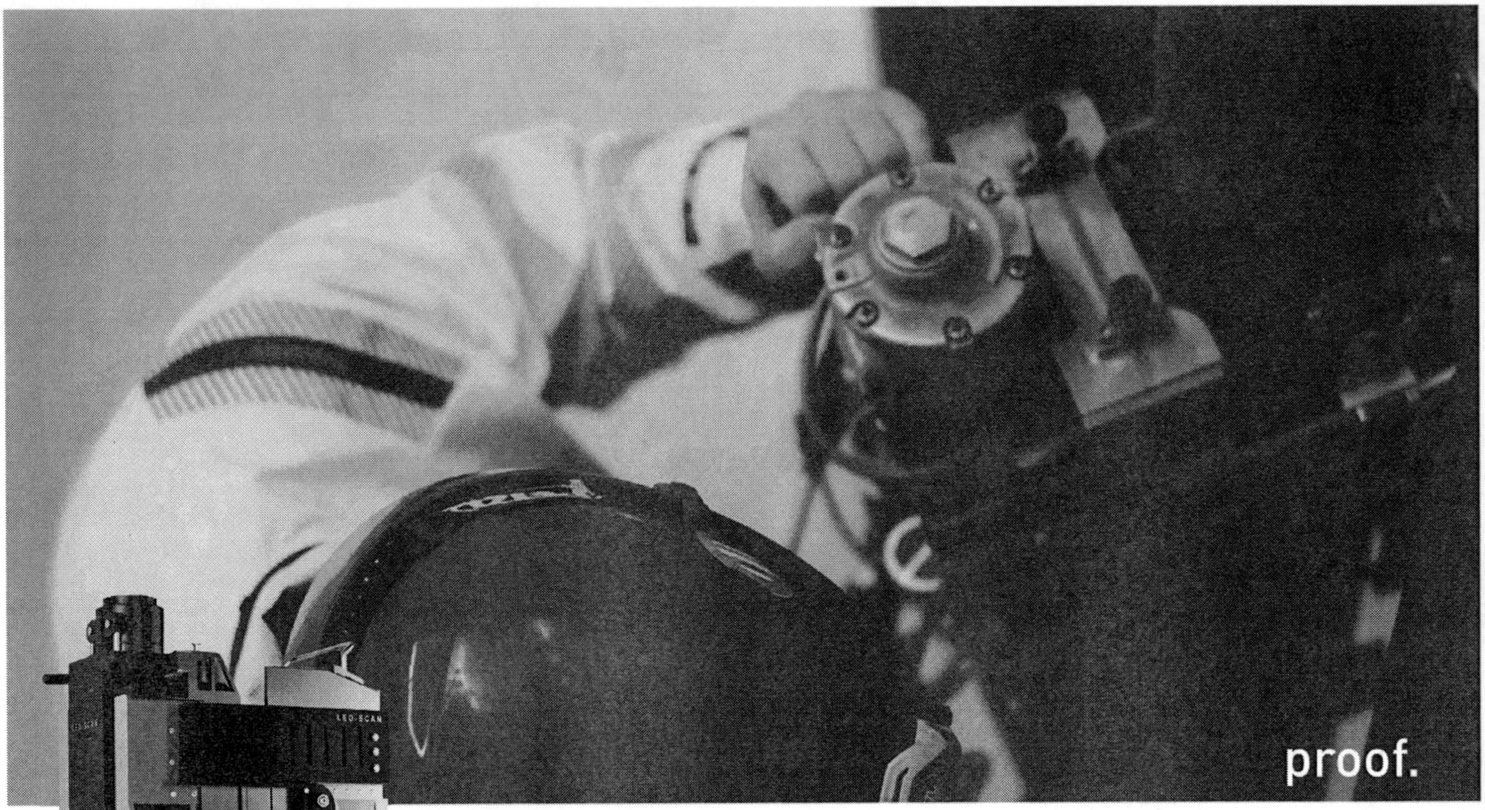
proof.

AIChE
120 Wall Street, FL 23
New York, NY 10005-4020

ISBN 978-1-7138-4502-7

9 781713 845027

CURRAN ASSOCIATES INC.
proceedings.com